Chapman & Hall/CRC Mathematical and Computational Biology Series

Computational Hydrodynamics of Capsules and Biological Cells

Edited by

C. Pozrikidis

CRC Press
Taylor & Francis Group
Boca Raton London New York

CRC Press is an imprint of the
Taylor & Francis Group an **informa** business
A CHAPMAN & HALL BOOK

CHAPMAN & HALL/CRC
Mathematical and Computational Biology Series

Aims and scope:
This series aims to capture new developments and summarize what is known over the entire spectrum of mathematical and computational biology and medicine. It seeks to encourage the integration of mathematical, statistical, and computational methods into biology by publishing a broad range of textbooks, reference works, and handbooks. The titles included in the series are meant to appeal to students, researchers, and professionals in the mathematical, statistical and computational sciences, fundamental biology and bioengineering, as well as interdisciplinary researchers involved in the field. The inclusion of concrete examples and applications, and programming techniques and examples, is highly encouraged.

Proposals for the series should be submitted to one of the series editors above or directly to:
CRC Press, Taylor & Francis Group
4th, Floor, Albert House
1-4 Singer Street
London EC2A 4BQ
UK

Published Titles

Algorithms in Bioinformatics: A Practical Introduction
Wing-Kin Sung

Bioinformatics: A Practical Approach
Shui Qing Ye

Biological Sequence Analysis Using the SeqAn C++ Library
Andreas Gogol-Döring and Knut Reinert

Cancer Modelling and Simulation
Luigi Preziosi

Cancer Systems Biology
Edwin Wang

Cell Mechanics: From Single Scale-Based Models to Multiscale Modeling
Arnaud Chauvière, Luigi Preziosi, and Claude Verdier

Combinatorial Pattern Matching Algorithms in Computational Biology Using Perl and R
Gabriel Valiente

Computational Biology: A Statistical Mechanics Perspective
Ralf Blossey

Computational Hydrodynamics of Capsules and Biological Cells
C. Pozrikidis

Computational Neuroscience: A Comprehensive Approach
Jianfeng Feng

Data Analysis Tools for DNA Microarrays
Sorin Draghici

Differential Equations and Mathematical Biology, Second Edition
D.S. Jones, M.J. Plank, and B.D. Sleeman

Engineering Genetic Circuits
Chris J. Myers

Exactly Solvable Models of Biological Invasion
Sergei V. Petrovskii and Bai-Lian Li

Gene Expression Studies Using Affymetrix Microarrays
Hinrich Göhlmann and Willem Talloen

Glycome Informatics: Methods and Applications
Kiyoko F. Aoki-Kinoshita

Handbook of Hidden Markov Models in Bioinformatics
Martin Gollery

Introduction to Bioinformatics
Anna Tramontano

An Introduction to Systems Biology: Design Principles of Biological Circuits
Uri Alon

Kinetic Modelling in Systems Biology
Oleg Demin and Igor Goryanin

Knowledge Discovery in Proteomics
Igor Jurisica and Dennis Wigle

Meta-analysis and Combining Information in Genetics and Genomics
Rudy Guerra and Darlene R. Goldstein

Modeling and Simulation of Capsules and Biological Cells
C. Pozrikidis

Niche Modeling: Predictions from Statistical Distributions
David Stockwell

Normal Mode Analysis: Theory and Applications to Biological and Chemical Systems
Qiang Cui and Ivet Bahar

Optimal Control Applied to Biological Models
Suzanne Lenhart and John T. Workman

Pattern Discovery in Bioinformatics: Theory & Algorithms
Laxmi Parida

Python for Bioinformatics
Sebastian Bassi

Spatial Ecology
Stephen Cantrell, Chris Cosner, and Shigui Ruan

Spatiotemporal Patterns in Ecology and Epidemiology: Theory, Models, and Simulation
Horst Malchow, Sergei V. Petrovskii, and Ezio Venturino

Stochastic Modelling for Systems Biology
Darren J. Wilkinson

Structural Bioinformatics: An Algorithmic Approach
Forbes J. Burkowski

The Ten Most Wanted Solutions in Protein Bioinformatics
Anna Tramontano

CRC Press
Taylor & Francis Group
6000 Broken Sound Parkway NW, Suite 300
Boca Raton, FL 33487-2742

Printed in the United States of America on acid-free paper
10 9 8 7 6 5 4 3 2 1

International Standard Book Number: 978-1-4398-2005-6 (Hardback)

Library of Congress Cataloging-in-Publication Data

Computational hydrodynamics of capsules and biological cells / editor, Constantine Pozrikidis.
p. ; cm. -- (Chapman and Hall/CRC mathematical and computational biology series)
Includes bibliographical references and index.
ISBN 978-1-4398-2005-6 (hardcover : alk. paper)
1. Body fluid flow--Mathematical models. 2. Fluid dynamics--Mathematical models. 3. Cells. I. Pozrikidis, C. II. Series: Chapman and Hall/CRC mathematical & computational biology series.
[DNLM: 1. Body Fluids--physiology. 2. Computational Biology--methods. 3. Biophysical Phenomena. 4. Blood Cells--physiology. 5. Cell Physiological Phenomena. 6. Hemorheology. QU 105 C738 2010]

QP517.R48C66 2010
612'.01522--dc22 2010014601

Visit the Taylor & Francis Web site at
http://www.taylorandfrancis.com

and the CRC Press Web site at
http://www.crcpress.com

Contents

Preface

Computational biofluid-dynamics addresses a diverse family of problems involving fluid flow inside and around living organisms, organs and tissue, biological cells, and other biological materials. Numerical methods combine aspects of computational mechanics, fluid dynamics, computational physics, computational chemistry and biophysics into a framework that integrates a broad range of scales. The goal of this edited volume is to provide a comprehensive, rigorous, and current introduction to the fundamental concepts, mathematical formulation, alternative approaches, and predictions of computational hydrodynamics of capsules and biological cells. The book is meant to serve both as a research reference and as a teaching resource.

Scope

The numerical methods discussed in the following eight chapters cover a broad range of possible formulations for simulating the motion of rigid particles (platelets) and the flow-induced deformation of liquid capsules and cells enclosed by viscoelastic membranes. Although some of the physical problems discussed in different chapters are similar or identical, the repetition is desirable so that solutions produced by different numerical approaches can be compared and the efficiency of alternative formulations can be assessed. The consistency of the results validates the procedures and offers several alternatives.

Boundary-integral formulations

The first three chapters present boundary-integral formulations. In Chapter 1, the editor of this volume discusses a boundary-element method for computing the flow-induced deformation of idealized two-dimensional red blood cells in Stokes flow and provides a pertinent MATLAB® code. In Chapter 2, Barthès-Biesel, Walter & Salsac discuss highly accurate boundary-element methods for simulating capsules with spherical unstressed shapes based on direct and variational formulations. Their work delineates the occurrence of membrane wrinkling due to local compression. In Chapter 3, Freund & Zhao present an advanced boundary-integral formulation based on Fourier expansions and discuss the results of simulations for cellular flow in domains with complex geometry. A snapshot of their simulations with thirty periodically repeated cells is featured on the cover of this volume. In the past, such depictions have been possible only by artist rendition or schematic illustration.

Immersed-boundary formulations

Chapters 4 and 5 present immersed-boundary methods. In Chapter 4, Zhang, Johnson & Popel discuss an immersed-boundary/lattice-Boltzmann formulation that takes into consideration cell aggregation. Results are presented for a broad range of conditions illustrating, among other effects, the significance of rouleaux formation. In Chapter 5, Bagchi discusses an immersed-boundary front-tracking method for computing the deformation of capsules and cells in dilute and dense suspensions. Combining the basic algorithm with a coarse-grain Monte-Carlo method for intermolecular forces allows the simulation of leukocyte rolling on an adhesive substrate under the influence of a shear flow.

Discrete models

In Chapter 6, Fedosov, Caswell & Kardiadakis present a discrete membrane model where a surface network of viscoelastic links emulates the spectrin network of the cytoskeleton. The motion of the internal and external fluids is computed by the method of dissipative particle dynamics. One important feature is that the network potential can be tuned to exhibit desired macroscopic properties without *ad-hoc* adjustment. The surface discretization scheme can be used as a module in other problems involving membrane–flow interaction. In Chapter 7, Secomb discusses a two-dimensional model of red and white blood cell motion. A novel feature of his approach is that the cell is represented by a network of membrane and interior viscoelastic elements mediating elastic response and viscous dissipation. Coupled with a finite-element method for computing the ambient plasma flow, the formulation yields a powerful technique for simulating cell motion in dilute and concentrated suspensions.

Platelet motion

In Chapter 8, Mody & King discuss the numerical simulation of platelet motion near a wall representing injured tissue. The computational model is based on the boundary-integral formulation for Stokes flow containing rigid particles with arbitrary shapes. Brownian motion and adhesive forces between platelets and a substrate are implemented in terms of surface bonds. The simulations furnish a wealth of information on microstructural platelet dynamics near an injured vascular wall.

Summary

The book is accompanied by an Internet site located at the Web address *http://dehesa.freeshell.org/CC2*, where computer codes, animations, and other information are provided.

The subject of computational cellular mechanics and biofluid-dynamics has emerged as an eminent topic, bridging biological, mathematical, computational, and engineering sciences. The authors of this volume conclude their chapters by pointing out venues for further work in the mathematical formulation and numerical implementation, and by identifying physiological problems to be addressed in future research. Their comments serve as a roadmap for students and researchers who wish to contribute to these efforts.

C. Pozrikidis

About the Editor

C. Pozrikidis is a Professor in the Department of Chemical Engineering at the University of Massachusetts, Amherst. He is the author of eight books and the editor of two volumes in fluid mechanics, computational fluid dynamics, biomechanics, boundary-element methods, finite and spectral-element methods, and scientific computing.

1 Flow-induced deformation of two-dimensional biconcave capsules

C. Pozrikidis

Department of Chemical Engineering
University of Massachusetts
Amherst

Numerical simulations of the flow-induced deformation of a two-dimensional biconcave capsule enclosed by an elastic membrane that exhibits flexural stiffness are conducted using the boundary-integral method for Stokes flow. The capsule represents an idealized red blood cell suspended in plasma. When the internal and external fluid viscosities are matched, a capsule immersed in infinite shear flow deforms to obtain a nearly steady elongated shape. A transition from flipping to tumbling is observed as a dimensionless shear rate, defined with respect to the membrane bending modulus, becomes smaller. When the internal fluid viscosity exceeds a critical threshold, the capsule engages in periodic tumbling motion as it stretches and compresses while rotating under the action of the shear flow. Simulations of the capsule motion in simple shear flow near a plane wall reveal that the capsule monotonically migrates away from the wall in the case of equal viscosities, and exhibits undulatory migration in the case of high interior viscosity. The behavior of two-dimensional capsules is qualitative similar to that exhibited by the mid-plane contour of three-dimensional biconcave capsules resembling red blood cells.

1.1 Introduction

Red blood cells are small liquid capsules enclosed by a biological membrane consisting of a lipid bilayer and an underlying protein cytoskeleton (e.g., Fung 1984). The interfacial layer exhibits a large modulus of dilatation that ensures surface incompressibility, and a small modulus of bending that prevents the development of large curvature leading to folds, cusps, and kinks. A complementary liquid-like behavior renders the membrane a viscoelastic, energy-dissipating medium endowed with dilatational and shear viscosity. The area incompressibility and resistance to bending are attributed to the membrane lipid bilayer, the elastic response in shearing deformation is attributed to the membrane cytoskeleton, and the viscous behavior is attributed to both (e.g., Skalak *et al.* 1973, Hochmuth 1982).

Numerous efforts have been made to describe the red blood cell motion and deformation in external and internal capillary flow, as reviewed by Keller & Skalak (1982) and Secomb (2003). Early models have neglected the cell deformability and modeled the red blood cells as rigid spherical or ellipsoidal particles. To account for the interfacial deformability, other authors proposed models of liquid drops. An improved model that couples the interior to the exterior flow by way of a prescribed tangential surface velocity was developed by Keller & Skalak (1982). In fact, the cell shape and surface velocity distribution must be found simultaneously as part of the solution to satisfy local interfacial force and torque balances involving interfacial elastic tensions, viscous tensions, and developing bending moments (e.g., Pozrikidis 2003).

With the availability of computing power and the development of advanced numerical methods, realistic simulations of the flow-induced deformation of capsules with spherical and spheroidal unstressed shapes and fluids with equal or different viscosities representing vesicles have become possible. Numerical simulations employ boundary-element, immersed-boundary, finite-element, and other methods (e.g., Lac *et al.* 2007, Zhang *et al.* 2008). For example, Doddi and Bagchi (2008) performed numerical simulations of the motion of a capsule with spherical unstressed shape in parabolic channel flow and observed migration toward the channel centerline (see also Chapter 5). As in the case of liquid drops, the higher the capsule deformability and interfacial mobility, the faster the migration velocity. Bagchi & Kalluri (2009) performed extensive simulations of capsules with oblate unstressed shape in simple shear flow and identified several modes of deformation determined by a dimensionless shear rate and fluid viscosities. Other simulation techniques are described in the following chapters.

Due to persistent computational challenges, only a few successful efforts have been made to describe the deformation of capsules with biconcave resting shapes and high interior fluid viscosity resembling red blood cells. Pozrikidis (2003, 2005) implemented a boundary-element method for simulating the deformation of a red blood cell in simple shear flow taking into consideration the membrane elastic response and flexural stiffness. Takagi *et al.* (2009) carried out successful simulations using the immersed-boundary method and demonstrated the significance of the membrane constitutive equation. Their results revealed that liquid liposomes enclosed by incompressible membranes deform differently than red blood cells. Doddi & Bagchi (2009) performed extensive simulations of solitary capsules and suspensions of biconcave capsules in channel flow. Efficient lattice-Boltzmann methods amenable to parallelization were implemented by Dupin *et al.* (2007) for simulating the deformation of a solitary three-dimensional capsule and the flow of three-dimensional suspensions (see also Chapter 4). Although the predictions of this novel approach are qualitatively similar to those obtained by traditional methods, some adjustable parameters must be chosen. Perhaps more important, a detailed description of the developed membrane tension field and perturbation flow is hard to extract from the numerical simulations. The effect of membrane viscosity was neglected in the computational model. Advanced simulation techniques are described in other chapters of this volume.

Computational difficulties for three-dimensional flow has motivated the development of two-dimensional models. Although such models clearly compromise the physics of the motion, they do provide us nevertheless with useful insights and often accurate predictions. Breyiannis & Pozrikidis (2000) applied the boundary-element method to investigate the motion of a non-dilute doubly periodic suspension of two-dimensional capsules whose viscosity is equal to that of the ambient fluid. Bagchi (2007) and Takagi *et al.* (2009) applied the immersed-boundary method to study the motion of a solitary capsule and investigated the channel flow of concentrated suspensions. Bagchi *et al.* (2005) and Zhang *et al.* (2008) presented simulations of two-dimensional cell aggregation and dissociation (see Chapters 4 and 5). Veerapaneni *et al.* (2009) developed an implicit boundary-integral method for simulating the motion of vesicles enclosed by inextensible membranes.

Secomb *et al.* (2007) proposed an interesting two-dimensional model of the motion and deformation of red blood cells in infinite shear flow and inside microvessels (see Chapter 7). The membrane of each cell is represented by an assemblage of discrete structural elements with defined mechanical properties. Internal viscous elements are introduced to mediate the effect of membrane energy dissipation. The formulation takes into consideration hydrodynamic interactions between cells and vessel walls and accounts for the lateral migration of the across flow streamlines. The results of their simulations are found to be in good agreement laboratory observations. Analytical and numerical studies of two-dimensional vesicle deformation and tumbling were presented by other authors (e.g., Rioual *et al.* 2004, Mishbah 2006).

Our goal in this chapter is to provide further documentation and physical insights into the motion and deformation of a red blood cell in infinite and wall-bounded shear flow, with emphasis on delineating the significance of the internal fluid viscosity. One particular issue of interest is the spontaneous cell migration away from a wall due to flow-induced deformation. Since the

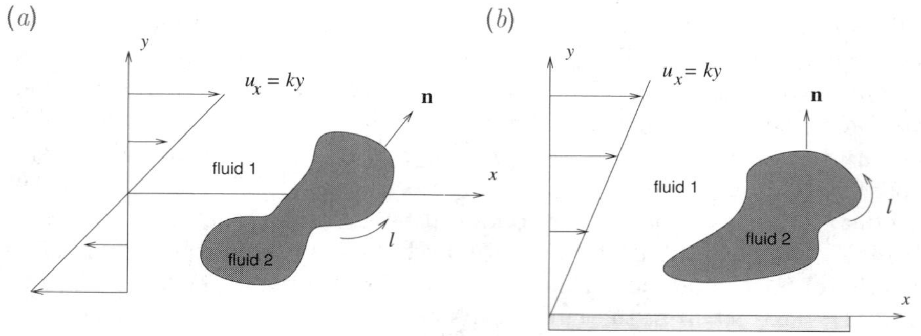

Figure 1.2.1 Illustration of a two-dimensional liquid capsule enclosed by an elastic membrane in (*a*) infinite shear flow, and (*b*) semi-infinite shear flow bounded by a plane wall.

simulation of three-dimensional capsules with biconcave unstressed shapes is prohibited by persistent computational challenges related to adaptive membrane discretization (e.g., Pozrikidis 2003), we consider to the more amenable but nonetheless nontrivial case of two-dimensional flow.

Two weaknesses of the two-dimensional model are that the effect of the membrane viscosity cannot be assessed and membrane flexural stiffness is necessary for the capsule to recover the unstressed shape after deformation. In spite of these limitations, the results are consistent with those reported previously for three-dimensional flow. The viability and longevity of the simulations hinge on an efficient regridding algorithm that allows us to accurately pursue the motion for an extended period of time. The MATLAB® programs developed to perform these simulations are available in the public domain for further research and education.

1.2 Mathematical framework

We consider the deformation of a two-dimensional liquid capsule enclosed by an elastic membrane in two-dimensional infinite or semi-infinite simple shear flow with shear rate k, as illustrated in figure 1.2.1. At low Reynolds numbers, the motion of the fluid outside the capsule is governed by (*a*) the Stokes equation expressing a differential force balance on an infinitesimal fluid parcel, and (*b*) the continuity equation ensuring mass conservation,

$$-\nabla p + \mu \nabla^2 \mathbf{u} = \mathbf{0}, \qquad \nabla \cdot \mathbf{u} = 0, \tag{1.2.1}$$

where $\mathbf{u}$ is the fluid velocity, p is the pressure, and μ is the ambient fluid viscosity. The motion of fluid inside the capsule is governed by the same set of equations, except that the viscosity is multiplied by the viscosity ratio, λ, corresponding to interior viscosity $\lambda\mu$. The viscosity ratio of the cytoplasma of healthy red blood cells *in vivo* is higher than that of the surrounding plasma due to the high concentration of cytoplasma proteins, $\lambda \simeq 5$. Ghost cells with unit viscosity ratio are readily produced in the laboratory.

The interfacial conditions require that the velocity, $\mathbf{u}$, is continuous across the capsule membrane. The hydrodynamic traction, $\mathbf{f} = \boldsymbol{\sigma} \cdot \mathbf{n}$, undergoes a discontinuity determined by the membrane load due to developing membrane tensions and bending moments,

$$\Delta\mathbf{f} \equiv (\boldsymbol{\sigma}^{(1)} - \boldsymbol{\sigma}^{(2)}) \cdot \mathbf{n}, \tag{1.2.2}$$

where $\boldsymbol{\sigma}$ is the Cauchy stress tensor and $\mathbf{n}$ is the unit vector normal to the membrane pointing outward (e.g., Pozrikidis 1997). The superscript 1 refers to the ambient fluid and the superscript 2 refers to the capsule fluid. In the case of semi-infinite flow bounded by a plane wall, no-slip and no-penetration conditions require the that the velocity is zero over the wall.

1.2.1 Membrane mechanics

In the absence of significant membrane mass and thus inertia, force equilibrium over an infinitesimal section of the membrane requires

$$\Delta\mathbf{f} = -\frac{\partial}{\partial l}\Big(\tau\,\mathbf{t} + q\,\mathbf{n}\Big) = -(\frac{\partial \tau}{\partial l} + \kappa\, q)\,\mathbf{t} + (\tau\,\kappa - \frac{\partial q}{\partial l})\,\mathbf{n}, \tag{1.2.3}$$

where l is the arc length measured in the direction of the unit tangent vector, $\mathbf{t}$, τ is the in-plane membrane tension, q is the transverse shear tension, and κ is the membrane curvature in the xy plane reckoned to be positive for a circular cell, as illustrated in figure 1.2.2. To derive the last expression in (1.2.3), we have used the Frenet relations

$$\frac{\mathrm{d}\mathbf{t}}{\mathrm{d}l} = -\kappa\,\mathbf{n}, \qquad \frac{\mathrm{d}\mathbf{n}}{\mathrm{d}l} = \kappa\,\mathbf{t}. \tag{1.2.4}$$

Torque equilibrium requires that

$$q = \frac{\partial m}{\partial l}, \tag{1.2.5}$$

where m is the membrane bending moment, as illustrated in figure 1.2.2. Substituting (1.2.5) in (1.2.3), we obtain

$$\Delta\mathbf{f} = -(\frac{\partial \tau}{\partial l} + \kappa\,\frac{\partial m}{\partial l})\,\mathbf{t} + (\tau\,\kappa - \frac{\partial^2 m}{\partial l^2})\,\mathbf{n}. \tag{1.2.6}$$

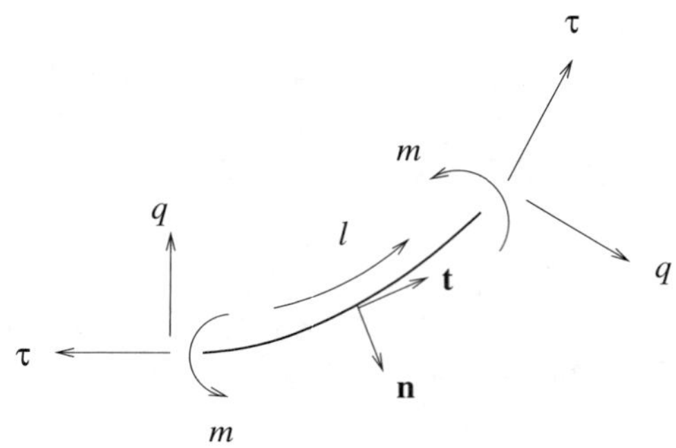

Figure 1.2.2 Illustration of a small section of a membrane showing the in-plane tension, τ, the transverse tension, q, and the bending moment, m. For the configuration shown, the membrane curvature is positive, $\kappa > 0$.

Next, we adopt a linear constitutive equation for the in-plane tension,

$$\tau = E_D\,(\lambda_s - 1), \tag{1.2.7}$$

where E_D is an elastic modulus of dilatation, $\lambda_s \equiv \partial l/\partial l_R$ is the stretch, and l_R is the membrane arc length in the relaxed configuration measured from an arbitrary origin defined by a designated point particle. An alternative constitutive equation based on the vanishing of the tension normal to the plane of the flow is $\tau = E_D\,(\lambda_s^3 - 1)/\lambda_s^{3/2}$ (e.g., Bagchi 2007). Other constitutive equations can be chosen to implement desired types of membrane response.

We complete the mathematical formulation by adopting a linear constitutive equation for the bending moment,

$$m = E_B\,(\kappa - \kappa_R), \tag{1.2.8}$$

where E_B is the bending modulus, κ is the curvature of the membrane in the xy plane at the position of a point particle, and κ_R is the resting curvature at the position of the same point particle.

Substituting (1.2.7) and (1.2.8) in (1.2.6) we find that

$$\Delta \mathbf{f} = -\Big(E_D\,\frac{\partial \lambda_s}{\partial l} + E_B\,\kappa\,\frac{\partial(\kappa - \kappa_R)}{\partial l}\Big)\,\mathbf{t} + \Big(E_D\,(\lambda_s - 1)\,\kappa - E_B\,\frac{\partial^2(\kappa - \kappa_R)}{\partial l^2}\Big)\,\mathbf{n}. \tag{1.2.9}$$

The accurate computation of the second derivative of the curvature requires special attention to prevent numerical instability in the numerical implementation.

Importance of flexural stiffness

Accounting for the flexular stiffness in terms of bending moments is imperative for a realistic description of capsule deformation in two-dimensional flow. The reason is that, in the absence of bending moments determined by the reference and the instantaneous membrane curvature, a deformed membrane will not necessarily return to the unstressed shape when the flow is stopped. To explain this, we may assume that a capsule is enclosed by an inextensible membrane, such as paper tissue, and deform the initial shape into a new shape. Because the membrane stretch remains constant at the value of unity around the membrane perimeter, elastic tensions do not develop. In contrast, a three-dimensional or axisymmetric capsule enclosed by an inextensible or extensible membrane returns to the reference shape when an external load is removed even in the absence of flexural stiffness. The reason is the principal stretches do not remain constant at the value of unity after deformation.

1.2.2 Boundary-integral formulation

The boundary-integral formulation provides us with an integral representation for the velocity at a chosen point, as well as with an integral equation for the velocity around the cell contour in the xy plane. The disturbance velocity at a point $\mathbf{x}_0$ in the ambient fluid due to the presence of the capsule is

$$u'_j(\mathbf{x}_0) = -\frac{1}{4\pi\mu} \oint_C \mathcal{G}_{ji}(\mathbf{x}_0, \mathbf{x})\, f_i'^{(1)}(\mathbf{x})\, \mathrm{d}l(\mathbf{x}) + \frac{1}{4\pi} \oint_C u_i'^{(1)}(\mathbf{x})\, \mathcal{T}_{ijk}(\mathbf{x}, \mathbf{x}_0)\, n_k(\mathbf{x})\, \mathrm{d}l(\mathbf{x}), \qquad (1.2.10)$$

where a prime denotes a perturbation variable, the integration is performed around the membrane contour, $\mathcal{G}_{ij}(\mathbf{x}, \mathbf{x}_0)$ is the Green's function of Stokes flow for the velocity, $\mathcal{T}_{ijk}(\mathbf{x}, \mathbf{x}_0)$ is the corresponding Green's function for the stress field, and $\mathbf{n}$ is the unit normal vector pointing into the ambient fluid (e.g., Pozrikidis 2002). The two integrals on the right-hand side of (1.2.10) represent the single- and double-layer potential of Stokes flow.

The Green's function represents the flow due to a point force in an infinite or bounded domain of flow. In the case of flow in free space, the Green's function is given by the Stokeslet,

$$\mathcal{G}_{ij}(\mathbf{x}, \mathbf{x}_0) = -\delta_{ij}\, \ln r + \frac{\hat{x}_i \hat{x}_j}{r^2}, \qquad \mathcal{T}_{ijk}(\mathbf{x}, \mathbf{x}_0) = -4\, \frac{\hat{x}_i \hat{x}_j \hat{x}_k}{r^4}, \qquad (1.2.11)$$

also known as the Oseen tensor, where $\hat{\mathbf{x}} = \mathbf{x} - \mathbf{x}_0$, $r = |\hat{\mathbf{x}}|$, and δ_{ij} is the Kronecker delta representing the identity matrix. Computer programs that evaluate the Green's functions for several boundary geometries, including periodic and nonperiodic flows bounded by an infinite plane wall, are available in the public library BEMLIB (Pozrikidis 2002).

The reciprocal theorem applied at the same point, $\mathbf{x}_0$, for the interior flow provides us with the identity

$$0 = -\frac{1}{4\pi\lambda\mu} \oint_C \mathcal{G}_{ji}(\mathbf{x}_0, \mathbf{x})\, f_i^{(2)}(\mathbf{x})\, \mathrm{d}l(\mathbf{x}) + \frac{1}{4\pi} \oint_C u_i^{(2)}(\mathbf{x})\, \mathcal{T}_{ijk}(\mathbf{x}, \mathbf{x}_0)\, n_k(\mathbf{x})\, \mathrm{d}l(\mathbf{x}). \qquad (1.2.12)$$

Combining (1.2.10) and (1.2.12) to formulate the difference between the disturbance exterior traction and the total interior traction, we find that

$$u_j'(\mathbf{x}_0) = -\frac{1}{4\pi\mu} \oint_C \mathcal{G}_{ji}(\mathbf{x}_0, \mathbf{x})\, [f_i'^{(1)} - f_i^{(2)}](\mathbf{x})\, \mathrm{d}l(\mathbf{x}) + \frac{1}{4\pi} \oint_C [u_i'^{(1)} - \lambda\, u_i^{(2)}](\mathbf{x})\, \mathcal{T}_{ijk}(\mathbf{x}, \mathbf{x}_0)\, n_k(\mathbf{x})\, \mathrm{d}l(\mathbf{x}). \qquad (1.2.13)$$

The reciprocal theorem applied at the same point, $\mathbf{x}_0$, for the incident simple shear flow, $\mathbf{u}^\infty = (ky, 0)$, provides us with the identity

$$0 = -\frac{1}{4\pi\mu} \oint_C \mathcal{G}_{ji}(\mathbf{x}_0, \mathbf{x})\, f_i^\infty(\mathbf{x})\, \mathrm{d}l(\mathbf{x}) + \frac{1}{4\pi} \oint_C u_i^\infty(\mathbf{x})\, \mathcal{T}_{ijk}(\mathbf{x}, \mathbf{x}_0)\, n_k(\mathbf{x})\, \mathrm{d}l(\mathbf{x}). \qquad (1.2.14)$$

Combining (1.2.14) with (1.2.13), demanding that the velocity is continuous across membrane, $\mathbf{u}^{(2)} = \mathbf{u}^{(1)} + \mathbf{u}^\infty = \mathbf{u}$, and adding the incident velocity to both sides, we obtain the desired integral representation,

$$u_j(\mathbf{x}_0) = u_j^\infty(\mathbf{x}_0) - \frac{1}{4\pi\mu} \oint_C \mathcal{G}_{ji}(\mathbf{x}_0, \mathbf{x})\, \Delta f_i(\mathbf{x})\, \mathrm{d}l(\mathbf{x}) + \frac{1-\lambda}{4\pi} \oint_C u_i(\mathbf{x})\, \mathcal{T}_{ijk}(\mathbf{x}, \mathbf{x}_0)\, n_k(\mathbf{x})\, \mathrm{d}l(\mathbf{x}). \qquad (1.2.15)$$

To derive an integral equation, we let the point $\mathbf{x}_0$ approach the capsule contour from the outside and use the identity

$$\lim_{\mathbf{x}_0 \to C} \oint_C u_i(\mathbf{x})\, \mathcal{T}_{ijk}(\mathbf{x}, \mathbf{x}_0)\, n_k(\mathbf{x})\, \mathrm{d}l(\mathbf{x}) \qquad (1.2.16)$$

$$= \oint_C^{PV} u_i(\mathbf{x})\, \mathcal{T}_{ijk}(\mathbf{x}, \mathbf{x}_0)\, n_k(\mathbf{x})\, \mathrm{d}l(\mathbf{x}) + 2\pi\, \mathbf{u}(\mathbf{x}_0),$$

where PV denotes the principal value of the double-layer potential. Rearranging, we obtain the integral equation

$$\frac{1+\lambda}{2}\, u_j(\mathbf{x}_0) = u_j^\infty(\mathbf{x}_0) - \frac{1}{4\pi\mu}\oint_C \mathcal{G}_{ji}(\mathbf{x}_0,\mathbf{x})\,\Delta f_i(\mathbf{x})\,\mathrm{d}l(\mathbf{x})$$
$$+\frac{1-\lambda}{4\pi}\oint_C^{PV} u_i(\mathbf{x})\,\mathcal{T}_{ijk}(\mathbf{x},\mathbf{x}_0)\,n_k(\mathbf{x})\,\mathrm{d}l(\mathbf{x}). \qquad (1.2.17)$$

Simplifications occur in the case of equal viscosities, $\lambda = 1$, as the interfacial velocity is given by an integral representation instead of satisfying an integral equation.

1.3 Numerical method

The numerical method involves solving the integral equation (1.2.17) for the interfacial velocity and then evolving the membrane contour from a specified initial configuration by time integration. The details of the numerical procedure play an important role in the overall performance of the numerical method.

We begin by describing the capsule contour with a set of Lagrangian point particles with a permanent identity parametrized by the arc length in the resting configuration, l_R. The membrane shape, normal vector, and curvature are reconstructed by periodic cubic spline interpolation of the x and y coordinates. The interpolation variable is the instantaneous polygonal arc length, η, computed by connecting pairs of successive marker points with straight lines.

The derivatives $\mathrm{d}l/\mathrm{d}l_R$, $\mathrm{d}m/\mathrm{d}l$, $\mathrm{d}\tau/\mathrm{d}l$, and $\mathrm{d}q/\mathrm{d}l$, are computed by second-order central difference approximations for arbitrarily spaced abscissas (e.g., Pozrikidis 2008). An alternative implementation where these derivatives are computed by cubic spline interpolation provides an entry for numerical instability in the case of significantly deformed membrane shapes.

Smoothing

Smoothing of the membrane tension field with respect to arc length using the five-point formula was enabled in some cases to ensure the longevity of the simulations (Longuet-Higgins & Cokelet 1976). Unwanted numerical oscillations are filtered out using the averaging formula

$$f_i \leftarrow \frac{1}{16}\,(f_{i-2} + 4f_{i-1} + 10f_i + 4f_{i+1} - f_{i+2}), \qquad (1.3.1)$$

where f is a function of interest, and i is an integer label of the interfacial point particles.

1.3.1 Solution of the integral equation

The single- and double-layer Stokes flow potentials over the individual interpolating cubic spline segments were computed using the six-point Gauss-Legendre quadrature. In performing the integration, the velocity components are assumed to vary linearly with respect to the polygonal arc length connecting two consecutive point particles, η. Formally, the interfacial velocity field is expressed in the form

$$\mathbf{u}(\eta) = \sum_{i=1}^{N} \mathbf{u}_i \phi_i(\eta), \tag{1.3.2}$$

where N is the number of point particles, $\mathbf{u}_i$ are the nodal velocities, and ϕ_i are linear, global, tent-like, cardinal interpolation functions. Over the ith element,

$$\mathbf{u}(\eta) = \mathbf{u}_i \phi_1(\omega) + \mathbf{u}_{i+1} \phi_2(\omega), \tag{1.3.3}$$

where $\omega = (\eta - \eta_i)/(\eta_{i+1} - \eta_i)$ is a scaled element variable and ϕ_1 and ϕ_2 are linear element interpolation functions. The first element function ϕ_1 drops linearly from unity at the ith node to zero at the $i+1$ node. The second element function ϕ_2 rises linearly from zero at the ith node to unity at the $i+1$ node.

In the case of two-dimensional flow presently considered, because the integrand of the double-layer potential exhibits a discontinuity across the nodal evaluation point, $\mathbf{x}_0$, special accommodations are not necessary. In contrast, the single-layer potential exhibits a logarithmic singularity that can be subtracted out and then integrated analytically with respect to the polygonal arc length, η.

To demonstrate the desingularization of the single-layer integral, we write

$$I_j^s(\mathbf{x}_0) \equiv \oint_C \mathcal{G}_{ji}(\mathbf{x}_0, \mathbf{x})\, \Delta f_i(\mathbf{x})\, \mathrm{d}l(\mathbf{x})$$
$$= \oint_C \mathcal{G}_{ji}(\mathbf{x}_0, \mathbf{x})\, \Delta f_i(\mathbf{x})\, h(\mathbf{x})\, \mathrm{d}\eta(\mathbf{x}), \tag{1.3.4}$$

where

$$h = \frac{\mathrm{d}l}{\mathrm{d}\eta} = \Big[\Big(\frac{\mathrm{d}x}{\mathrm{d}\eta}\Big)^2 + \Big(\frac{\mathrm{d}y}{\mathrm{d}\eta}\Big)^2 \Big]^{1/2} \tag{1.3.5}$$

is an arc length metric coefficient. Now we write

$$I_j^s(\mathbf{x}_0) = \oint_C \Big(\mathcal{G}_{ji}(\mathbf{x}_0, \mathbf{x})\, \Delta f_i(\mathbf{x})\, h(\mathbf{x}) + \Delta f_j(\mathbf{x}_0)\, h(\mathbf{x}_0)\, \ln|\eta - \eta(\mathbf{x}_0)| \Big)\, \mathrm{d}\eta(\mathbf{x})$$
$$-\Delta f_j(\mathbf{x}_0)\, h(\mathbf{x}_0) \oint_C \ln|\eta - \eta_0|\, \mathrm{d}\eta(\mathbf{x}). \tag{1.3.6}$$

The first integral on the right hand side is nonsingular and can be computed by the Gauss-Legendre quadrature. The second integral is computed by elementary analytical methods.

In the case of fluids with different viscosities, the numerical formulation produces a dense system of linear equations,

$$\mathbf{A} \cdot \mathbf{w} = \mathbf{r}, \tag{1.3.7}$$

where the vector $\mathbf{w}$ contains the x and y components of the point particle velocities, the coefficient matrix $\mathbf{A}$ encapsulates the x and y components of the double-layer potential, and the vector $\mathbf{r}$ encapsulates the x and y components of the single-layer potential. The linear system is solved using the linear solver embedded in MATLAB®.

Adaptive point redistribution

To resolve the membrane shape with adequate accuracy, interfacial point particles are inserted or removed at each time step to ensure that the local aperture angle and distance between successive points lie inside a specified window. In most calculations, the maximum and minimum point separation is set to $0.2a$ and $0.1a$, respectively, and the aperture angle is set to $\pi/4$, where a is the equivalent areal radius defined in Section 1.3. When new points are introduced, the unstressed arc length and curvature are computed by fourth-order local interpolation with respect to the polygonal arc length, η, implemented in terms of Lagrange polynomials.

In the absence of point redistribution, the numerical method is able to describe the capsule evolution only for a limited time due to the onset of numerical instability stemming from inadequate or excessive spatial resolution. In the simulations discussed later in this chapter, the number of point particles ranges from thirty-two to over one hundred.

Membrane point particle advancement

Once the integral equation has been solved, the interfacial point particle position, $\mathbf{X}$, is advanced by integrating in time the equation of point particle motion using the second-order Runge–Kutta method (RK2) with a constant time step (e.g., Pozrikidis 2008)

$$\frac{\mathrm{d}\mathbf{X}}{\mathrm{d}t} = \mathbf{u}(\mathbf{X}). \tag{1.3.8}$$

In the second sub-step of the RK2, the geometry and membrane load are re-evaluated, but point redistribution is disabled to ensure that membrane point particles retain their identity. The time step must be sufficiently small to prevent numerical instability. The maximum time step is determined by the number of point particles and the interface elastic and bending moduli.

1.3.2 MATLAB® code rbc_2d

The MATLAB® code is available in directory `rbc_2d` inside directory `stokes` of the boundary-element software library BEMLIB (Pozrikidis 2002). The main program residing in file `rbc_2d` performs the simulation and simultaneously animates the cell motion. An option is provided for infinite flow, semi-infinite flow bounded by a plane wall, and periodic semi-infinite flow bounded by a plane wall. The type of flow selected affects only the choice of the Green's function in the mathematical formulation.

The code contains the following files encapsulating same-named functions, listed in alphabetical order below:

1. `gauss_leg`
   Base points and weights for the Gauss-Legendre quadrature.

2. `interpolate`
   Four-point Lagrange interpolation used in point redistribution.

3. `prd_2d`
   Adaptive point redistribution around the capsule contour.

4. `rbc_df`
   Computation of the jump in traction across the membrane.

5. `rbc_input`
   Input module for restarting a simulation.

6. `rbc_sdlp_spline`
   Computation of the single- and double-layer potential over a cubic spline segment.

7. `rbc_slp_spline`
   Computation of the single-layer potential over a cubic spline segment.

8. `rbc_velocity`
   Solution of the integral equation for the membrane velocity.

9. `sgf_2d_1p_w`
   Periodic Green's function of Stokes flow for a semi-infinite domain bounded by a plane wall.

10. `sgf_2d_fs`
    Free-space Green's function of Stokes flow.

11. `sgf_2d_w`
    Green's function of Stokes flow for a semi-infinite domain bounded by a plane wall.

12. `smooth`
    Interfacial smoothing.

13. `splc_geo`
    Computation of geometrical variables using cubic spline interpolation.

14. `splc_pr`
    Cubic spline interpolation with periodic boundary conditions.

All physical and flow parameters are set in the main code residing in file `rbc_2d`. The output is recorded in file `rbc_2d.out`.

1.4 Cell shapes and dimensionless numbers

In the absence of flow, normal red blood cells suspended in an isotonic solution assume the shape of biconcave disks. The average cell contour in the first two quadrants of the xy plane is described in parametric form by the equations

$$\frac{x}{a} = c \cos\phi,$$
$$\frac{y}{a} = \frac{c}{2} \sin\phi \; (0.207 + 2.003 \cos^2\phi - 1.123 \cos^4\phi), \tag{1.4.1}$$

where the parameter ϕ varies in the range $0 \le \phi \le 2\pi$, a is the equivalent cell radius defined by the equation $V_c = \frac{4\pi}{3} a^3$, and V_c is the cell volume (e.g., Evans & Fung, 1972). The dimensionless coefficient $c = 1.3858189$ is the ratio of the maximum radius of the biconcave disk to the equivalent cell radius, a. For a normal red blood cell whose mean value of the maximum diameter is approximately 7.82μm, the equivalent cell radius is $a \simeq 2.82\mu$m.

Nondimensionalization

Nondimensionalizing all variables using as characteristic length the equivalent cell radius, a, as characteristic velocity ka, and as characteristic stress μk, we find that the cell deformation depends on the viscosity ratio, λ, and the dimensionless groups

$$G_D = \frac{\mu k a}{E_D}, \qquad G_B = \frac{\mu k a^3}{E_B} = \frac{G_D}{\tilde{E}_B}. \tag{1.4.2}$$

The dimensionless ratio

$$\tilde{E}_B = \frac{E_B}{a^2 E_D} \tag{1.4.3}$$

is a property parameter determining the relative importance of bending and extensional deformation. Physically, the value of $\tilde{E}_B$ defines a threshold for the minimum membrane radius of curvature developing during the deformation.

In the simulations presented in this chapter, we study the capsule deformation as a function of the viscosity ratio, λ, dimensionless shear rate, G_D, and dimensionless bending parameter, $\tilde{E}_B$.

Physical properties

Because the membrane of a real three-dimensional red blood cell is nearly incompressible, the modulus of dilatation is much larger than the shearing modulus. Simulations of three-dimensional cell deformation have shown that the perimeter of the membrane contour and the cross-sectional area in the midplane that is perpendicular to the vorticity vector of the simple shear flow decrease in the course of the deformation in simple shear flow (e.g., Pozrikidis 2003). Physically, stretching of the membrane along the vorticity vector causes a reduction in the circumferential membrane arc length and cross-sectional area. Unless an unphysical elongational flow is artificially imposed driving fluid away from the xy plane, this behavior cannot be reproduced in a two-dimensional model.

It is sensible to set the modulus of dilatation of a two-dimensional membrane equal to a value that is comparable to the shear modulus of a three-dimensional membrane. Substituting in (1.4.3) $E_D = 4.2 \times 10^{-3}$ dyn/cm, $E_B = 1.8 \times 10^{-12}$ dyn cm, and $a \simeq 2.82\mu$m, we find that $\tilde{E}_B = 0.0054$, and conclude that $\tilde{E}_B$ is on the order of 10^{-3}. Unless otherwise specified, results presented in this chapter were obtained with $\tilde{E}_B = 0.001$. For plasma viscosity $\mu = 1$ cp, as the shear rate k increases from 0 to 100 s^{-1}, the dimensionless group G_D increases from 0 to 0.1.

Importance of flexural stiffness

We have mentioned that flexural stiffness is necessary for a deformed two-dimensional capsule to return to the unstressed biconcave shape in the absence of flow. To demonstrate this explicitly, we consider the relaxation of a biconcave capsule deformed into a perfect circle in an otherwise quiescent infinite fluid.

Figure 1.4.1 shows a sequence of evolving profiles during relaxation for $\lambda = 1$ and dimensional bending modulus $\tilde{E} = 0$ and 0.005. These simulations were conducted with a time step $\Delta t E_D/(\mu a) = 0.5$. Profiles are shown at equal time intervals after every 25 time steps. In the absence of flexural stiffness, the capsule evolution is highly sensitive to the number of points

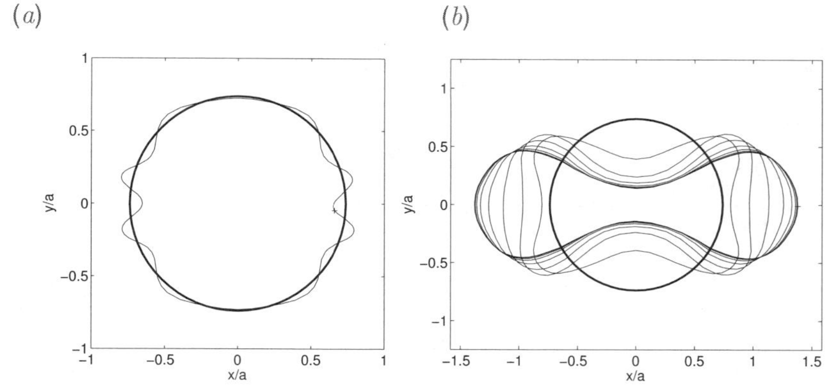

Figure 1.4.1 Relaxation of a biconcave capsule deformed into a circle for dimensionless bending modulus (a) $\tilde{E}_B = 0$ and (b) 0.005. A two-dimensional cell recovers the biconcave shape only in the presence of flexural stiffness.

distributed around the contour, and the membrane evolves to obtain a convoluted shape. Including a small amount of flexural stiffness allows the capsule to return to the biconcave shape through a sequence of interesting transient profiles.

1.5 Capsule deformation in infinite shear flow

We begin by discussing the deformation of a capsule with biconcave unstressed shape in infinite shear flow. Unless otherwise specified, the simulations were conducted using dimensionless time step $k\Delta t = 0.02$.

Effect of the viscosity ratio

Figure 1.5.1(a, b) illustrates the transient deformation of a capsule initially inclined at an angle 45° with respect to the x axis for dimensionless shear rate $G_D = 1/20$ and viscosity ratio $\lambda = 1$ and 5. Membrane profiles are shown at equal time intervals every 50 time steps for $\lambda = 1$, corresponding to scaled time interval $kt = 1.0$, and every 100 time steps for $\lambda = 5$, corresponding to scaled time interval $kt = 2.0$. For clarity, successive profiles have been shifted along the x axis by the arbitrary distance $3a$.

When $\lambda = 1$, the capsule deforms to obtain a nearly steady shape and then exhibits small-amplitude wobbling and flipping fluctuations as the dimple of the unstressed biconcave disk rotates around the perimeter of the

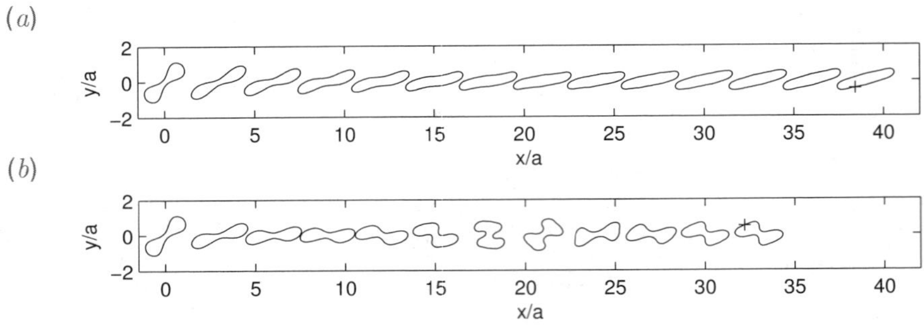

Figure 1.5.1 Deformation of a biconcave capsule initially inclined at 45° with respect to the x axis in infinite simple shear flow for $G_D = 1/20$, $\tilde{E}_B = 0.001$, and viscosity ratio (a) $\lambda = 1$ and (b) 5. Shifted profiles are shown at equal time intervals of $kt = 1.0$ for $\lambda = 1$ and $kt = 2.0$ for $\lambda = 5$. (c).

deformed membrane. When $\lambda = 5$, the capsule performs a complete rotation while undergoing significant changes in shape, and then engages in periodic motion. The evolving shapes shown in figure 1.5.1(b) for $\lambda = 5$ are similar to those of the trace of a three-dimensional biconcave cell in the midplane for the same viscosity ratio and dimensionless shear rate (Pozrikidis 2003). This comparison supports, at least in a qualitative sense, the physical relevance of the two-dimensional model.

Figure 1.5.2(a–d) illustrates the distribution of the membrane curvature, tangential velocity, in-plane tension, and transverse tension around the capsule contour at the end of the simulations presented in figure 1.5.1. The first interfacial marker point defining the origin of arc length around the capsule perimeter at the end of the simulation is marked with a cross in the last profile shown in figure 1.5.1(a, b). The solid and broken lines in figure 1.5.2(a–d) correspond to $\lambda = 1$ and 5. The distribution of curvature around the capsule perimeter depicted in figure 1.5.2(a) is consistent with the shapes shown in figure 1.5.1(a, b). High curvature occurs at the tips and low curvature occurs over the main body of the capsule. A double peak is observed for $\lambda = 5$ corresponding to the sigmoidal shape where regions of negative curvature develop inside the pronounced dimple. The tangential velocity distribution shown in figure 1.5.2(b) is nearly constant for $\lambda = 1$ and shows significant fluctuations for $\lambda = 5$.

The in-plane tension distribution shown in figure 1.5.2(c) can take positive values, corresponding to extension, and negative values, corresponding to compression. Pronounced negative values occur inside the dimple for $\lambda = 5$.

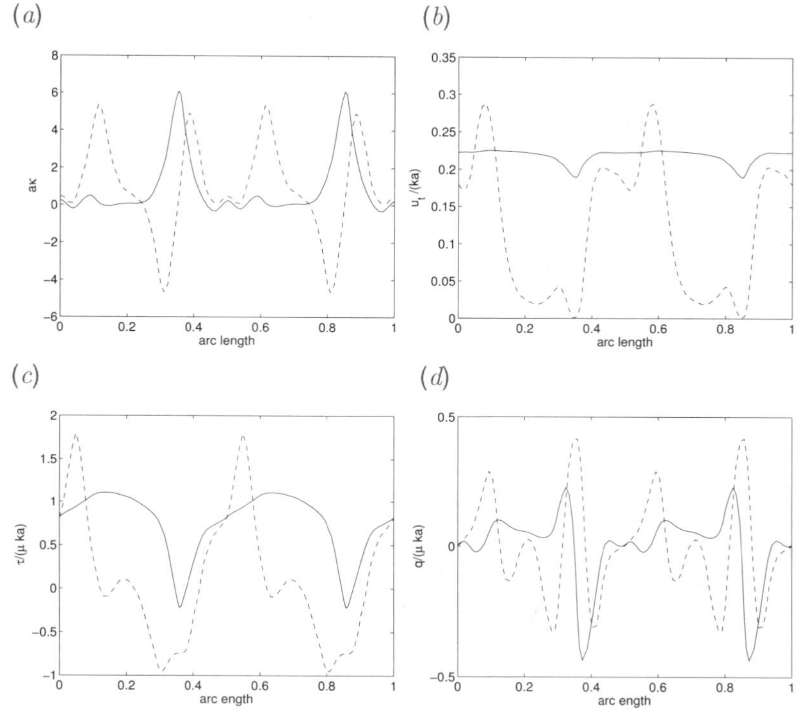

Figure 1.5.2 Distribution of the (a) membrane curvature, (b) tangential velocity, (c) in-plane tension, and (d) transverse tension around the capsule at the end of the simulations shown in figure 1.5.1, plotted with respect to membrane arc length scaled by the capsule perimeter. The solid lines correspond to viscosity ratio $\lambda = 1$ and the dashed lines correspond to viscosity ratio $\lambda = 5$.

Compression induces buckling, but the viscosity of the interior and exterior fluids prevents immediate wrinkling. The slope of the transverse shear tension distribution shown figure 1.5.2(d) fluctuates to give positive and negative bending moments.

The thin lines in figure 1.5.3 describe the evolution of the membrane arc length scaled by the initial value corresponding to the biconcave disk for the simulations described in figure 1.5.1. The medium-thickness lines describe the evolution of the capsule aspect ratio and the heavy lines describe the evolution of the capsule inclination angle divided by π. The solid lines correspond to $\lambda = 1$, and the broken lines correspond to $\lambda = 5$. The capsule aspect ratio is quantified by the ratio of the eigenvalues of the second surface moment of

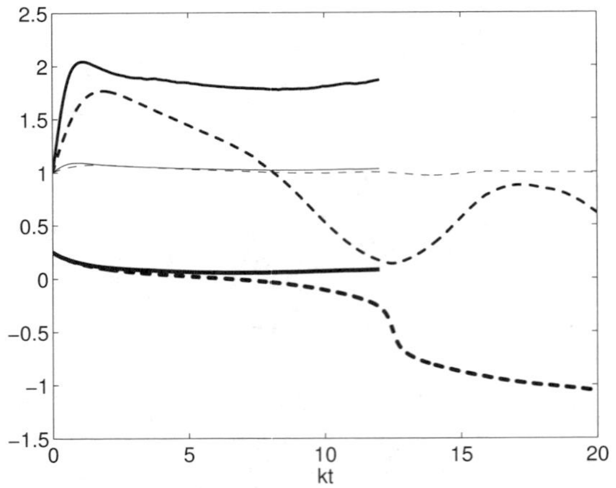

Figure 1.5.3 Evolution of the total membrane length scaled by the initial value (thin lines), capsule aspect ratio scaled by the aspect ratio of the biconcave disk (medium-thickness lines), and capsule inclination angle scaled by π (heavy lines). The solid lines correspond to viscosity ratio $\lambda = 1$ and the broken lines correspond to viscosity ratio $\lambda = 5$.

inertia tensor, and the inclination angle is quantified by the direction of the corresponding orthogonal eigenvectors. The results show that the membrane perimeter increases, reaches a maximum, and then declines while remaining near unity due to the large elastic modulus of dilatation. For $\lambda = 1$, the capsule aspect ratio increases, reaches a maximum, and then slowly fluctuates around a mean value that is higher than unity. For $\lambda = 5$, the aspect ratio varies periodically in a range between zero and unity.

The overall capsule rotation for $\lambda = 5$ is reflected in the sigmoidal shape of the thick broken line in figure 1.5.3, describing the capsule inclination. The scaled period of rotation, kT, is estimated to lie in the range 25–30, which is somewhat higher than that reported for a three-dimensional biconcave capsule, lying in the range 20–25 (Pozrikidis 2003). The longer period is attributed to stronger viscous effects in two-dimensional flow. In the limit of infinite viscosity ratio, the capsule rotates as a rigid particle independent of the interfacial elastic and bending modulus.

Effect of the initial inclination

Further simulations have shown that the capsule motion is insensitive to the initial inclination after an initial start-up period. Figures 1.5.4 and

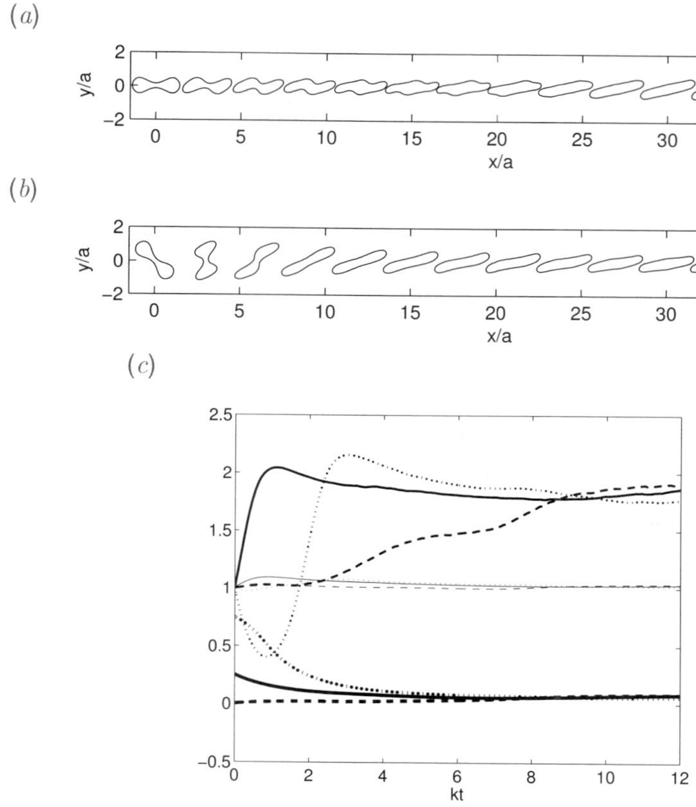

Figure 1.5.4 (*a, b*) Deformation of a biconcave capsule initially parallel to the x axis or inclined at 135° with respect to the x axis for $G_D = 1/20$, $\tilde{E}_B = 0.001$, and $\lambda = 1$. (*c*) Flow diagnostics similar to those presented in figure 1.5.3; the solid, dashed, and dotted lines correspond to initial inclination 45°, 0°, and 135°.

1.5.5 show results for a capsule with viscosity ratio $\lambda = 1$ or 5, respectively, initially aligned with the x axis or inclined by 135° with respect to the x axis. In frames (*a, b*), shifted profiles are shown at a sequence of evenly spaced time intervals corresponding to those shown in figure 1.5.1(*a, b*). The insensitivity of the motion to the initial orientation after an initial transient period is clearly manifested in the flow diagnostics presented in figure 1.5.4(*c*) and 1.5.5(*c*). The high-viscosity capsule described in figure 1.5.5 rotates slowly and quickly tumbles in the vertical orientation where the inclination angle is an odd integer multiple of $\pi/2$. Variations in the capsule shape for a certain

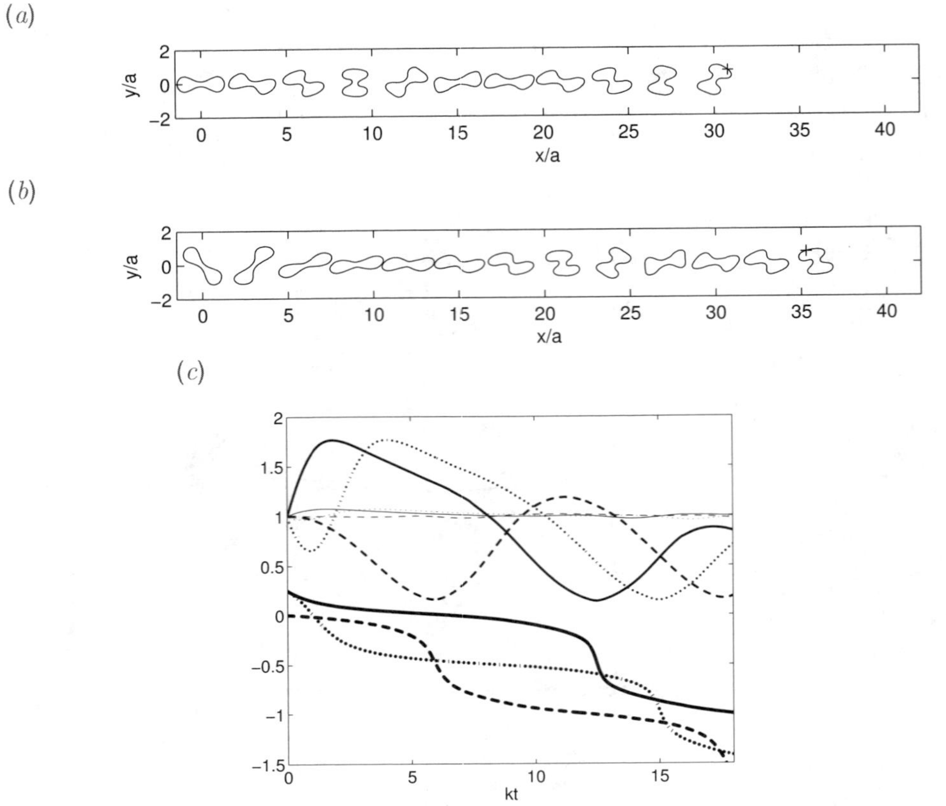

Figure 1.5.5 Same as figure 1.5.4 but for a higher viscosity ratio, $\lambda = 5$.

orientation are attributed to differences in the position of the dimple around the unstressed shape.

It is interesting to compare the transient profiles shown in figure 1.5.1(b) for $\lambda = 5$ to those shown in figure 1.5.4(a) for $\lambda = 1$. In both cases, a traveling hump is observed on the upper and lower surface of the capsule in the horizontal orientation. In the case of high capsule viscosity, the amplitude of the hump amplifies to give the impression of tumbling. In the case of low capsule viscosity, the amplitude of the hump diminishes due to stretching under the action of the shear flow. This observation suggests that capsule kinematics can be described in terms of interfacial wave motion. In this interpretation, a wave with two peaks traveling around a circle with undiminished intensity describes tumbling, whereas a modulated traveling wave describes flipping.

Effect of the shear rate

All simulations presented thus far correspond to dimensionless shear rate $G_D = 1/20$. Figure 1.5.6(*a, b*) shows a sequence of profiles during the deformation of a capsule for a higher and a lower dimensionless shear rate, $G_D = 1/5$ and $1/50$, and equal viscosities, $\lambda = 1$. The three panels in figure 1.5.6(*b*) constitute a continued time series of the same simulation. As in previous figures, successive profiles have been shifted for clarity along the x axis by the arbitrary distance $3a$. Figure 1.5.6(*c*) shows corresponding flow diagnostics illustrating the history of the capsule deformation and inclination.

Increasing or decreasing the shear rate raises or lowers the capsule deformation but does not drastically affect the nature of the motion. Specifically, raising G_D by a factor of four from $1/20 = 0.05$ to $1/5 = 0.2$–solid to dashed lines in figure 1.5.6(*c*)–causes the capsule aspect ratio to increase at long times from around 1.8 to nearly 2.5 times the value corresponding to the biconcave disk, but has a small effect on the capsule orientation. Decreasing the shear rate from $G_D = 0.05$ to 0.02–solid to dotted lines in figure 1.5.6(*c*)–significantly reduces the cell deformation and introduces noticeable oscillations in the capsule inclination, leading to a mild flipping behavior depicted in figure 1.5.6(*b*). We conclude that, as the shear rate is lowered, the capsule motion undergoes a mild qualitative transition.

Figure 1.5.7 shows corresponding results for $\lambda = 5$. Decreasing the dimensionless shear rate G_D to $1/50$ allows the capsule to maintain a nearly biconcave shape throughout the rotation, as shown in figure 1.5.7(*a*). Increasing G_D to $1/5$ causes the capsule contour to be squeezed in the vertical orientation into a double pumpkin or peanut shape. Further increasing G_D by a factor of ten to 2.0 causes the capsule to become significantly elongated in the horizontal orientation. Inspecting the flow diagnostics in figure 1.5.7(*d*) shows that raising the shear rate considerably increases the period of rotation.

The orientation curve for $G_D = 1/5$ in figure 1.5.7(*d*) seemingly indicates a change in the character of the motion from rotation to flipping, in that the inclination angle returns to the initial value after a time period approximately equal to $15/k$. However, because the distinction between rotation and flipping depends on the particular method used to quantify the capsule orientation and is blurred by the convoluted membrane shape, we must rely on a visual impression for an appropriate classification.

Effect of the membrane bending stiffness

All simulations presented thus far correspond to dimensionless bending modulus $\tilde{E}_B = 0.001$. Increasing the bending modulus allows the capsule to retain the unstressed biconcave shape and preserve the dimple throughout the

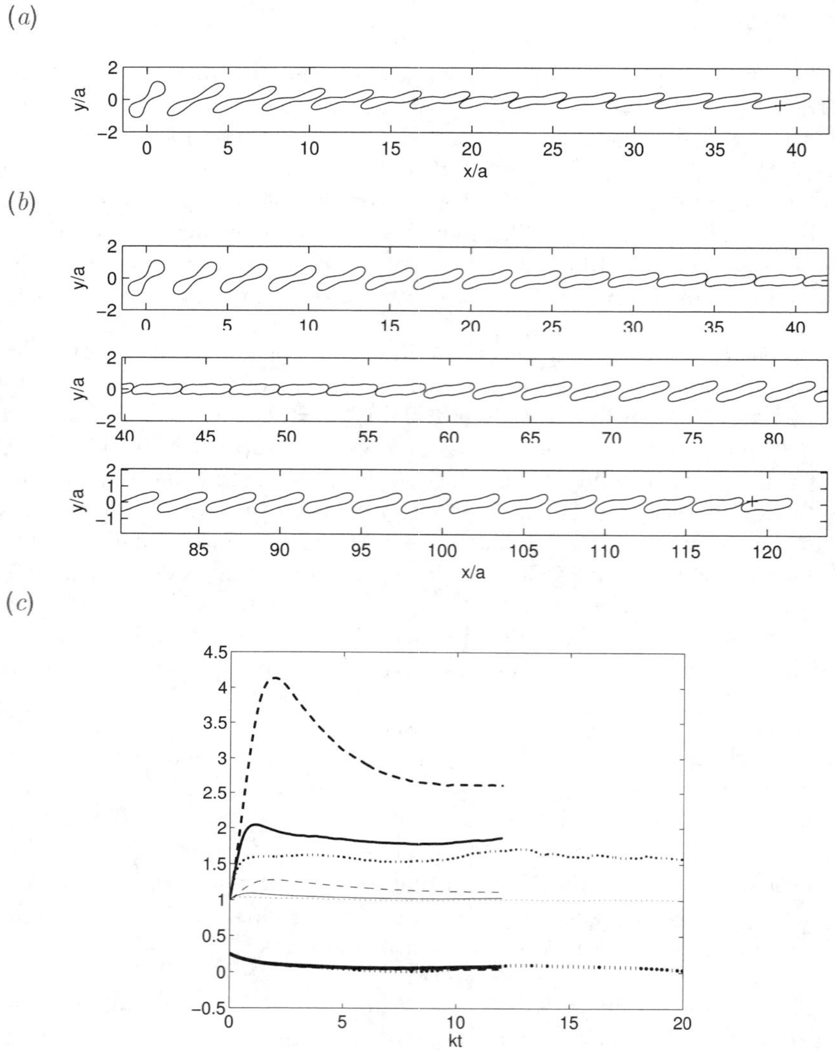

Figure 1.5.6 Stages in the deformation of an inclined biconcave capsule for viscosity ratio $\lambda = 1$ and $\tilde{E}_B = 0.001$, at (*a*) a high dimensionless shear rate, $G_D = 1/5$, and (*b*) a low dimensionless shear rate, $G_D = 1/50$. (*c*) Flow diagnostics similar to those presented in figure 1.5.3. The solid lines correspond to $G_D = 1/20$, the dashed lines correspond to $G_D = 1/5$, and the dotted lines correspond to $G_D = 1/50$.

(*a*)

(*b*)

(*c*)

(*d*)

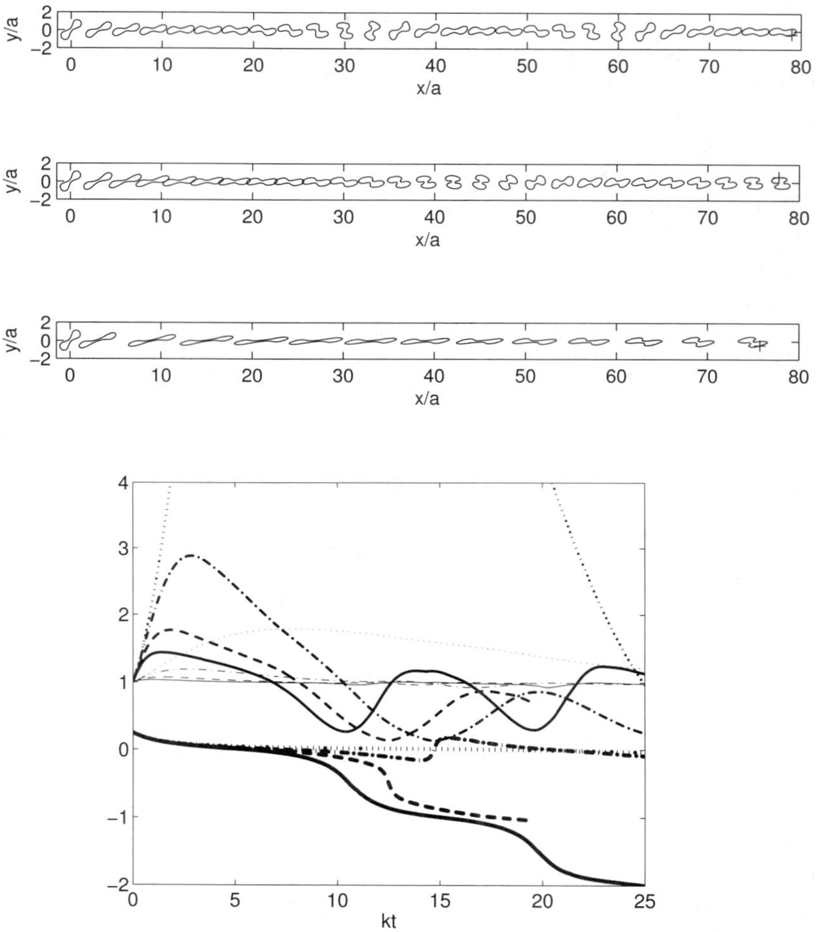

Figure 1.5.7 Stages in the deformation of an inclined biconcave capsule for $\lambda = 5$, $\tilde{E}_B = 0.001$, and dimensionless shear rates increasing by a factor of ten: (*a*) $G_D = 1/50$, (*b*) $1/5$, and (*c*) 2. (*d*) Flow diagnostics similar to those presented in figure 1.5.3; the solid lines correspond to $G_D = 1/20$, the dashed lines correspond to $G_D = 1/5$, the dot-dashed lines correspond to $G_D = 1/5$, and the dotted lines correspond to $G_D = 2$.

motion. Figure 1.5.8 shows simulations for $\lambda = 1$, $G_D = 1/50$, $\tilde{E}_B = 0.01$ and 0.05, corresponding to $G_B = 2$ and 2/5, conducted with time step $k\Delta t = 0.01$. A simulation for a lower bending modulus, $\tilde{E}_B = 0.001$ corresponding to $G_B = 20$, was shown in figure 1.5.1(*b*). Comparing the results shown in these figures, we conclude that increasing the bending modulus causes a radical change in the character of the motion. Lowering the reduced shear rate, G_B, below a threshold that lies somewhere in the range $(2, 20)$ causes a transition from flipping to tumbling.

Bagchi (2007) performed simulations of two-dimensional capsule motion in pressure-driven parabolic channel flow for viscosity $\lambda = 5$. Evolving profiles shown in his figure 2(*a*, *b*) for a single capsule and $\tilde{E}_B = 0.5 \times 10^{-3}$ are strikingly similar to those shown in the present figures 1.5.1(*b*) and 1.5.5(*a*, *b*). Increasing the bending modulus reduces the capsule deformation, in agreement with the present simulations. Bagchi (2007) found that, when the flexural stiffness is sufficiently small, increasing the dimensionless shear rate, E_D, causes a transition from tumbling to tank-treading. Although this behavior is consistent with the trends documented in figure 1.5.8, we have been unable to reproduce the shapes published in Bagchi's figures 2(*c*). It is possible that the particular form of the membrane constitutive equation and the parabolic profile in channel flow are instrumental in enabling this transition; the latter is the most likely cause.

Discussion

Experiments have shown that vesicles and red blood cells suspended in simple shear flow may exhibit three types of motion depending on the shear rate: flipping, tumbling, and an intermittent behavior where the two modes occur in an alternating fashion (Walter *et al.* 2001, Abkarian & Viallat 2008). Tumbling of low-viscosity capsules can be induced by decreasing the shear rate. Skotheim & Secomb (2007) theoretically analyzed the capsule motion and established thresholds for transition between these modes in terms of the intensity of the in-plane membrane rotation. In their analysis, the membrane tank-treading motion is determined by an energy balance that accounts for membrane viscous dissipation. The present simulations show that, when the bending modulus is low compared to the modulus of elasticity, that is, $\tilde{E}_B$ is sufficiently small, the character of the motion is insensitive to the shear rate. However, as $\tilde{E}_B$ is raised, a transition from flipping to tumbling occurs at sufficiently low dimensionless shear rates, G_B. In spite of differences in the underlying physical mechanism, these results are consistent with previous experimental observations and theoretical deductions.

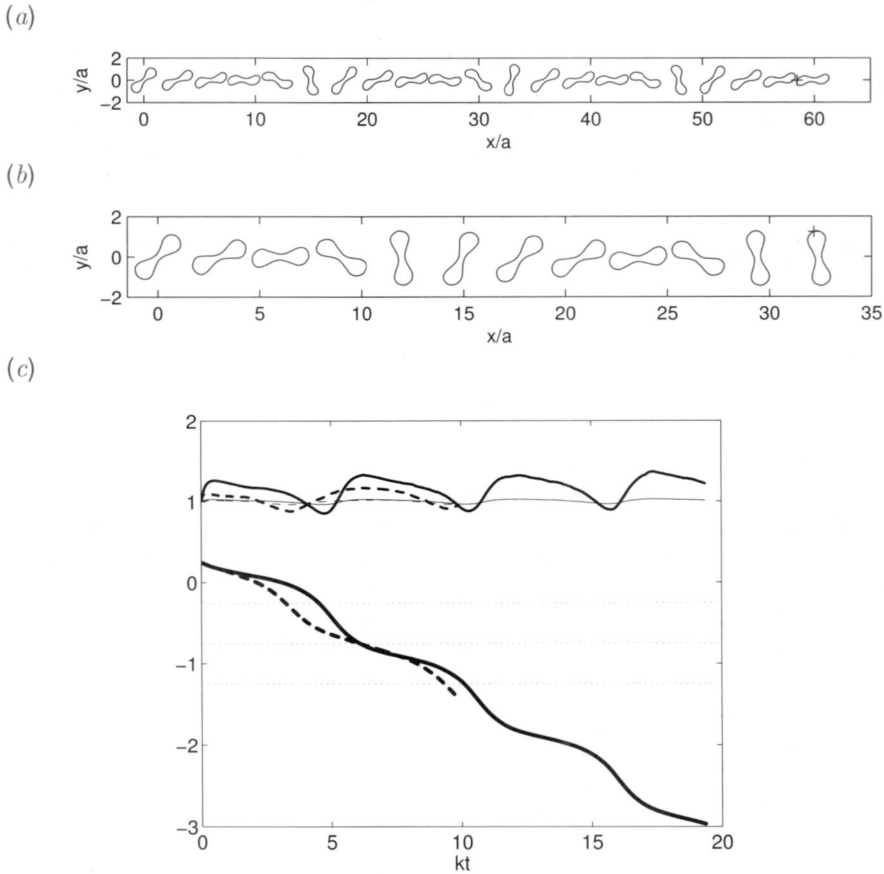

Figure 1.5.8 Capsule rotation for $\lambda = 1$ for a small dimensionless shear rate, $G_D = 1/50$, and a high dimensionless membrane bending modulus (*a*) $\tilde{E}_B = 0.01$ and (*b*) 0.05. In the second case, the capsule rotates nearly undeformed. (*c*) Flow diagnostics similar to those presented in figure 1.5.3; the solid lines correspond to $\tilde{E}_B = 0.01$ and the dashed lines correspond to $\tilde{E}_B = 0.05$. The horizontal dotted lines are separated by $\pi/2$.

1.6 Capsule motion near a wall

In this section, we turn to discussing the capsule motion near an infinite plane wall under the action of simple shear flow, as illustrated in figure 1.2.1(*a*).

Equal fluid viscosities

Figure 1.6.1 shows results of simulations for $\lambda = 1$, dimensionless shear rate $G_D = 1/20$, and dimensionless bending modulus $\tilde{E}_B = 0.001$. At the initial instant, the capsule center is placed at the origin of the x axis at a distance $y_c/a = 1.25$ or 3.0 above the wall. The biconcave shape is initially inclined at an angle $\pi/4$ with respect to the x axis. Figure 1.6.1(a, b) presents sequences of evolving capsule profiles drawn at equal time intervals, $2/k$ or $1/k$. The results show that the lower portion of the capsule initially tends to stick to the wall, but then it separates while the membrane deforms as in the case of infinite shear flow.

Cursory inspection reveals that the capsule migrates away from the wall, and the migration velocity decreases monotonically as the capsule moves farther from the wall. The overall capsule motion is described in figure 1.6.1(c) where graphs of the transverse scaled capsule center position, y_c/a, and scaled streamwise center position reduced by time elapsed since the beginning the simulation, $x_c/(kat)$, are shown. Because the capsule is initially convected under the action of the simple shear flow in the absence of a perturbation flow due to the equal fluid viscosities, the ratio $x_c/(kat)$ is equal to y_c/a at the beginning of the simulation. The graphs reveal that the capsule center position exhibits an initial decline due to a relaxation of the biconcave shape. This is followed by an overall increase, accompanied by small-amplitude fluctuations as the capsule moves away from the wall. Because the velocity of the simple shear flow increases linearly with y_c, the graph of $x_c/(kat)$ mimics that of y_c/a.

The capsule motion is illustrated directly in graphs of the migration velocity, computed by numerically differentiating the capsule center position, against the capsule center position shown in figure 1.6.1(d). The dimensionless migration velocity shown in this figure is scaled by ka. In spite of significant numerical noise, the data clearly indicate an initial decline followed by a significant increase and then an undulating decline. The results in the late part of the simulation for the remote capsule are consistent with those for the capsule near the wall, in that the two plots merge into a universal shape. The data suggest that the migration velocity scales with the inverse of the capsule center from the wall, $\mathrm{d}y_c/\mathrm{d}t \sim 1/y_c$. A two-dimensional force-free capsule induces a disturbance flow that decays like $1/r$, where r is the distance from the particle center. This flow is reflected by the wall to push the capsule away from the wall. In the case of a three-dimensional cell, we expect a weaker rate of migration due to the faster decay of the disturbance flow, $\mathrm{d}y_c\mathrm{d}t \sim y_c^2$.

Figure 1.6.2(a, b) describes the capsule motion for $\lambda = 1$, $\tilde{E}_B = 0.001$, and a higher dimensionless shear rate, $G_D = 1/5$. Two initial capsule center

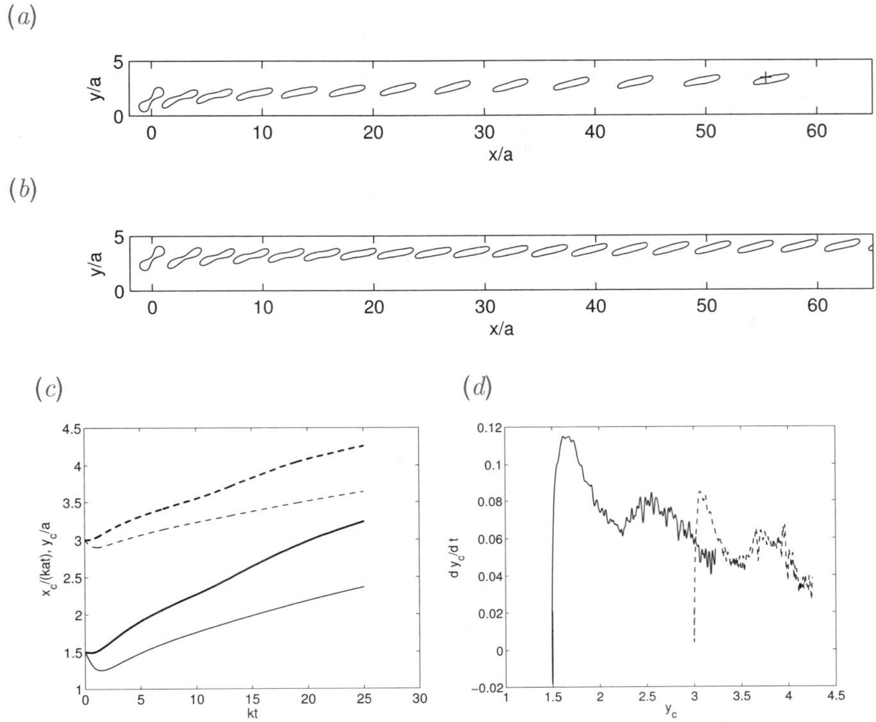

Figure 1.6.1 $(a,\ b)$ Migration of a biconcave capsule in simple shear flow above a plane wall located at $y = 0$ for $G_D = 1/20$, $\tilde{E}_B = 0.001$, $\lambda = 1$, initial capsule inclination $\pi/4$ and center position $y_c/a = 0.15$ and 0.3. (c) Evolution of the scaled capsule center distance from the wall y_c/a (bold lines), and scaled streamwise position, $x_c/(kat)$. (d) Dependence of the migration velocity on the distance from the wall. In frames (c) and (d), the solid lines correspond to initial capsule center position $y_c/a = 0.15$ and the dashed lines correspond to $y_c/a = 0.3$.

positions are considered. Graphs of the transverse capsule center position, y_c/a, axial center position reduced by time, $x_c/(kat)$, and particle center velocity, are presented in figure 1.6.2$(c,\ d)$. The results show that increasing the shear rate reduces the migration velocity after an initial period of adjustment where the capsule loses the biconcave shape. Because of the pronounced membrane stretching, strong oscillations in the migration velocity are not observed. Physically, as the shear rate increases, the capsule elongates to obtain a flat shape which is then convected mostly parallel to the wall.

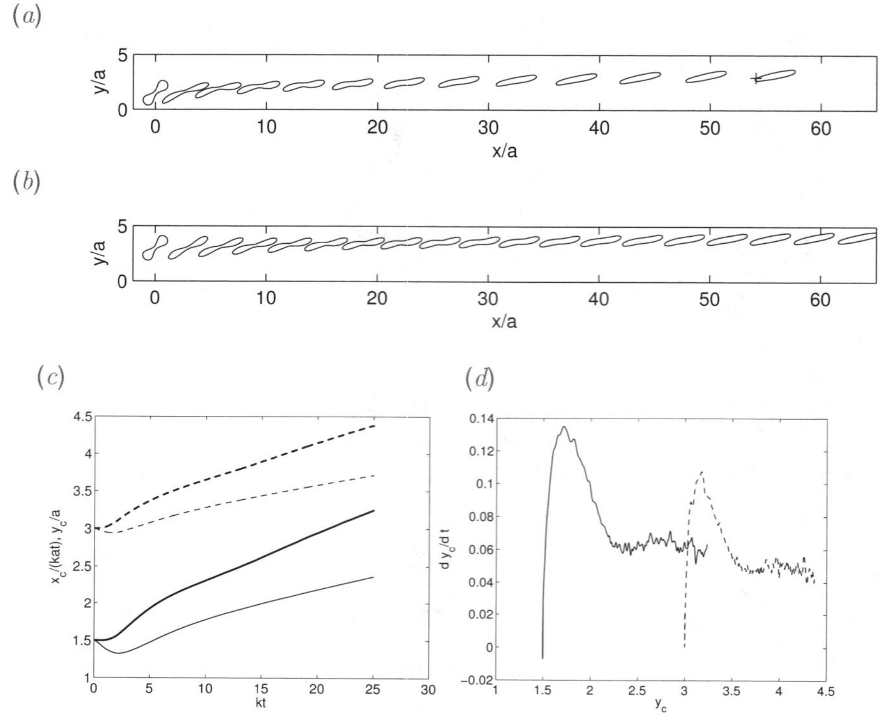

Figure 1.6.2 Same as figure 1.6.1 but for a higher dimensionless shear rate, $G_D = 1/5$.

Effect of the capsule viscosity

Simulations for non-unit viscosity ratio in wall-bounded flow are challenging. Difficulties are traced to the need to solve an integral equation of the second kind for the interfacial velocity, which introduces a reflection flow that amplifies the numerical error. A successful simulation requires careful selection of the number of interfacial marker points, point redistribution parameters, and time step to prevent the onset of numerical instability. In some cases, smoothing of the interfacial profiles is necessary after several hundred time steps in order to filter out numerical oscillations and effectively restart the simulation.

Figure 1.6.3(a) illustrates the motion of a capsule for viscosity ratio $\lambda = 5$, $G_D = 1/20$, and $\tilde{E}_B = 0.005$. At the initial instant, the capsule center is placed at the origin of the x axis at a distance $y_c/a = 1.5$ above

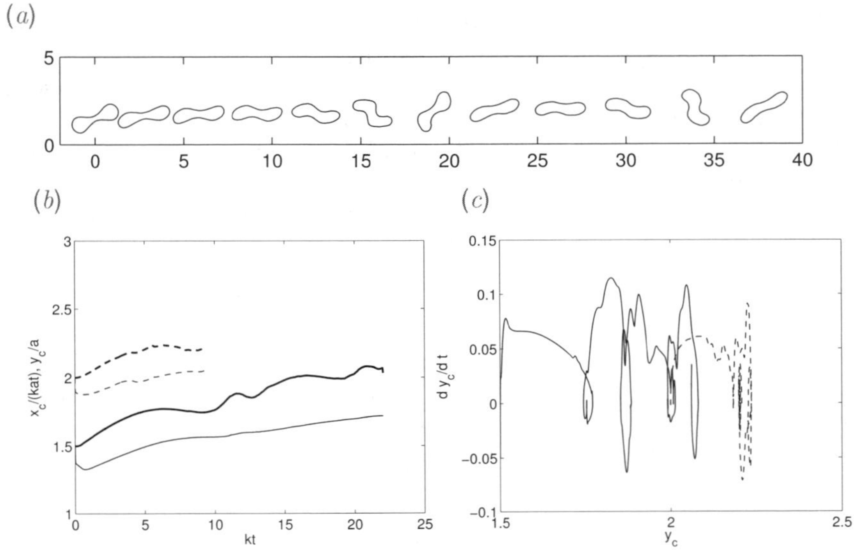

Figure 1.6.3 (*a*) Migration of a biconcave capsule in simple shear flow above a plane wall located at $y = 0$ for $\lambda = 5$, $G_D = 1/20$, $\tilde{E}_B = 0.005$, and initial capsule center position $y_c/a = 0.150$. (*b*) Evolution of the capsule center distance from the wall, y_c/a, (bold lines) and scaled streamwise position, $x_c/(kat)$. (*c*) Dependence of the migration velocity on the distance from the wall. In (*c*) and (*d*), the solid line corresponding to initial capsule center position $y_c/a = 0.150$ and the dashed line to $y_c/a = 2.0$.

the wall. The capsule contour is initially inclined at an angle of $\pi/8$ with respect to the x axis. The simulation requires nearly two days of CPU time on a high-end workstation. The results show that the capsule rotates as in the case of infinite flow discussed earlier in this chapter, while exhibiting a net migration away from the wall. The developing capsule profiles are similar to those of the midplane of a three-dimensional capsule computed by a boundary-element method with corresponding physical and geometrical parameters (Pozrikidis 2003). In the absence of regriding, the simulation of the three-dimensional motion terminates when numerical instability due to insufficient spatial resolution arises.

The motion of the capsule center is described in figure 1.6.3(*c*, *d*). Because of the different fluid viscosities, the ratio x_c/t is less than ky_c at the initial instant. The results shown in figure 1.6.3(*c*) clearly demonstrate a net migration away from the wall due to the capsule deformability, in agreement with Goldsmith's (1971) glass tube observations. The migration velocity becomes negative during periods of time when the capsule quickly rotates under the action of the simple shear flow. Although significant numerical noise is present in the graph of the capsule migration velocity shown in figure 1.6.3(*d*), the data clearly demonstrate systematic fluctuations according to the capsule orientation. Results for initial position $y_c/a = 2$ represented by the dashed lines are consistent with those for the smaller initial distance from the wall.

Bagchi (2007) simulated the migration of a two-dimensional capsule with viscosity ratio $\lambda = 5$ in pressure-driven parabolic channel flow. Results presented in his figure 3 demonstrate that the capsule migrates away from the walls toward the centerline, in agreement with the present predictions. Fluctuations similar to those shown in the present figure 1.6.3(*c, d*) are observed in Bagchi's simulations due to tumbling. The magnitude of the migration velocity is comparable to that predicted by the present simulations.

1.7 Discussion

We have developed and implemented a boundary-element method for simulating the motion and deformation of two-dimensional capsules resembling red blood cells in infinite and semi-infinite shear flow. The formulation takes into consideration the elasticity and flexural stiffness of the capsule membrane but disregards the effect of the membrane viscosity. The numerical results have confirmed that the viscosity ratio is an important parameter for infinite and wall-bounded flow. In the case of wall-bounded flow, the viscosity ratio determines the instantaneous and overall capsule migration velocity over a time interval that is long compared to the capsule period of rotation.

The occurrence of flow-induced migration is consistent with the well-documented formation of a cell-free zone near blood vessel walls responsible for the Fåhraeus-Lindqvist effect describing the dependence of the whole blood viscosity on the vessel diameter. Capsule migration is also responsible for plasma skimming at side branches and bifurcations (Goldsmith 1971). Unfortunately, laboratory data that would allow us to quantitatively confirm the theoretical predictions on the migration velocity in the case of solitary blood cells are not available.

The two-dimensional capsule model discussed in this chapter can be criticized regarding physiological consistency and relevance. First, in the absence of bending moments, a two-dimensional capsule does not return to the reference shape in the absence of flow. Second, the membrane viscosity is irrelevant, as an inextensible two-dimensional membrane cannot produce energy dissipation. A three-dimensional capsule dissipates energy due to the interfacial velocity variations from the midplane to the tips along the z axis, normal to the xy plane. In spite of these shortcomings, comparison with results of three-dimensional simulations have confirmed consistency and good qualitative agreement on several aspects of the motion.

Acknowledgment

This work was supported by the National Science Foundation.

References

ABKARIAN, M. & VIALLAT, A. (2008) Vesicles and red blood cells in shear flow. *Soft Matter.* **4**, 653–657.

BAGCHI, P. (2007) Mesoscale simulation of blood flow in small vessels. *Biophys. J.* **92**, 1858–1877.

BAGCHI, P. & KALLURI, R. M. (2009) Dynamics of nonspherical capsules in shear flow. *Biophys. J.* **80**, 016307.

BAGCHI, P., JOHNSON, P. C. & POPEL, A. S. (2005) Computational fluid dynamic simulation of aggregation of deformable cells in a shear flow. *J. Biomech. Eng.* **127**, 1070–1080.

BREYIANNIS, G. & POZRIKIDIS, C. (2000) Simple shear flow of suspensions of elastic capsules. *Theor. Comp. Fluid Dyn.* **13**, 327–347.

DODDI, S. K. & BAGCHI, P. (2008) Lateral migration of a capsule in a plane Poiseuille flow in a channel. *Int. J. Multiph. Flow* **34**, 966–986.

DODDI, S. K. & BAGCHI, P. (2009) Three-dimensional computational modeling of multiple deformable cells flowing in microvessels. *Phys. Rev. E* **79**, 046318.

DUPIN, M. M., HALLIDAY, I., CARE, C. M., ALBOUL, L. & MUNN, L. L. (2007) Modeling the flow of dense suspensions of deformable particles in three dimensions. *Phys. Rev. E* **75**, 066707.

EGGLETON, C. D. & POPEL, A. S. (1998) Large deformation of red blood cell ghosts in simple shear flow. *Phys. Fluids* **10**, 1834–1845.

EVANS, E. A. & FUNG, Y. C. (1972) Improved measurements of the erythrocyte geometry. *Macrovasc. Res.* **4**, 335–347.

FUNG, Y. C. (1984) *Biodynamics: Circulation.* Springer–Verlag, New York.

GOLDSMITH, H. L. (1971) Red cell motions and wall interactions in tube flow. *Fed. Proc.* **30**, 1578–1590.

HOCHMUTH, R. M. (1982) Solid and liquid behavior of red cell membrane. *Ann. Rev. Biophys. Bioeng.* **11**, 43–55.

KELLER, S. R. & SKALAK, R., (1982) Motion of a tank-treading ellipsoidal particle in a shear flow. *J. Fluid Mech.* **120**, 27–47.

LAC, E., MOREL, A. & BARTHÈS-BIESEL, D. (2007) Hydrodynamic interaction between two identical capsules in shear flow. *J.Fluid Mech.* **573**, 149–169.

LONGUET-HIGGINS, M. S. & COKELET, E. D. (1976) The deformation of steep surface waves on water: I–A numerical method of computation. *Proc. R. Soc. Lond. A* **350**, 1–26.

MISBAH, C. (2006) Vacillating, breathing and tumbling of vesicles under shear flow. *Phys. Rev. Lett.* **96**, 028104.

POZRIKIDIS, C. (1992) *Boundary Integral and Singularity Methods for Linearized Viscous Flow.* Cambridge University Press, New York.

POZRIKIDIS, C. (1997) *Introduction to Theoretical and Computational Fluid Dynamics.* Oxford University Press, New York.

POZRIKIDIS, C. (2002) *A Practical Guide to Boundary Element Methods with the Software Library BEMLIB.* Chapman & Hall/CRC, Boca Raton.

POZRIKIDIS, C. (2003) Numerical simulation of the flow-induced deformation of red blood cells. *Ann. Biomed. Eng.* **31**, 1194–1205.

POZRIKIDIS, C. (2005) Axisymmetric motion of a file of red blood cells through capillaries. *Phys. Fluids* **17**, 031503.

POZRIKIDIS, C. (2008) *Numerical Computation in Science and Engineering.* Second Edition, Oxford University Press, New York.

RIOUAL, F., BIBEN, T. & MISBAH, C. (2004) Analytical analysis of a vesicle tumbling under a shear flow. *Phys. Rev. E* **69**, 061914.

SECOMB, T. W. (2003) Mechanics of red blood cells and blood flow in narrow tubes. In: *Modeling and Simulation of Liquid Capsules and Biological Cells*, Pozrikidis, C. (Ed.), Chapman & Hall/CRC, Boca Raton.

SECOMB, T. W., STYP-REKOWSKA, B. & PRIES, A. R. (2007) Two-dimensional simulation of red blood cell deformation and lateral migration in microvessels. *Ann. Biomed. Eng.* **35**, 755–765.

SKALAK R., TOZEREN A., ZARDA R. P. & CHIEN S. (1973) Strain energy function of red blood cell membrane. *Biophys. J.* **13**, 245–264.

SKOTHEIM, J. M. & SECOMB, T.W. (2007) Red blood cells and other non-spherical capsules in shear flow: Oscillatory dynamics and the tank-treading-to-tumbling transition. *Phys. Rev. Lett.* **98**, 078301.

TAKAGI, S., YAMADA, T., GONG, X. & MATSUMOTO, Y. (2009) The deformation of a vesicle in a linear shear flow. *J. Appl. Mech.* **76**, 021207.

VEERAPANENI, S. K., GUEYFFIER, D., ZORIN, D. & BIROS, G. (2009) A boundary integral method for simulating the dynamics of inextensible vesicles suspended in a viscous fluid in 2D. *J. Comp. Phys.* **228**, 2334–2353.

WALTER, A., REHAGE, H. & LEONHARD, H. (2001) Shear induced deformation of micro-capsules: shape oscillations and membrane folding. *Colloids Surf. A* **123**, 183–185.

ZHANG, J., JOHNSON, P. C. & POPEL, A. S. (2008) Red blood cell aggregation and dissociation in shear flows simulated by lattice Boltzmann method. *J. Biomech.* **41**, 47–55.

2 Flow-induced deformation of artificial capsules

D. Barthès-Biesel, J. Walter, A.-V. Salsac

Biomécanique et Bioingénierie, UMR CNRS 6600
Université de Technologie de Compiègne
Compiègne, France

Two numerical methods are presented for modeling the deformation of a microcapsule suspended in an external flow. The capsule is enclosed by a zero-thickness hyperelastic membrane obeying different non-linear constitutive laws that can be either strain-hardening or strain-softening. In the absence of fluid inertia, a boundary-integral method is used to compute the internal and external Stokes flow for the case of matched fluid viscosities. The force exerted by the capsule membrane on the fluid is obtained from a membrane equilibrium equation with negligible bending stiffness. In the first method, the local membrane equilibrium equation is solved, and the membrane geometry is reconstructed by means of projections on B-spline functions. In the second method, a variational formulation for the membrane equilibrium is developed based on a finite-element implementation. Results obtained by the two methods are compared and discussed for simple shear and planar hyperbolic flows.

2.1 Introduction

A capsule consists of an internal medium enclosed by a semi-permeable membrane that controls the exchange between the internal content and the ambient environment, and thus plays a protective role. Examples of natural capsules include cells, bacteria, and eggs. Artificial capsules are widely used in the pharmaceutical, cosmetic, and food industry to control the release of active principles, aromas and flavors. Capsules are used in bioengineering applications for drug targeting and encapsulation of cell culture for artificial organs (e.g., Kühtreiber, Lanza & Chick 1999).

Artificial capsules can be produced by interfacial polymerization of liquid droplets. The approximately spherical particles obtained using this process are enclosed by a thin polymerized membrane whose mechanical properties depend on the fabrication procedure. In biological applications, membranes typically consist of natural or synthetic polymers such as poly-L-lysine, alginate, and polyacrylates. These capsules are different from vesicles that are enclosed by phospholipid bilayers with nearly constant surface area. A vesicle with a non-spherical initial shape is endowed with excess surface area and can thus easily deform to obtain different geometrical shapes. However, deformation of an initially spherical capsule requires membrane stretching at some elastic energy cost.

In many situations of interest, capsules are suspended in another liquid and may be subjected to hydrodynamic forces in flow. The motion of the suspending and internal fluids generates viscous stresses and surface tractions on either side of the membrane that may lead to capsule breakup. Controlling this process is essential for successfully designing artificial capsules and ensuring that the capsules remain intact. Models of the underlying mechanics are difficult to develop because of the involved complex fluid structure interaction on an interface with unknown geometry. The capsule membrane is subjected not only to large displacement, but also to large deformation requiring a nonlinear constitutive law.

For simplicity, we consider artificial capsules with a Newtonian liquid core enclosed by a thin hyperelastic membrane freely suspended in another Newtonian liquid. The flow-induced capsule deformation has been studied extensively by a number of techniques in the past three decades. In the absence of fluid inertia, the Stokes velocity field at any point in the fluid can be evaluated in terms of surface integrals defined on the flow domain boundaries, including the capsule membrane (e.g., Li, Barthès-Biesel & Helmy 1988, Ramanujan & Pozrikidis 1998, Lac *et al.* 2004, Dodson & Dimitrakopoulos 2008). This approach reduces the geometric dimension of the problem with respect to the physical dimension by one unit. Efficient, accurate, and stable numerical methods have been developed based on this formulation to describe the deformation of capsules subjected to shear flows.

Immersed-boundary methods have been developed to solve the fluid flow problem at nonzero Reynolds numbers (e.g., Eggleton & Popel 1998, Doddi & Bagchi 2008a, Li & Sarkar 2008). In the numerical implementation, a stationary three-dimensional grid is introduced for the fluid flow, and a moving two-dimensional boundary grid is set for the interface. The forces exerted by the membrane on the fluid and the flow velocity advecting the membrane are transferred between the two grids by smoothed Dirac delta functions (Chapter 5). This introduces some fuzziness in the membrane position that may cause the loss of precision during Lagrangian interfacial tracking. A significant advantage of the immersed-boundary method compared to the boundary-integral method is that it is able to describe nonzero Reynolds number and non-Newtonian flow.

Two approaches are available for modeling the capsule membrane mechanics. In the first approach, the equations of force equilibrium on the capsule wall are enforced locally at each point (strong form). In the second approach, the equilibrium equations are multiplied by a test function and then integrated over the capsule surface to give a variational formulation (weak form). Most capsule studies have been conducted based on the strong form.

Numerical simulations have been performed for axisymmetric configurations using a collocation technique to determine the unknown forces and velocities at discrete locations in elongational flow (Li, Barthès-Biesel & Helmy 1988, Pozrikidis 1990) and pore flow (Leyrat-Maurin & Barthès-Biesel 1994, Quéguiner & Barthès-Biesel 1997, Diaz & Barthès-Biesel 2002). The full three-dimensional problem has been solved for infinite-flow configurations. Capsules in simple shear flow have been considered by Pozrikidis (1995) and Ramanujan & Pozrikidis (1998) who computed the membrane load as a piecewise constant function. More recently, Li & Sarkar (2008) also used this method. Lac *et al.* (2004) and Lac & Barthès-Biesel (2005) implemented instead a projection technique on bicubic B-spline functions in order to compute with high accuracy the membrane loads. Dodson & Dimitrakopoulos (2008) chose a spectral discretization of the interface.

An alternative is to use a finite-element method based on the weak formulation of the membrane equilibrium equations. Two groups have implemented a finite-element method coupled with an immersed-boundary method to compute the motion of a capsule in shear flow (Eggleton & Popel 1998, Doddi & Bagchi 2008a, 2008b). However, their implementation employs the linear interpolation functions of Charrier *et al.* (1989) and lacks generality. Recently, Walter *et al.* (2010) developed a novel method where the boundary-integral formulation of the Stokes flow equations is coupled with a finite-element formulation of the membrane. The combined numerical technique was shown to be both precise and stable.

In this chapter, we present the salient features of the mechanical behavior of a capsule deforming under the influence of viscous forces in infinite flow, assuming negligible inertia forces for the fluid and membrane motion. The flow of the internal and external liquids is described by the boundary-integral form of the Stokes equation. Two coupling strategies are presented based on the strong or weak formulation of the membrane equations. In the first approach, the membrane is discretized into a structured mesh and all quantities

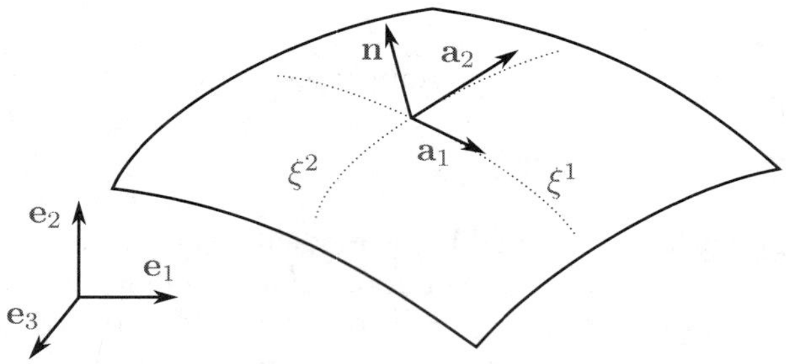

Figure 2.2.1 Illustration of a local covariant base $(\mathbf{a}_1, \mathbf{a}_2, \mathbf{n})$ following the surface deformation. The Cartesian frame $(\mathbf{e}_1, \mathbf{e}_2, \mathbf{e}_3)$ is fixed.

of interest are projected onto B-spline functions. In the second approach, an unstructured mesh is used together with a finite-element method to compute the load exerted by the membrane on the fluids.

We first present the mechanics of thin membranes undergoing large displacements and deformations. Different nonlinear constitutive laws will be used to describe the rheological behavior of the membrane. We then consider the fluid-structure coupled problem when a single capsule is freely suspended in a linear shear flow. The two numerical models are compared and their advantages and disadvantages are discussed for the specific case of an initially spherical capsule suspended in linear shear flow.

2.2 Membrane mechanics

When the thickness of a capsule membrane is small compared to the capsule dimensions and typical radius of curvature, the membrane can be modeled as a hyperelastic *surface* devoid of bending resistance (e.g., Skalak *et al.* 1973). Even with this simplification, quantifying the capsule deformation is a complicated geometrical problem involving the description of curved surfaces and their deformation. We will only outline the basic concepts and provide references to monographs for mathematical details (e.g., Green & Adkins 1960, Crisfield 1997). Throughout the chapter, summation is performed over repeated indices. Greek indices take values 1 or 2, and Latin indices take values 1, 2 or 3.

2.2.1 Membrane deformation

The position of a point P on the membrane surface S is determined by two surface curvilinear coordinates, (ξ^1, ξ^2). It is convenient to use two coordinate bases, as depicted in figure 2.2.1. The first is a fixed Cartesian base,

$(\mathbf{e}_1, \mathbf{e}_2, \mathbf{e}_3)$, corresponding to position $\mathbf{x}(\xi^1, \xi^2)$. The second is a local covariant base, $(\mathbf{a}_1, \mathbf{a}_2, \mathbf{a}_3)$, following the surface deformation. The first two base vectors $(\mathbf{a}_1, \mathbf{a}_2)$ are tangential to lines of constant ξ^α,

$$\mathbf{a}_\alpha = \mathbf{x}_{,\alpha}, \quad \alpha = 1, 2, \tag{2.2.1}$$

where the notation $\cdot_{,\alpha}$ denotes a derivative with respect to ξ^α. The third base vector $\mathbf{a}_3 = \mathbf{n}$ is the unit normal vector pointing outward. The associated contravariant base, $(\mathbf{a}^1, \mathbf{a}^2, \mathbf{a}^3)$, is defined by $\mathbf{a}^\alpha \cdot \mathbf{a}_\beta = \delta^\alpha_\beta$, where δ^α_β is the Kronecker tensor. A vector $\mathbf{v}$ can then be described as

$$\mathbf{v} = v^i \, \mathbf{a}_i = v_i \, \mathbf{a}^i = v_{X_i} \mathbf{e}_i \,, \tag{2.2.2}$$

where X_i are the Cartesian components. The covariant and contravariant metric tensors on S are

$$a_{\alpha\beta} = \mathbf{a}_\alpha \cdot \mathbf{a}_\beta, \qquad a^{\alpha\beta} = \mathbf{a}^\alpha \cdot \mathbf{a}^\beta. \tag{2.2.3}$$

A differential surface element can be expressed in the form

$$\mathrm{d}S = \sqrt{|a_{\alpha\beta}|} \, \mathrm{d}\xi^1 \mathrm{d}\xi^2, \tag{2.2.4}$$

where $|a_{\alpha\beta}| = a_{11} a_{22} - a_{12}^2$ is the determinant of $a_{\alpha\beta}$. The curvature tensor is $b_{\alpha\beta} = \boldsymbol{n} \cdot \boldsymbol{a}_{\alpha,\beta}$.

A membrane material point is identified by its coordinates (ξ^1, ξ^2) corresponding to the position $\mathbf{X}(\xi^1, \xi^2)$ in the reference state and to the position $\mathbf{x}(\xi^1, \xi^2, t)$ in the deformed state. By convention, all quantities in the reference state are denoted by capital letters. Assuming negligible bending stiffness, deformation occurs only in the plane of the membrane. A material vector normal to the surface remains normal to the surface after deformation. The gradient of the transformation $\mathbf{F}$ is defined by

$$\mathrm{d}\mathbf{x} = \mathbf{F} \cdot \mathrm{d}\mathbf{X}, \tag{2.2.5}$$

and is given by

$$\mathbf{F} = \mathbf{a}_\alpha \otimes \mathbf{A}^\alpha, \tag{2.2.6}$$

where $\mathbf{A}^\alpha$ are the curvilinear base vectors in the undeformed state. The local deformation of the surface can be measured by the Cauchy–Green dilation tensor

$$\mathbf{C} = \mathbf{F}^T \cdot \mathbf{F}, \tag{2.2.7}$$

or by the Green–Lagrange strain tensor

$$\mathbf{e} = \frac{1}{2}(\mathbf{C} - \mathbf{I}), \tag{2.2.8}$$

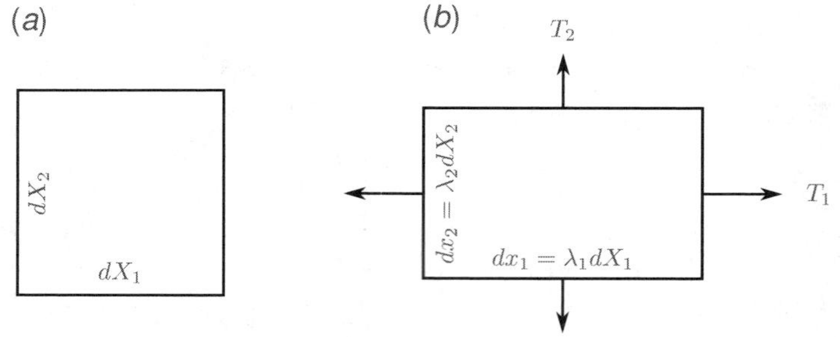

Figure 2.2.2 Deformation of a membrane element along the principal axes. (*a*) Reference state with dimensions d$X_1 \times$ dX_2; (*b*) deformed state with extension ratios λ_1 and λ_2 and principal tensions T_1 and T_2.

where **I** is the two-dimensional identity tensor. The invariants of the transformation are

$$I_1 = \text{tr}\,\mathbf{C} - 2 = A^{\alpha\beta} a_{\alpha\beta} - 2, \quad I_2 = \det\,\mathbf{C} - 1 = |A^{\alpha\beta}||a_{\alpha\beta}| - 1\,. \tag{2.2.9}$$

The membrane deformation can also be quantified by the principal dilation ratios λ_1 and λ_2 in its plane, as shown in figure 2.2.2. The deformation invariants are then

$$I_1 = \lambda_1^2 + \lambda_2^2 - 2, \qquad I_2 = \lambda_1^2\lambda_2^2 - 1 = J_s^2 - 1. \tag{2.2.10}$$

The Jacobian, $J_s = \det(\mathbf{F}) = \lambda_1\lambda_2$, expresses the ratio of the deformed to the undeformed surface areas.

2.2.2 Membrane constitutive laws and equilibrium

Elastic stresses in an infinitely thin membrane are replaced by elastic tensions corresponding to forces per unit arc length measured in the plane of the membrane. Deformation is measured by the relative extension and distortion of lines in the plane of the membrane, as discussed in the previous section. Assuming that the membrane is a two-dimensional isotropic material, we relate the Cauchy tension tensor, **T**, to a strain energy function per unit area of undeformed membrane, $w_s(I_1, I_2)$, by the equation

$$\mathbf{T} = \frac{1}{J_s}\mathbf{F} \cdot \frac{\partial w_s}{\partial \mathbf{e}} \cdot \mathbf{F}^T\,. \tag{2.2.11}$$

Using the chain rule, we obtain the following expression for the contravariant representation of **T** (Green & Adkins 1960)

$$T^{\alpha\beta} = \frac{2}{J_s}\frac{\partial w_s}{\partial I_1} A^{\alpha\beta} + 2J_s \frac{\partial w_s}{\partial I_2} a^{\alpha\beta}\,. \tag{2.2.12}$$

A number of laws are available for modeling thin hyperelastic membranes (e.g., Oden 1972). Different material behavior can be described for large deformation, including either the strain-softening behavior of gelled membranes exhibiting rubber-like elasticity, or the strain-hardening behavior exhibited by membranes made of a polymerized network with strong covalent links. Simple laws with constant material coefficients are considered in our analysis.

In the limit of small deformation, all laws reduce to the two-dimensional Hooke's law (H)

$$w_s^H = G_s \left(\mathrm{tr}(\boldsymbol{\epsilon}^2) + \frac{\nu_s}{1-\nu_s} (\mathrm{tr}\boldsymbol{\epsilon})^2 \right) , \qquad (2.2.13)$$

where $\boldsymbol{\epsilon}$ is the two-dimensional linearized Green–Lagrange strain tensor and G_s is the surface shear elastic modulus. The surface Poisson ratio, $\nu_s \in]-1,+1[$, is related to the area dilation modulus by $K_s = G_s(1+\nu_s)/(1-\nu_s)$. An area-incompressible membrane thus corresponds to $\nu_s = 1$.

The widely used neo-Hookean law (NH) describes the behavior of an infinitely thin sheet of a three-dimensional isotropic and incompressible material,

$$w_s^{NH} = \frac{G_s^{NH}}{2} \left(I_1 - 1 + \frac{1}{I_2+1} \right) . \qquad (2.2.14)$$

Because of volume incompressibility, area dilation is balanced by membrane thinning. The area dilation modulus K_s^{NH} is then shown to be $3G_s^{NH}$ (Barthès-Biesel, Diaz & Dhenin 2002). Another law (SK) has been derived by Skalak *et al.* (1973) for two-dimensional materials with independent surface shear and area dilation moduli,

$$w_s^{SK} = \frac{G_s^{SK}}{4} \left(I_1^2 + 2I_1 - 2I_2 + CI_2^2 \right) , \quad C > -1/2 . \qquad (2.2.15)$$

The area dilation modulus is $K_s^{SK} = G_s^{SK}(1+2C)$. The SK law was initially designed to model the area-incompressible membrane of biological cells, such as red blood cells corresponding to $C \gg 1$. However, the law is very general and can be used to also model other types of membranes for which the shear and area dilation moduli are of the same order of magnitude, as in the case of alginate membranes (Carin *et al.* 2002).

Equivalence between the laws requires

$$G_s = G_s^{NH} , \qquad \nu_s = \frac{1}{2} \qquad (2.2.16)$$

for the neo-Hookean law, and

$$G_s = G_s^{SK}, \qquad \nu_s = \frac{C}{1+C} \tag{2.2.17}$$

for the Skalak law. When $C = 1$, the NH and SK laws predict the same behavior in small deformation of a membrane with $K_s = 3G_s$, corresponding to $\nu_s = 1/2$. However, different nonlinear tension–strain relations are obtained under large deformations. In particular, it is easily checked that the NH law is strain-softening under uniaxial stretching, whereas the SK law is strain-hardening (Barthès-Biesel, Diaz & Dhenin 2002).

Membrane equilibrium

For an infinitely thin membrane, inertia can be neglected and the membrane mechanics is governed by the local equilibrium equation

$$\nabla_s \cdot \mathbf{T} + \mathbf{q} = \mathbf{0}, \tag{2.2.18}$$

where $\mathbf{q}$ is the external load and $\nabla_s \cdot$ is the surface divergence operator in the deformed configuration.

In surface curvilinear coordinates, the load vector is given by $\boldsymbol{q} = q^\beta \boldsymbol{a}_\beta + q^n \mathbf{n}$. The equilibrium equation (2.2.18) is resolved into its tangential and normal components,

$$\begin{cases} \dfrac{\partial T^{\alpha\beta}}{\partial \xi^\alpha} + \Gamma_{\alpha\,\lambda}^{\,\alpha} T^{\lambda\beta} + \Gamma_{\alpha\,\lambda}^{\,\beta} T^{\alpha\lambda} + q^\beta = 0, \qquad \beta = 1, 2, \\ \\ T^{\alpha\beta} b_{\alpha\beta} + q^n = 0, \end{cases} \tag{2.2.19}$$

where $T^{\alpha\beta}$ is the twice-contravariant representation of the tension tensor and $\Gamma_{\alpha\,\lambda}^{\,\beta} = \boldsymbol{a}_{\alpha,\lambda} . \boldsymbol{a}^\beta$ are the Christoffel symbols.

Equation (2.2.18) can be written in a weak form using the virtual work principle: for any virtual displacement field $\hat{\mathbf{u}}$, the virtual work of external and internal forces balances to zero,

$$\iint_S \hat{\mathbf{u}} \cdot \mathbf{q}\, \mathrm{d}S - \iint_S \hat{\epsilon}(\hat{\mathbf{u}}) : \mathbf{T}\, \mathrm{d}S = 0, \tag{2.2.20}$$

where $\hat{\epsilon}(\hat{\mathbf{u}}) = \frac{1}{2}\left(\nabla_s \hat{\mathbf{u}} + \nabla_s \hat{\mathbf{u}}^T\right)$ is the virtual deformation tensor.

2.2.3 Osmotic effects and prestress

A positive pressure difference may exist between the capsule interior and exterior, particularly in bioengineering applications involving semipermeable membranes; that is, membranes that are permeable to small molecules, such as water or small ions, but impermeable to large molecules (e.g., Kühtreiber, Lanza & Chick 1999). The pressure difference established on either side of the membrane is then due to osmotic effects. In the case of a simple capsule consisting of a droplet of saline solution enclosed by an alginate membrane, partial dissolution of the membrane occurs leading to an unknown concentration of large molecules trapped inside the capsule (Sherwood *et al.* 2003). The internal concentration is then usually underestimated. When the capsule is suspended in a saline solution whose concentration is nominally the same as that of the internal medium, a concentration jump is often established across the membrane leading to osmotic pressure.

If a positive osmotic pressure difference $p^{(0)}$ exists between the internal and external phase, the membrane of an initially spherical capsule is prestressed by an isotropic elastic tension $T^{(0)}$ given by the Laplace law,

$$T_1 = T_2 = T^{(0)} = \frac{1}{2} a p^{(0)}, \tag{2.2.21}$$

where a is the radius of the inflated capsule. The membrane is then stretched with an initial isotropic elongation $\lambda_1 = \lambda_2 = a/a_0 = 1 + \alpha$, where a_0 is the capsule radius in the unstressed configuration. The relation between $T^{(0)}$ and α depends on the membrane constitutive law (Lac & Barthès-Biesel 2005). In the limit of small inflation, $T^{(0)} = 6\alpha\, G_s$.

2.3 Capsule dynamics in flow

Consider a capsule with typical dimension a, filled with an incompressible Newtonian liquid and enclosed by an infinitely thin membrane with surface shear elastic modulus G_s and areal dilation modulus K_s. The capsule is freely suspended in another incompressible Newtonian liquid undergoing shear flow with velocity $\mathbf{v}^\infty(\mathbf{x})$, as shown in figure 2.3.1. The external (superscript 1) and internal (superscript 2) liquids are Newtonian with viscosity μ and $\lambda\mu$, respectively, and equal density ρ. Gravitational effects are thus neglected as the capsule is assumed to be neutrally buoyant. The characteristic velocity and shear rate are V^∞ and $\dot{\gamma}$, respectively. The Reynolds number of the flow based on the capsule dimension is defined as $\mathrm{Re} = \rho V^\infty a/\mu$.

At low Reynolds numbers, $\mathrm{Re} \ll 1$, the motion of the internal and external liquids is governed by the equations of Stokes flow. In the laboratory Cartesian frame of reference $(\mathbf{e}_1, \mathbf{e}_2, \mathbf{e}_3)$ centered at the capsule center of mass, as shown in figure 2.3.1, the governing equations are

$$\nabla p^{(k)} = \mu^{(k)} \nabla^2 \mathbf{v}^{(k)}, \qquad \nabla \cdot \mathbf{v}^{(k)} = 0, \quad k = 1, 2, \tag{2.3.1}$$

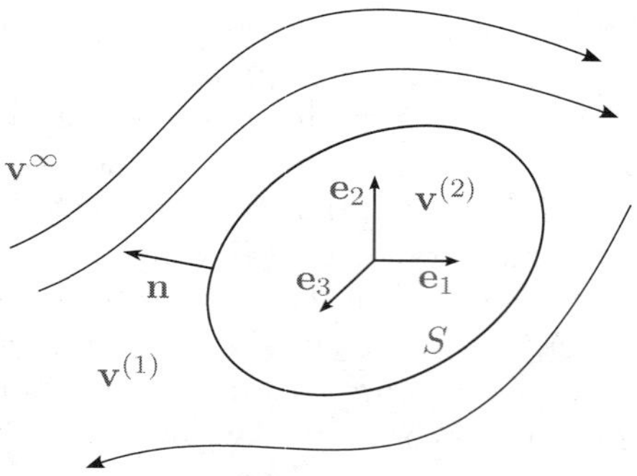

Figure 2.3.1 Schematic illustration of an isolated capsule freely suspended in simple shear flow.

Far from the capsule, the disturbance velocity vanishes,

$$\mathbf{v}^{(1)}(\mathbf{x}) \to \mathbf{v}^\infty \quad \text{as} \quad |\mathbf{x}| \to \infty. \tag{2.3.2}$$

An alternative to (2.3.1) and (2.3.2) is to express the interfacial velocity in terms of an integral equation defined over the deformed capsule surface,

$$\frac{1+\lambda}{2}\,\mathbf{v}(\mathbf{x}) = \mathbf{v}^\infty(\mathbf{x}) - \frac{1}{8\pi\mu}\iint_S \boldsymbol{\mathcal{G}}(\mathbf{x},\mathbf{y})\cdot[\boldsymbol{\sigma}(\mathbf{y})]\cdot\mathbf{n}(\mathbf{y})\,\mathrm{d}S(\mathbf{y})$$
$$+\frac{1-\lambda}{8\pi}\iint_S^{PV} \mathbf{v}(\mathbf{y})\cdot\boldsymbol{\mathcal{T}}(\mathbf{x},\mathbf{y})\cdot\mathbf{n}(\mathbf{y})\,\mathrm{d}S(\mathbf{y}). \tag{2.3.3}$$

where $\boldsymbol{\sigma}$ is the stress tensor, $\mathbf{n}$ is the outward unit normal vector to the membrane surface, S, PV denotes the principal value, and

$$[\boldsymbol{\sigma}(\mathbf{x})]\cdot\mathbf{n} = [\boldsymbol{\sigma}^{(1)}(\mathbf{x}) - \boldsymbol{\sigma}^{(2)}(\mathbf{x})]\cdot\mathbf{n} \tag{2.3.4}$$

is the jump in traction across the interface (e.g., Pozrikidis 1992). The Oseen tensor $\boldsymbol{\mathcal{G}}$ and associated stress tensor $\boldsymbol{\mathcal{T}}$ are defined as

$$\mathcal{G}_{ij}(\mathbf{x},\mathbf{y}) = \frac{\delta_{ij}}{|r|} + \frac{r_i\,r_j}{|r|^3}, \qquad \mathcal{T}_{ijk} = 6\,\frac{r_i r_j r_k}{|r|^5}, \tag{2.3.5}$$

where $\mathbf{r} = \mathbf{x} - \mathbf{y}$ and $r = \|\mathbf{r}\|$, and integration is performed over the instantaneous deformed capsule surface S.

It is henceforth assumed that the fluid viscosities are equal, $\lambda = 1$. The integral equation for the interfacial velocity simplifies into the integral representation

$$\mathbf{v}(\mathbf{x}) = \mathbf{v}^\infty(\mathbf{x}) - \frac{1}{8\pi\mu}\iint_S \boldsymbol{\mathcal{G}}(\mathbf{x},\mathbf{y})\cdot[\boldsymbol{\sigma}(\mathbf{y})]\cdot\mathbf{n}(\mathbf{y})\,\mathrm{d}S(\mathbf{y}). \tag{2.3.6}$$

Two fluid-structure coupling conditions are introduced next.

A kinematic condition requires continuity of the membrane velocity and of the internal and external fluid velocities at the interface,

$$\mathbf{v}^{(1)}(\mathbf{x},t) = \mathbf{v}^{(2)}(\mathbf{x},t) = \mathbf{v}(\mathbf{x},t) = \frac{\partial \mathbf{x}(\mathbf{X},t)}{\partial t}, \quad \mathbf{x} \in S, \tag{2.3.7}$$

where $\mathbf{x}$ is the current position of a membrane material point and $\mathbf{X}$ is the corresponding position in the reference state. A dynamic condition requires that the load $\mathbf{q}$ exerted on the membrane is balanced by the viscous traction jump across the interface,

$$[\sigma(\mathbf{x})] \cdot \mathbf{n} = \mathbf{q} = -\nabla_s . \mathbf{T}, \quad \mathbf{x} \in S. \tag{2.3.8}$$

An important parameter of the problem is the ratio of viscous and elastic forces expressed by the dimensionless group

$$\mathrm{Ca} = \frac{\mu V^{\infty}}{G_s}, \quad \text{or} \quad \mathrm{Ca} = \frac{\mu \dot{\gamma} a}{G_s}, \tag{2.3.9}$$

playing the role of a capillary number where surface tension is replaced by the membrane shear elastic modulus. For a given capsule, Ca may also be viewed as a non-dimensional shear rate.

Because of the strong coupling between fluid and solid mechanics for large deformations, the solution of (2.3.1) – (2.3.8) subject to a constitutive law discussed in Section (2.2.2) is difficult to obtain. Another difficulty arises from the simultaneous use of Lagrangian ($\mathbf{X}$) and Eulerian ($\mathbf{x}$) descriptions for the membrane and fluid motion. The two descriptions are related by (2.3.7).

2.3.1 Instability due to compression

In the absence of fluid inertia, the capsule deformation at any time is governed by a balance between the membrane elastic tensions and the viscous stresses exerted by the flowing liquids. However, the equilibrium shape found by solving the steady equations is not necessarily stable. Since the membrane bending modulus has been neglected, the capsule wall should be everywhere under tension; otherwise, the membrane may buckle locally in the regions where the elastic tensions are compressive. Surface wrinkling is well known for thin elastic sheets (e.g., Cerda & Mahadevan 2003, Luo & Pozrikidis 2007, Pozrikidis & Luo 2010). When compressive tensions develop, a full shell model including bending moments and transverse shear forces is necessary to properly describe the mechanics of the capsule wall.

2.3.2 Numerical procedure

To describe the deformation of large capsules, we resort to numerical methods. In the numerical procedure, an undeformed capsule is injected in the flow field and its motion and deformation are followed numerically in time. At any given time, the position $\mathbf{x}(\mathbf{X}, t)$ of a membrane material point is known. The deformation of the capsule membrane may then be computed from (2.2.5) and (2.2.7) or (2.2.8) by comparison with the initial reference state. The elastic tensions $\mathbf{T}$ follow from (2.2.12), where the strain energy function w_s is given by a chosen membrane constitutive law. The solution of the membrane equilibrium equation leads to the values of the load $\mathbf{q}$ and traction jump $[\boldsymbol{\sigma}].\mathbf{n}$ on the membrane. Solving the Stokes equations (2.3.1) or (2.3.3) produces the velocity $\mathbf{v}(\mathbf{x})$ of the membrane points. Time integration of the kinematic condition (2.3.7) leads to the new position of membrane material points.

In the framework of the integral representation, the fluid mechanics equation (2.3.3) and membrane equation (2.3.8) can be solved on the same grid, thus reducing the geometric dimension of the problem by one unit. A main difficulty is determining the load $\mathbf{q}$ on the membrane (or equivalently the traction jump) from knowledge of the material point displacement $\mathbf{U} = \mathbf{x} - \mathbf{X}$ at each time step. This can be achieved by enforcing the membrane equilibrium equation (2.2.18) at each grid point. Projection on functions with continuous second-order derivatives at each grid point is then required. Bicubic B-spline functions are well suited for this purpose.

Another approach is to compute $\mathbf{q}$ by solving a variational problem corresponding to equation (2.2.20) using a finite-element method. The two approaches are presented in detail in the next two sections for the case of an initially spherical capsule and unit viscosity ratio, $\lambda = 1$. In that case, the Stokes integral equation simplifies to the integral representation (2.3.6).

2.4 B-spline projection

The projection of the membrane shape on B-spline functions was originally used to compute the axisymmetric deformation of a capsule in an elongational flow (Diaz, Pelekasis & Barthès-Biesel 2000, Diaz & Barthès-Biesel 2001) and in a small cylindrical channel (Diaz & Barthès-Biesel 2002). The technique was then extended to three-dimensional flow (Lac *et al.* 2004).

Mesh generation

A structured mesh is first generated based on a parametric description (ξ^1, ξ^2) of the surface S. In the case of a spherical capsule, ξ^1 and ξ^2 correspond to the azimuthal and meridional angles θ and ϕ (figure 2.4.1). The initial shape is discretized into $n \times m$ elements corresponding to equal intervals

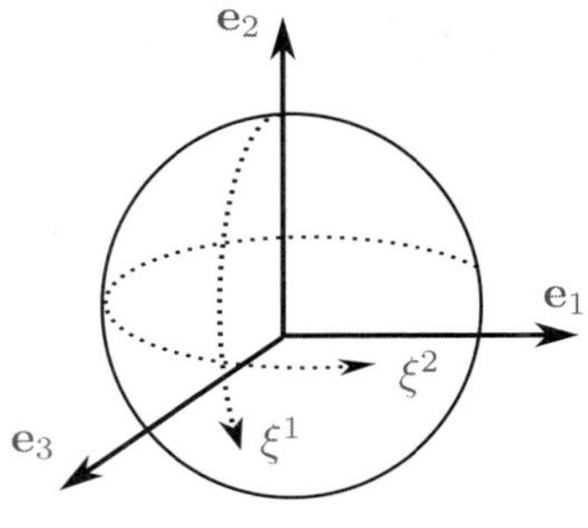

Figure 2.4.1 Surface coordinates on an initially spherical capsule. At time $t = 0$, the curvilinear coordinates ξ^1 and ξ^2 correspond to the azimuthal and meridional angles θ and ϕ in spherical coordinates.

of $\xi^1 = \theta$ and $\xi^2 = \phi$. The intersections of azimuthal curves, parametrized by i, and meridian curves, parametrized by j, define grid nodes (ξ_i^1, ξ_j^2),

$$\xi_i^1 = \frac{i\pi}{n}, \quad \xi_j^2 = \frac{2j\pi}{m}, \quad \text{for } i \in [0, n], \ \ j \in [0, m]. \tag{2.4.1}$$

Since the capsule surface is closed, the first ($j = 0$) and last ($j = m$) meridians coincide, and the nodal points (ξ_i^1, ξ_0^2) and (ξ_i^1, ξ_m^2) are identical for all i. Two singular points arise at the poles, $\xi^1 = 0, \pi$. Consequently, there is a total of $m(n-1)+2$ distinct nodes.

B-spline projection

A function $f(\xi^1, \xi^2)$ defined over the surface is projected on a base of regular B-spline functions,

$$f(\xi^1, \xi^2) = \sum_{k,l} \bar{f}_{kl}\, B_k^n(\xi^1)\, B_l^m(\xi^2), \tag{2.4.2}$$

where B_k^n is a B-spline piecewise cubic polynomial taking nonzero values in the interval $[\xi_{k-2}^1, \xi_{k+2}^1]$, and B_l^m is a corresponding polynomial taking nonzero values in the interval $[\xi_{l-2}^2, \xi_{l+2}^2]$. The superscripts n and m indicate the size of the constant intervals between the nodes: $\Delta\xi = \pi/n$ for B_k^n, and $\Delta\xi = 2\pi/m$ for B_k^m. A regular B-spline function $B_i(\xi)$ is then defined in the interval $[\xi_{i-2}, \xi_{i+2}]$ by

$$
\begin{aligned}
B_i(\xi) &= \frac{1}{6(\Delta\xi)^3}(\xi-\xi_{i-2})^3 \quad \text{for} \quad \xi \in [\xi_{i-2},\xi_{i-1}],\\
B_i(\xi) &= \frac{1}{6(\Delta\xi)^3}\Big[-3(\xi-\xi_{i-1})^3 + 3\Delta\xi(\xi-\xi_{i-1})^2\\
&\qquad +3(\Delta\xi)^2(\xi-\xi_{i-1}) + (\Delta\xi)^3\Big] \quad \text{for} \quad \xi \in [\xi_{i-1},\xi_i],\\
B_i(\xi) &= \frac{1}{6(\Delta\xi)^3}\left[3(\xi-\xi_i)^3 - 6\Delta\xi(\xi-\xi_i)^2 + 4(\Delta\xi)^3\right]\\
&\qquad \text{for} \quad \xi \in [\xi_i,\xi_{i+1}],\\
B_i(\xi) &= \frac{1}{6(\Delta\xi)^3}(\xi-\xi_{i+1})^3 \quad \text{for} \quad \xi \in [\xi_{i+1},\xi_{i+2}],\\
B_i(\xi) &= 0 \quad \text{otherwise.}
\end{aligned}
\tag{2.4.3}
$$

It is easily checked that both families of polynomials B_k^n and B_l^m have continuous first and second derivatives at the nodes.

A set of $(n+3)(m+3)$ spline coefficients $\bar{f}_{kl}$ are computed from the values f_{ij} of f at the nodes,

$$
f_{ij} = \sum_{k,l} \bar{f}_{kl}\, B_k^n(\xi_i^1)\, B_l^m(\xi_j^2), \tag{2.4.4}
$$

for $k \in [i-1, i+1]$ and $l \in [j-1, j+1]$. Conditions (2.4.4) provide us with $m(n-1)+2$ relations. An additional set of $3n+4m+7$ conditions arise by requiring the following:

1. *Periodicity of f along ξ^2* expressed by

$$
f(\xi_i^1,\xi_0^2) = f(\xi_i^1,\xi_m^2), \quad \forall i \in 0,\dots,n, \tag{2.4.5}
$$

leading to $3(n+3)$ equations

$$
\bar{f}_{k,m+l} = \bar{f}_{k,l}, \qquad \text{for} \quad l \in \{-1,0,1\} \quad \text{and} \quad \forall k \in \{-1,\dots,n+1\}. \tag{2.4.6}
$$

2. *Multiple value of f at the poles* expressed by

$$
f(\xi_i^1,\xi_0^2) = f(\xi_i^1,\xi_1^2) = \cdots = f(\xi_i^1,\xi_{m-1}^2), \quad i = 0, n, \tag{2.4.7}
$$

leading to $2(m-1)$ equations,

$$
\sum_{k=i-1}^{i+1}\sum_{l=j-1}^{j+1} \bar{f}_{kl}\, B_k^n(\xi_0^1) B_l^m(\xi_j^2) = \sum_{k=i-1}^{i+1}\sum_{l=j}^{j+2} \bar{f}_{kl}\, B_k^n(\xi_0^1) B_l^m(\xi_j^2),
$$
$$
\text{for} \quad j \in 0,\dots,m-2. \tag{2.4.8}
$$

3. *Periodicity of f along ξ^1* requiring that the derivatives of f with respect to ξ^1 be continuous at the poles:

$$\frac{\partial f}{\partial \xi^1}(\xi_0^1, \xi_j^2) = -\frac{\partial f}{\partial \xi^1}(\xi_0^1, \xi_j^2 + \pi), \quad \forall j \in \{0, \ldots, m-1\}, \tag{2.4.9}$$

leading to $2m$ equations

$$\bar{f}_{k,l+m/2} = \bar{f}_{k,l}, \tag{2.4.10}$$

$$\text{for} \quad k \in \{-1, 1, n-1, n+1\} \quad \text{and} \quad \forall l \in \{2, \ldots, m/2+1\}.$$

Determining the spline coefficients requires solving a linear system of $(n+3)(m+3)$ equations, $\mathcal{B} \cdot \bar{\mathcal{F}} = \mathcal{F}$, where the matrix $\mathcal{B}$ contains the values of the B-spline functions evaluated at the nodes, $\bar{\mathcal{F}}$ is a vector containing the unknown spline coefficients, and $\mathcal{F}$ is a vector containing the known values f_{ij} at the nodes. The spline coefficients are then given by

$$\bar{\mathcal{F}} = \mathcal{B}^{-1} \cdot \mathcal{F}. \tag{2.4.11}$$

Because the matrix $\mathcal{B}$ depends only on the node distribution, it needs to be inverted only once. After the spline coefficients $\bar{f}_{kl}$ have been determined, the value of the function f at an element defined by $\xi^1 \in [\xi_i^1, \xi_{i+1}^1]$ and $\xi^2 \in [\xi_i^2, \xi_{j+1}^2]$ is obtained by B-spline projection in four adjacent elements,

$$f(\xi^1, \xi^2) = \sum_{k=i-1}^{i+2} \sum_{l=j-1}^{j+2} \bar{f}_{kl} \, B_k^n(\xi^1) \, B_l^m(\xi^2). \tag{2.4.12}$$

The value of f at the node is identical to that given in (2.4.4).

Computation of surface properties

The position of a membrane material point, $\mathbf{x}(t)$, after inception of the flow is found by projecting on the B-spline functions, yielding

$$\mathbf{x}(\xi^1, \xi^2, t) = \sum_{k,l} \bar{\mathbf{x}}_{kl}(t) \, B_k^n(\xi^1) \, B_l^m(\xi^2). \tag{2.4.13}$$

Since the projection guarantees second-order continuity of derivatives at the nodes, the surface metrics are easily obtained. For example, the covariant basis vector $\mathbf{a}_1$ is given by

$$\mathbf{a}_1 = \frac{\partial \mathbf{x}}{\partial \xi^1} = \sum_{k,l} \bar{\mathbf{x}}_{kl}(t) \, \frac{\mathrm{d}B_k^n(\xi^1)}{\mathrm{d}\xi^1} \, B_l^m(\xi^2). \tag{2.4.14}$$

Second derivatives of the position vector are necessary for the evaluation of the Christoffel symbols, $\Gamma_{\alpha\lambda}^{\beta}$, and curvature tensor, $b_{\alpha\beta}$. The elastic tensions

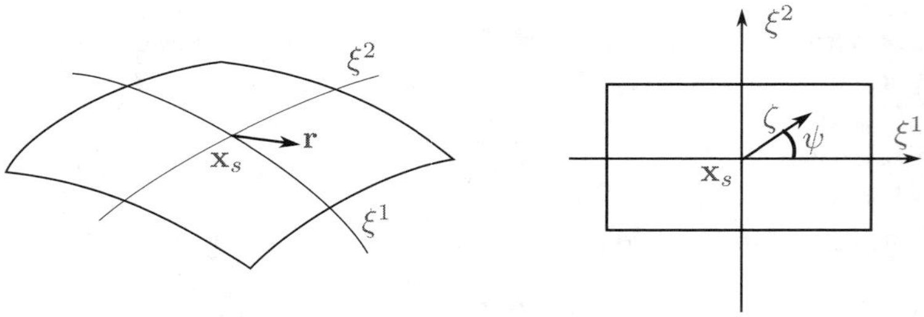

Figure 2.4.2 Polar coordinates (ζ, ψ) are used to calculate the $1/r$ integral at the node x over four neighboring elements.

are computed from (2.2.12) for a given constitutive law. Finally, the load is computed from the surface gradient of $T^{\alpha\beta}$ in (2.2.19).

In summary, the method allows the direct computation of a load $\mathbf{q}$ that is continuous over the interface, except at the poles where singularities occur.

2.4.1 Computation of boundary integrals

The velocity at a given node point $\mathbf{x}$ is obtained by evaluating the integral in (2.3.6) using a Gaussian quadrature over elements where the kernel $\boldsymbol{\mathcal{G}}$ is nonsingular. However, each node, $\mathbf{x}$, is shared by four elements where the Euclidean distance $|r|$ may vanish and where the kernel $\boldsymbol{\mathcal{G}}$ becomes singular, as shown in figure 2.4.2. Even if the quadrature points where the integrand is calculated do not coincide with the nodes themselves, the distance $|r|$ becomes small enough to generate large errors in the calculation of the improper integral.

To compute the singular integrals, polar coordinates (ζ, ψ) centered at the singular point denoted $\mathbf{x}_s$ are introduced, as shown in figure 2.4.2, such that

$$\xi^1 = \xi^1_s + \zeta \cos\psi, \qquad \xi^2 = \xi^2_s + \zeta \sin\psi. \tag{2.4.15}$$

The surface element $\mathrm{d}S$ can be expressed as

$$\mathrm{d}S = \sqrt{|a_{\alpha\beta}|}\mathcal{J}\, \mathrm{d}\zeta\, \mathrm{d}\psi. \tag{2.4.16}$$

The Jacobian $\mathcal{J} = \zeta$ tends to zero as fast as the Euclidean distance, thus eliminating the singularity in the evaluation of the integral.

After the velocity of each node has been computed, the nodes are convected by integrating (2.3.7) using the explicit second-order Runge-Kutta

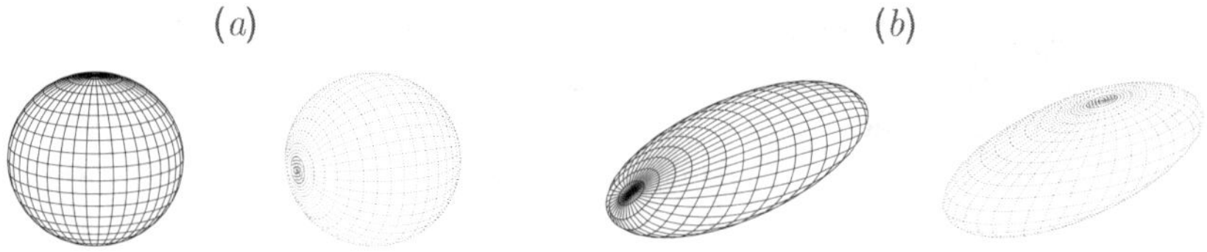

Figure 2.4.3 Illustration of the two-grid method. The solid lines show grid (1) and the dotted lines show grid (2). Grids are shown (*a*) at the origin of computational time and (*b*) at a later time. The second grid arises by rotating the initial first grid by $\pi/2$ around the $\mathbf{e}_3$ axis. From Lac, Morel & Barthès-Biesel (2007).

method. Since the time integration is explicit, stability requires a small time step Δt,

$$\dot{\gamma}\Delta t < O(h\mathrm{Ca}), \tag{2.4.17}$$

where h is the characteristic size of a mesh.

Because of the fundamental properties of the Green's function kernel, $\mathcal{G}$, the velocity computed by (2.3.6) is divergence-free for any surface force distribution, $\mathbf{q} = [\boldsymbol{\sigma}] \cdot \mathbf{n}$. One way of assessing the accuracy of the integration procedure combined with the spline representation of the surface is to monitor the capsule volume during deformation. With this technique, the maximum relative volume change is typically 10^{-4} for a 20×40 grid, and 10^{-6} for a 30×60 grid.

2.4.2 Two-grid method

The dominant error of the numerical scheme comes from the determination of $\mathbf{q}$ at the two poles. When the steady capsule shape is obtained after a small or moderate number of iterations, the polar values of the load can be extrapolated at each time step without introducing a large error (Lac *et al.* 2004, Lac & Barthès-Biesel 2005). However, in long simulations, if the pole value of the load is extrapolated, the accumulated error may become large and eventually cause the numerical method to fail.

In order to calculate the surface load with improved accuracy, the capsule surface is described by *two* grids whose poles are orthogonal at the initial instant, as shown in figure 2.4.3(*a*). Grid (1) was described in Section 2.4. Grid (2) is obtained by a similar $n \times m$ partition of the initial surface, but with poles orthogonal to those of grid (1), i.e., located on points $\xi^1 = \pi/2$;

$\xi^2 = 0, \pi$ of grid (1). The secondary grid coordinates on the primary grid, (ξ^1, ξ^2), are used in (2.4.13) to determine the position of the nodes of grid (2) on the deformed surface (figure 2.4.3(*b*), and then repeat the load computation of (2.2.19) on grid (2). This procedure yields two load distributions, $\mathbf{q}_{(1)}$ and $\mathbf{q}_{(2)}$, corresponding to grids (1) and (2), excluding the respective poles. The final load is given by

$$\mathbf{q} = w\,\mathbf{q}_{(1)} + (1 - w)\,\mathbf{q}_{(2)}, \tag{2.4.18}$$

where $w(\xi^1, \xi^2)$ is a weight function tending to zero at the poles of grid (1) and to unity at the poles of grid (2). When the capsule is spherical, the singularity in equation (2.2.19) behaves like $\sin^{-2}\xi^1$ in the vicinity of $\xi^1 = 0, \pi$. We can therefore choose

$$w = \frac{\sin^2 \xi^1_{(1)}}{\sin^2 \xi^1_{(1)} + \sin^2 \xi^1_{(2)}}, \tag{2.4.19}$$

where $\xi^\alpha_{(1)}$ are the coordinates on the first grid and $\xi^\alpha_{(2)}$ are the coordinates on the second grid. The usefulness and improved stability of the two-grid method has been demonstrated by Lac *et al.* (2007).

2.5 Coupling finite elements and boundary elements

An alternative strategy for the membrane solver is to turn the virtual work principle expressed by (2.2.20) into a variational formulation where $\mathbf{q}$ is an unknown, and then produce a solution by a finite-element method. The surface of the capsule is discretized into an unstructured triangular mesh, and isoparametric interpolation is used where both the element shape and field quantities are interpolated with the same functions, called shape functions.

2.5.1 Isoparametric interpolation

The position of a point particle on an element is determined by coordinates intrinsic to the element, (η^1, η^2), defined such that η^1, η^2, and $1-\eta^1-\eta^2$ lie in the interval $[0, 1]$, as shown in figure 2.5.1. Flat P_1 elements with three nodes–one at each vertex–and linear shape functions (figure 2.5.1(*a*)), and curved P_2 elements with six nodes–one at each vertex and in the middle of each side–and quadratic shape functions (figure 2.5.1(*b*)) are employed. The method presented in this section is general and can be used for any interpolation choice.

The shape functions N depend on the chosen element type. For P_1 elements, their expression is

$$N^{(1)}(\eta^1, \eta^2) = 1 - \eta^1 - \eta^2, \qquad N^{(2)}(\eta^1, \eta^2) = \eta^1, \qquad N^{(3)}(\eta^1, \eta^2) = \eta^2. \tag{2.5.1}$$

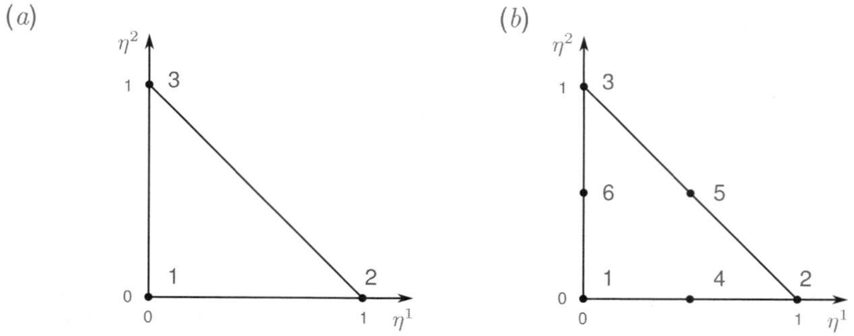

Figure 2.5.1 Reference coordinate system and node numbering for (*a*) P_1 and (*b*) P_2 elements.

For P_2 elements, their expression is

$$\begin{aligned} N^{(1)}(\eta^1,\eta^2) &= (1-\eta^1-\eta^2)(1-2\eta^1-2\eta^2)\,, \\ N^{(2)}(\eta^1,\eta^2) &= \eta^1(2\eta^1-1)\,, \\ N^{(3)}(\eta^1,\eta^2) &= \eta^2(2\eta^2-1)\,, \\ N^{(4)}(\eta^1,\eta^2) &= 4\eta^1(1-\eta^1-\eta^2)\,, \\ N^{(5)}(\eta^1,\eta^2) &= 4\eta^1\eta^2\,, \\ N^{(6)}(\eta^1,\eta^2) &= 4\eta^2(1-\eta^1-\eta^2)\,. \end{aligned} \tag{2.5.2}$$

These formulas can be used to interpolate any field quantity **f**,

$$\mathbf{f}(\eta^1,\eta^2) = \sum_{p=1}^{n_n} f_{Xj}^{(p)} N^{(p)}(\eta^1,\eta^2)\,\mathbf{e}_j\,, \tag{2.5.3}$$

where $N^{(p)}(\eta^1,\eta^2)$ is the shape function associated with node $p \in \{1\ldots n_n\}$, and nn is the number of nodes of the element. The nodal value $f_{Xj}^{(p)}$ is the jth Cartesian component of the vectorial quantity **f** at node p.

2.5.2 Mesh generation

A spherical surface is meshed by first introducing an inscribed icosahedron, that is, a regular polyhedron with twenty triangular faces (Ramanujan & Pozrikidis 1998). The mesh is generated by placing a new node in the middle of each edge, and dividing each element into four new elements. The new nodes are projected onto the sphere and the procedure is repeated until a desired number of elements has been attained. Figure 2.5.2 shows undeformed and deformed meshes for P_1 elements. The P_2 elements are obtained from the P_1 elements by dividing the edges in half in the last step, and projecting the

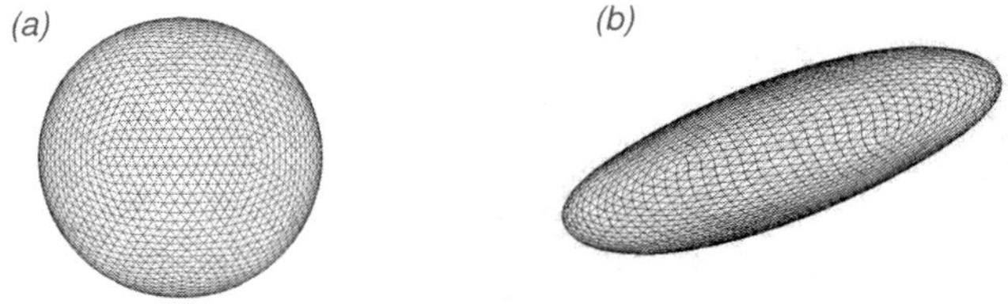

Figure 2.5.2 Typical capsule shape meshed with flat triangles (P_1 elements) (*a*) at rest, and (*b*) under deformation, for $N_E = 5120$, $N_N = 2562$.

new nodes onto the sphere. The total number of elements and nodes are, respectively, N_E and N_N.

2.5.3 Membrane finite-element formulation

Given the instantaneous capsule shape, the membrane problem is solved using the virtual work principle expressed by (2.2.20), treating $\mathbf{q}$ as an unknown. In order to develop the finite-element formulation, the domain is discretized into a finite-element space, $\mathcal{V}_h$, based on the mesh described in Section 2.5.2. Since isoparametric elements are used, $\hat{\mathbf{u}}$, $\mathbf{x}$ and $\mathbf{q}$ are discretized in the same space.

The membrane mechanics problem can be stated as follows: find $\mathbf{q} \in \mathcal{V}_h$, such that

$$\iint_S \hat{\mathbf{u}} \cdot \mathbf{q} \, \mathrm{d}S = \iint_S \hat{\epsilon}(\hat{\mathbf{u}}) : \mathbf{T} \, \mathrm{d}S \tag{2.5.4}$$

for $\forall \hat{\mathbf{u}} \in \mathcal{V}_h$. The objective is to restate the membrane problem as a linear system of algebraic equations involving the nodal values of the unknown function, $\mathbf{q}$, and test function, $\hat{\mathbf{u}}$. The integrals over S are computed as sums of integrals over the individual elements. When calculating the metrics inside a given element, the coordinates (ξ^1, ξ^2) are replaced by the intrinsic coordinates in an isoparametric triangular element, (η^1, η^2). For example,

$$\mathbf{a}_\alpha = \frac{\partial N^{(p)}}{\partial \eta^\alpha} \mathbf{x}^{(p)} . \tag{2.5.5}$$

The metric tensor and Christoffel symbols can then be obtained from this expression.

The left-hand side of (2.5.4) becomes

$$\iint_S \hat{\mathbf{u}} \cdot \mathbf{q} \, \mathrm{d}S = \sum_{el} \hat{u}_{Xj}^{(p)} \left(\int_0^1 \int_0^{1-\eta^2} N^{(p)} N^{(q)} \sqrt{|a_{\alpha\beta}|} \, \mathrm{d}\eta^1 \mathrm{d}\eta^2 \right) q_{Xj}^{(q)} , \tag{2.5.6}$$

or in vector form,

$$\iint_S \hat{\mathbf{u}} \cdot \mathbf{q} \, \mathrm{d}S = \sum_{el} \{\hat{u}_{el}\}^T [M_{el}] \{q_{el}\} \, . \tag{2.5.7}$$

The vectors $\{q_{el}\}$ and $\{\hat{u}_{el}\}$, of size $3n_n$, contain, respectively, the Cartesian components of the discrete load and of the virtual displacement at the element nodes,

$$\begin{aligned} \{q_{el}\} &= \left\{ q_{X1}^{(1)}, q_{X2}^{(1)}, q_{X3}^{(1)} \quad \cdots \quad q_{X1}^{(n_n)}, q_{X2}^{(n_n)}, q_{X3}^{(n_n)} \right\} , \\ \{\hat{u}_{el}\} &= \left\{ \hat{u}_{X1}^{(1)}, \hat{u}_{X2}^{(1)}, \hat{u}_{X3}^{(1)} \quad \cdots \quad \hat{u}_{X1}^{(n_n)}, \hat{u}_{X2}^{(n_n)}, \hat{u}_{X3}^{(n_n)} \right\} . \end{aligned} \tag{2.5.8}$$

Since the matrix $[M_{el}]$ corresponding to the integral of equation (2.5.6) involves the Jacobian $\sqrt{|a_{\alpha\beta}|}$ to account for the deformed local surface area, it must be recomputed at each time step.

The elementary vectors and matrices are assembled into their global counterparts $\{q\}$, $\{\hat{u}\}$, and $[M]$ of size $3N_N$ for the vectors and size $3N_N \times 3N_N$ for the matrix (Crisfield 1997). Note that $[M]$ is a sparse matrix and should be treated as such to reduce computation time.

The left-hand side of (2.5.4) thus becomes

$$\iint_S \hat{\mathbf{u}} \cdot \mathbf{q} \, \mathrm{d}S = \{\hat{u}\}^T [M] \{q\} \, . \tag{2.5.9}$$

Using a similar procedure, the right-hand side of equation (2.5.4) becomes

$$\iint_S \hat{\boldsymbol{\epsilon}}(\hat{\mathbf{u}}) : \mathbf{T} \, \mathrm{d}S = \sum_{el} \int_0^1 \int_0^{1-\eta^2} \hat{\epsilon}_{\alpha\beta} T^{\alpha\beta} \sqrt{|a_{\alpha\beta}|} \, \mathrm{d}\eta^1 \mathrm{d}\eta^2 \, . \tag{2.5.10}$$

The tensions $T^{\alpha\beta}$ are calculated directly from the metric tensor of the deformed configuration using equation (2.2.12).

The tensor $\hat{\epsilon}_{\alpha\beta}$ is related to the covariant representation of $\hat{\mathbf{u}}$ by

$$\hat{\epsilon}_{\alpha\beta} = \frac{1}{2} \left(\hat{u}_{\alpha,\beta} + \hat{u}_{\beta,\alpha} - 2\Gamma^i_{\alpha\beta} \hat{u}_i \right) , \tag{2.5.11}$$

where $\Gamma^3_{\alpha\beta}$ is the curvature tensor $b_{\alpha\beta}$ introduced in (2.2.19). The covariant component of the virtual displacement vector $\hat{\mathbf{u}}$ must now be expressed in terms of its Cartesian components that appear in equation (2.5.8)

$$\hat{u}_i = a_i^{Xj} \hat{u}_{Xj} = N^{(p)} a_i^{Xj} \hat{u}_{Xj}^{(p)} , \tag{2.5.12}$$

where a_i^{Xj} are the Cartesian components of $\mathbf{a}_i$. The virtual strain tensor can therefore be written as

$$\hat{\epsilon}_{\alpha\beta} = \hat{u}_{Xj}^{(p)} \chi_{\alpha\beta}^{(p)j} , \tag{2.5.13}$$

where

$$\chi_{\alpha\beta}^{(p)j} = \frac{1}{2} N_{,\beta}^{(p)} a_\alpha^{Xj} + \frac{1}{2} N_{,\alpha}^{(p)} a_\beta^{Xj} + N^{(p)} a_{\alpha,\beta}^{Xj} - \Gamma_{\alpha\beta}^i N^{(p)} a_i^{Xj} . \tag{2.5.14}$$

After extracting the nodal values of $\hat{\mathbf{u}}$ from the various terms in the integral, the right-hand side of (2.5.4) becomes

$$\iint_S \hat{\epsilon}(\hat{\mathbf{u}}) : \mathbf{T} \, \mathrm{d}S = \sum_{el} \hat{u}_{Xj}^{(p)} \left(\int_0^1 \int_0^{1-\eta^2} \chi_{\alpha\beta}^{(p)j} T^{\alpha\beta} \sqrt{|a_{\alpha\beta}|} \, d\eta^1 d\eta^2 \right) . \tag{2.5.15}$$

The vector form of this equation is

$$\iint_S \hat{\epsilon}(\hat{\mathbf{u}}) : \mathbf{T} \, \mathrm{d}S = \sum_{el} \{\hat{u}_{el}\}^T \{R_{el}\} = \{\hat{u}\}^T \{R\} . \tag{2.5.16}$$

Because the vector $\{R\}$ depends on the metrics, it must also be computed and assembled at each time step.

The discrete solid mechanics problem finally reads

$$\{\hat{u}\}^T [M] \{q\} = \{\hat{u}\}^T \{R\}, \quad \forall \{\hat{u}\}^T \in \mathcal{V}_h. \tag{2.5.17}$$

Since $\{q\}$ satisfies equation (2.5.17) for any virtual displacement $\{\hat{u}\} \in \mathcal{V}_h$,

$$[M]\{q\} = \{R\}. \tag{2.5.18}$$

In the numerical implementation, surface integration is performed to produce $[M]$ and $\{R\}$ using six Hammer points on each element (Hammer, Marlowe & Stroud 1956). Equation (2.5.18) is then solved using the sparse solver Pardiso (Schenk & Gärtner 2004, 2006).

It should be noted that the implemented finite-element method is unconventional. In the classical formulation, finite-element analysis produces a solution for the displacement. In the case of capsules, the membrane displacement is imposed by the fluids, and the solid solver determines the membrane reaction forces exerted on the fluid. Although this unconventional use of the finite-element method prevents us from using commercial and legacy finite-element codes, it does allow for simplifications. For example, in the standard approach, the displacement and therefore the current configuration are unknown, thus complicating the calculation of the surface integrals. In our method, integration is performed on a known deformed state.

2.5.4 Computation of boundary integrals

As in the B-spline method, the velocity is obtained from the boundary-integral representation (2.3.6). For triangular elements, numerical integration is performed using Hammer points. The procedure detailed in Section 2.4.1 is used to perform the integration over singular elements. The new position is calculated with an explicit second-order Runge-Kutta method.

Monitoring the capsule volume reveals that the finite-element discretization is slightly less accurate than the B-spline discretization. For $N_N = 2562$, the maximum volume change is typically 10^{-4} for P_1 and P_2 elements.

2.6 Capsule deformation in linear shear flow

We will discuss the deformation of a spherical capsule of radius a enclosed by a membrane obeying the NH or SK ($C = 1$) law, with shear modulus G_s and area dilation modulus K_s. The capsule is suspended in a Newtonian liquid with the same viscosity ($\lambda = 1$) undergoing a linear shear flow with velocity

$$\mathbf{v}^{\infty} = \dot{\gamma}\,(\mathbf{e} + \boldsymbol{\omega}) \cdot \mathbf{x}, \tag{2.6.1}$$

where $\mathbf{e}$ and $\boldsymbol{\omega}$ are the rate-of-deformation and vorticity tensors.

The capsule deformation has been studied previously for membranes obeying the NH law (Li, Barthès-Biesel & Helmy 1988, Ramanujan & Pozrikidis 1998, Diaz, Pelekasis & Barthès-Biesel 2000, Eggleton & Popel 1998, Lac *et al.* 2004, Doddi & Bagchi 2008a, Walter *et al.* 2010) or the SK law (Lac *et al.* 2004, Li & Sarkar 2008, Walter *et al.* 2010). We will compare results obtained for the B-spline method using a typical 30×60 grid with those obtained for the finite-element formulation using 2562 nodes and 1280 P_2 elements.

2.6.1 Simple shear flow

In the first case study, we consider simple shear in the $(\mathbf{e}_1, \mathbf{e}_2)$ plane described by

$$\mathbf{v}^{\infty} = \dot{\gamma}\, x_2\, \mathbf{e}_1 \,. \tag{2.6.2}$$

The deformation of the capsule is measured by the Taylor parameter

$$D = \frac{L - B}{L + B}, \tag{2.6.3}$$

where L and B denote, respectively, the maximum and minimum diameters in the shear plane of the ellipsoid of inertia of the deformed capsule.

Deformation of an unstressed capsule

Both numerical methods predict the same qualitative behavior. The capsule deforms and inclines with respect to streamlines under the effect of viscous stresses. At long times, the capsule may reach a steady deformed shape where the membrane continuously rotates in a tank-treading mode. The stability of the equilibrium shape is determined by the sign of the principal tensions developing in the membrane. Three different situations exist depending on the value of the dimensionless shear rate, Ca. For flow strength lower than a critical threshold, $\text{Ca} < \text{Ca}_L$, the capsule appears to reach an equilibrium deformed state. However, the equilibrium is mechanically unstable due to the presence of negative principal tensions in the equatorial plane causing membrane buckling, as shown in figure 2.6.1(*a*). The numerical results are in agreement with laboratory experiments where membrane folds are observed (Walter, Rehage & Leonhard 2001). Negative tensions are also predicted by a small-deformation asymptotic analysis (Lac *et al.* 2004).

For $\text{Ca} > \text{Ca}_L$, the membrane deforms into a steady equilibrium shape, as shown in figure 2.6.1(*b*). Because the deformation and subsequent area dilation are now higher, the principal tensions are positive and the membrane undergoes stretching at every point. The equilibrium shape is then stable, and a membrane model without bending resistance is appropriate. The capsule deformation strongly depends on the membrane constitutive law, as illustrated in figure 2.6.1(*d*). For a given shear rate measured by Ca, a capsule with a NH membrane deforms more than a capsule with a SK membrane. Steady elongated capsule shapes with a tank-treading membrane have been observed in the laboratory (Chang & Olbricht 1993, Walter, Rehage & Leonhard 2000, Rehage, Husman & Walter 2002). For large shear rates, $\text{Ca} > \text{Ca}_H$, where Ca_H is a higher critical threshold, the capsule is highly elongated and has high curvature tips that are bent off the flow elongation direction by the vorticity of the incident flow, as shown in figure 2.6.1(*c*). Negative tension occurs near the tip area rendering the equilibrium state unstable. Similar tips have been observed experimentally for capsules with nylon membranes immediately before break-up, albeit for small values of the viscosity ratio (Chang & Olbricht 1993).

Because of the moderate deformation, the low threshold Ca_L depends weakly on the membrane constitutive law. As the membrane response changes from a strain-softening (NH) to strain-hardening (SK), Ca_L decreases from 0.45 to 0.4. Conversely, as the high threshold Ca_H occurs for large deformations, it depends on the membrane constitutive law and increases as the strain hardening effects in the membrane increase (figure 2.6.1(*d*)). Thus the value of Ca_H is larger for a SK membrane than for a NH membrane.

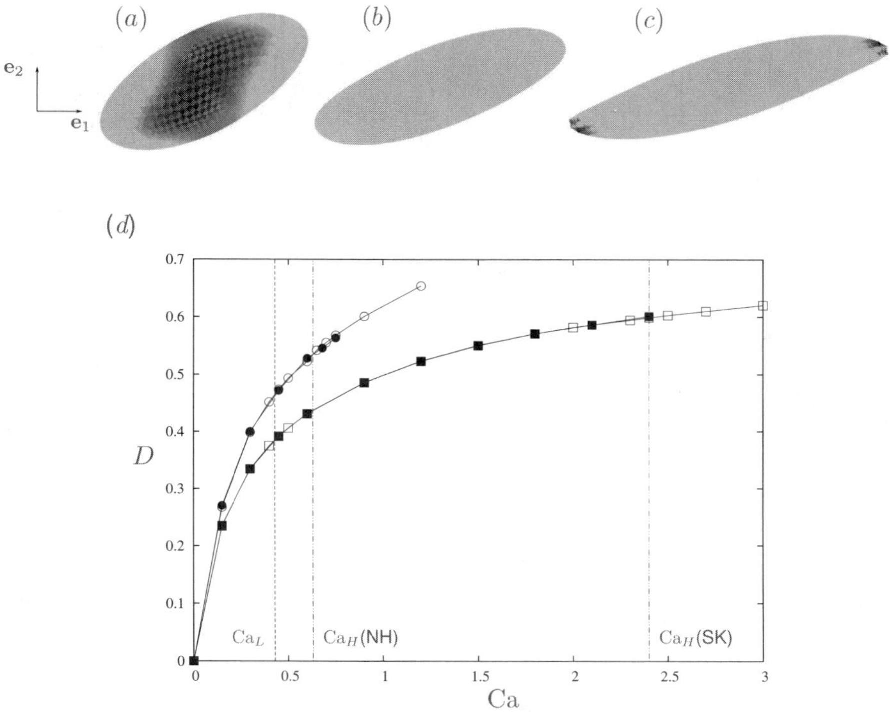

Figure 2.6.1 (*a–c*) Deformed profiles of a capsule with a NH membrane in simple shear flow; (*a*) at low shear rates, $\mathrm{Ca} = 0.3 < \mathrm{Ca}_L$, the membrane undergoes compression and buckles; (*b*) at medium shear rates, $\mathrm{Ca} = 0.6$, the equilibrium profile is stable and the membrane rotates around a steady shape; (*c*) at high shear rates, $\mathrm{Ca} = 1.2 > \mathrm{Ca}_H$, high curvature tips develop where the membrane is compressed. The gray scale measures the intensity of compression forces. (*d*) Effect of the membrane constitutive law on the steady-state deformation in the plane of shear: ○: NH law, □: SK law, filled symbols: B-spline method, open symbols: finite-element method.

Numerical models allow us to describe variables that are difficult to measure experimentally, such as the elastic tension developing in the membrane. The sign of the minimum tension is useful for assessing the stability of mechanical equilibrium (Lac *et al.* 2004). The numerical results show that the maximum tension in the membrane at steady state, T_{max}/G_s, sharply increases with Ca or with the strain hardening effect, as shown in figure 2.6.2. It is important to have an evaluation of the maximum tension value as it may allow to predict the occurrence of capsule burst under flow, provided a failure

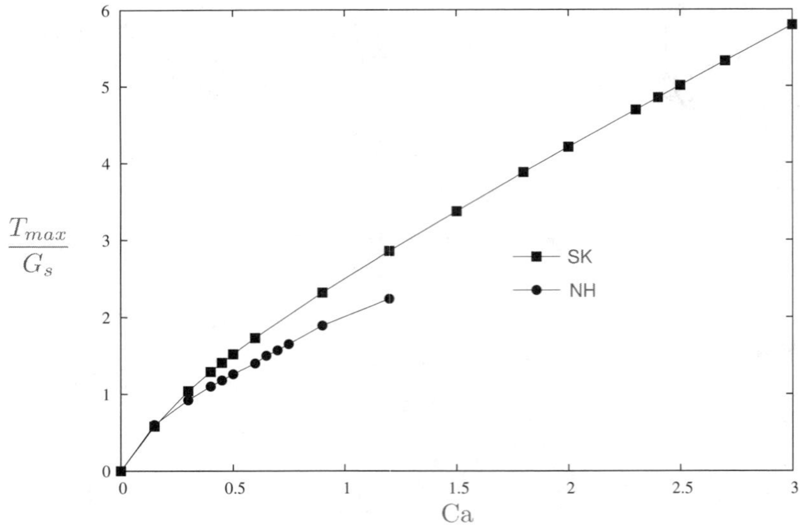

Figure 2.6.2 Maximum steady elastic tension developing in the membrane as a function of the shear rate.

criterion for the membrane has been determined (by means of experiments, usually).

Effect of osmotic pressure

When a capsule is subjected to a positive osmotic pressure difference between the internal and external phase, $p^{(0)}$, prestress can counteract the negative tension arising at low shear rates, and thereby prevent membrane buckling (Lac & Barthès-Biesel 2005). In the following discussion, the capillary number Ca is based on the measured inflated capsule radius, a, rather than on the unstressed capsule radius, a_0. Prestress decreases the capsule deformation for a given shear rate and increases the elastic tension over the membrane at a given level of deformation.

The effects of prestress and strain hardening combine to decrease the membrane deformation for a given flow strength, as shown in figure 2.6.3. For example, a capsule with a 10% pre-inflation and a SK membrane requires quite large values of Ca to reach a 60% deformation. Furthermore, prestress decreases significantly the value of Ca_L that can even vanish. Because the effect of the capsule initial state becomes relatively unimportant at large deformation, prestress increases only slightly the value of Ca_H.

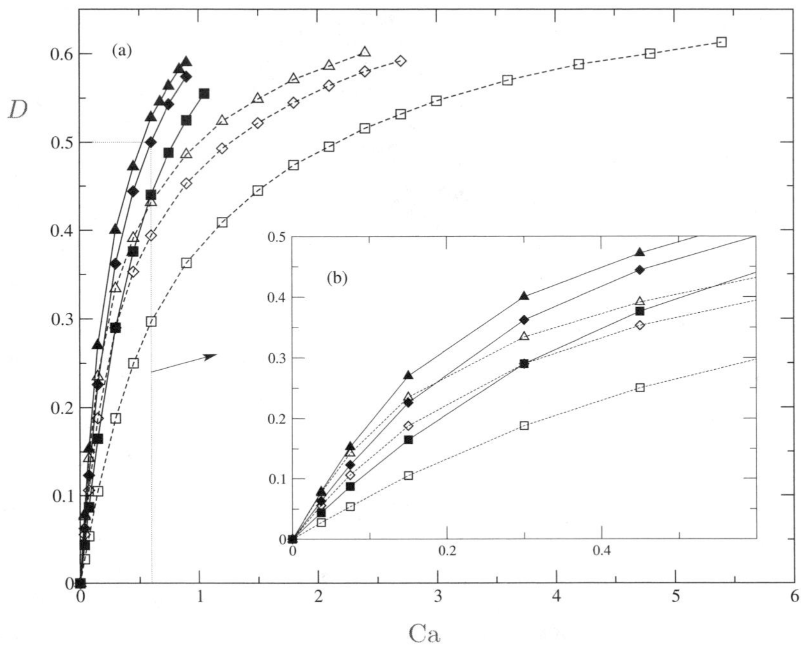

Figure 2.6.3 Effect of pre-inflation and membrane constitutive law on the steady deformation in the shear plane: $\triangle$: $\alpha = 0$; $\diamond$: $\alpha = 2.5\%$; $\square$: $\alpha = 10\%$; filled symbols: NH law; open symbols: SK law ($C = 1$). The capillary number Ca is based on the inflated capsule radius, a. *Reproduced from Lac & Barthès-Biesel (2005), ©2005, American Institute of Physics.*

Effect of the numerical method

When the membrane is everywhere under tension, ($\mathrm{Ca}_L < \mathrm{Ca} < \mathrm{Ca}_H$), the B-splines and finite-elements produce the same results within 1% for deformation, orientation, tank-treading frequency, and membrane tension (Walter *et al.* 2010). Furthermore, the two methods yield the same values for the critical capillary numbers between which all tensions are positive

$$\begin{aligned} &\mathrm{Ca}_L = 0.45\,, \quad \mathrm{Ca}_H = 0.63\,, \quad \text{for NH law,} \\ &\mathrm{Ca}_L = 0.4\,, \quad \mathrm{Ca}_H = 2.4\,, \quad \text{for SK law } (C = 1). \end{aligned} \tag{2.6.4}$$

However, the behavior of a capsule undergoing negative tension strongly depends on the method.

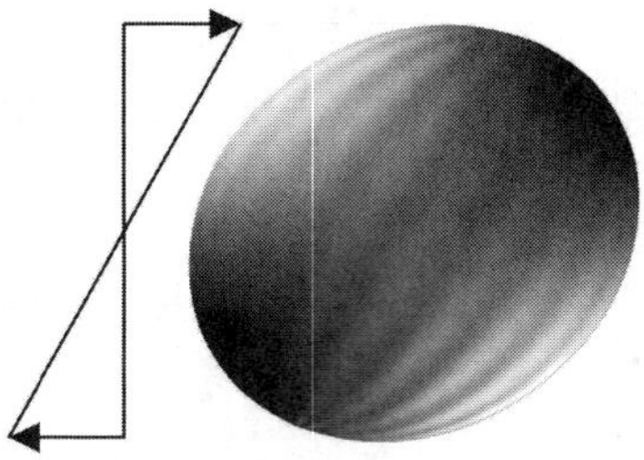

Figure 2.6.4 Normal load distribution for a NH membrane at $\mathrm{Ca} = 0.0375$. The oscillations near the equator have the same spatial period as the B-spline mesh. *Adapted from Lac et al. (2004).*

For $\mathrm{Ca} < \mathrm{Ca}_L$, negative tensions tend to reduce the distance between the mesh nodes as there is no bending stiffness to oppose this effect. With the B-spline method, the spline polynomials may oscillate as they must go through the nodes. This eventually causes the numerical method to fail. As an example, the steady state distribution of the normal load is shown in figure 2.6.4 for a capsule with a NH membrane at $\mathrm{Ca} = 0.0375 < \mathrm{Ca}_L$. The oscillations are clearly mesh-dependent and are thus an artefact of the numerical method (Lac *et al.* 2004). Similar behavior occurs at the tips where one tension becomes negative when $\mathrm{Ca} > \mathrm{Ca}_H$. Consequently, a numerically stable steady solution can be obtained with the B-spline method only when the membrane tension is positive over the whole capsule. As discussed previously, the numerical solution is then consistent with the assumption of negligible bending stiffness. When the capsule motion induces negative tension, a small amount of prestress can render all tensions positive and thereby remove the buckling instability. Stabilization by press-stress has been applied to study the hydrodynamic interaction between two identical capsules in simple shear flow (Lac, Morel & Barthès-Biesel 2007, Lac & Barthès-Biesel 2008).

Negative tensions are tolerated by the finite-element method. A numerically stable steady capsule deformation is obtained with membrane folding at the equator ($\mathrm{Ca} < \mathrm{Ca}_L$) or at the tips ($\mathrm{Ca} > \mathrm{Ca}_H$) (Walter *et al.* 2010). Figure 2.6.1(a) shows that, at low shear rates, the spatial size of the oscillations is comparable to the mesh size. Numerically stable solutions can then be obtained for values of Ca significantly larger than Ca_H, up to 1.2 for NH law, and up to 3.0 for the SK law, as shown in figure 2.6.1(d).

In summary, the finite-element method allows the study of the capsule deformation even when the membrane is under partial compression. Physically, the finite-element implementation introduces some *numerical* stiffness that prevents the overlapping of nodes and emulates mechanical bending stiffness. Such models, although not exact from the mechanics point of view, are nevertheless useful for describing a capsule with a wall that is mostly under extension, and subjected to slight compression in relatively small regions. Accounting for a finite bending stiffness of the membrane may remove the folds (depending on the bending modulus) and prevent the low shear instability. It must be understood though, that shell models are much more complicated than simple membrane models.

2.6.2 Plane hyperbolic flow

Next, we consider a capsule suspended in a plane hyperbolic flow in the $(\mathbf{e}_1, \mathbf{e}_2)$ plane with velocity

$$\mathbf{v}^\infty(\mathbf{x}) = \dot{\gamma}(x_1\,\mathbf{e}_1 - x_2\,\mathbf{e}_2)\,. \tag{2.6.5}$$

The capsule deformation has been studied with the B-spline method (Lac *et al.* 2004), the finite-element method (Walter *et al.* 2010), and more recently with a boundary-integral method coupled with a spectral element discretization (Dodson & Dimitrakopoulos 2008).

Capsule deformation

When placed in a hyperbolic flow described by (2.6.5), the capsule elongates in the $\mathbf{e}_1$ direction. During the transient phase, compressive tensions and folds appear in the equatorial plane orthogonal to the direction of elongation. For capillary numbers below a critical value Ca_L, the folds persist at steady state, as shown in figure 2.6.5(*a*). As in the case of simple shear flow, Ca_L is defined as the value below which negative tensions appear at steady state. For higher values, $\mathrm{Ca} > \mathrm{Ca}_L$, a steady state free of compressive tensions is reached, as shown in figure 2.6.5(*b*).

The steady deformation in the shear plane is shown in figure 2.6.5 as a function of Ca. As Ca is increased, capsule deformation depends strongly on the membrane constitutive law. With the strain-softening NH law, a critical capillary number Ca_∞ exists above which the capsule keeps extending without ever reaching a steady state. A capsule with a strain-hardening SK law always reaches a stationary deformed shape for all values of Ca considered.

The different high-Ca behaviors of NH and SK membranes can be explained with a simple order of magnitude estimate. Let L_1 and L_3 be the dimensions of the deformed capsule in the $\mathbf{e}_1$ and $\mathbf{e}_3$ directions. The internal liquid is at rest and exerts no stress at steady state. The viscous deforming force exerted by the external flow is of order $\mu\dot{\gamma}L_1L_3$. The main elastic deformation is an elongation in the $\mathbf{e}_1$ direction, leading to an elastic tension

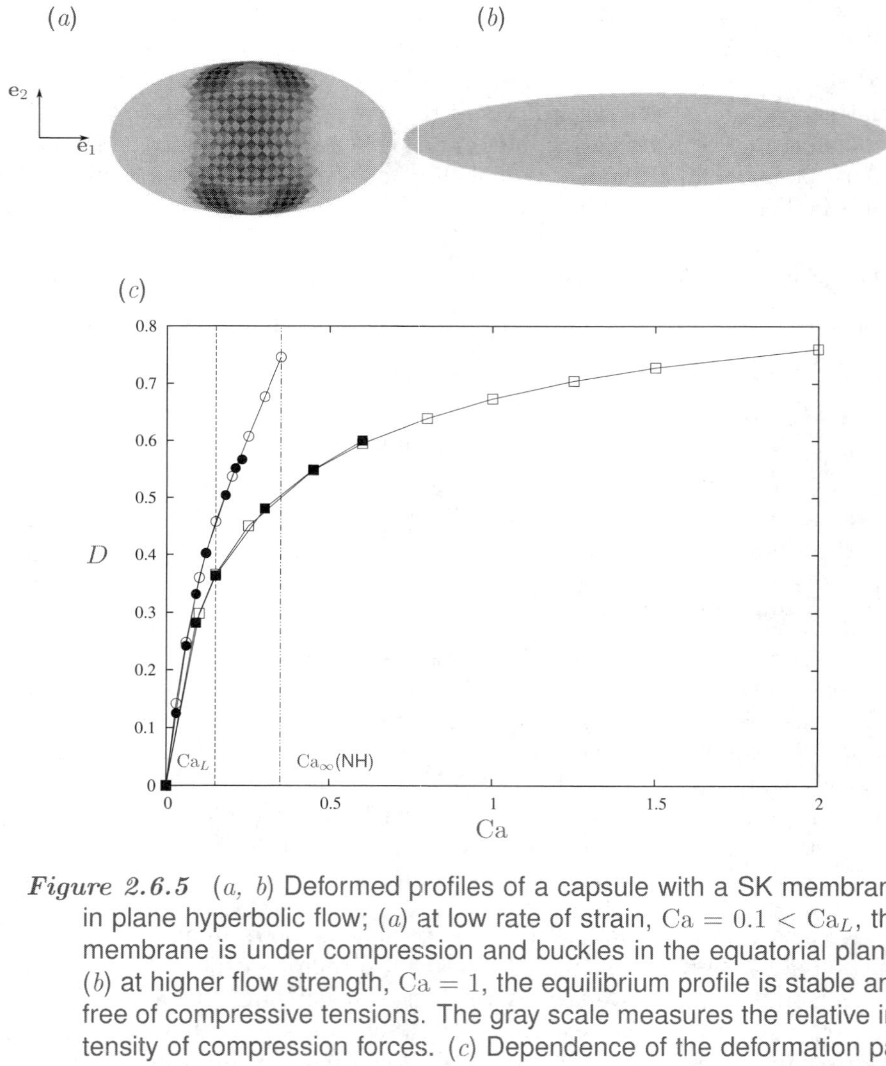

Figure 2.6.5 (*a, b*) Deformed profiles of a capsule with a SK membrane in plane hyperbolic flow; (*a*) at low rate of strain, $\mathrm{Ca} = 0.1 < \mathrm{Ca}_L$, the membrane is under compression and buckles in the equatorial plane; (*b*) at higher flow strength, $\mathrm{Ca} = 1$, the equilibrium profile is stable and free of compressive tensions. The gray scale measures the relative intensity of compression forces. (*c*) Dependence of the deformation parameter, D, on Ca for a capsule suspended in a planar hyperbolic flow. Same legend as in figure 2.6.1(*d*). The low critical capillary number Ca_L is roughly the same value for the two membrane laws considered. For $\mathrm{Ca} > \mathrm{Ca}_\infty$, a steady deformation with a NH membrane cannot be found. The B-spline method is unable to describe large deformations with high tip curvature, and eventually fails.

of order $G_s(L_1/a)^n$, where the exponent n depends on the law ($n > 2$ for a strain hardening law and $n < 2$ for a strain softening law; see Barthès-Biesel, Diaz & Dhenin 2002). The corresponding elastic force is thus of order $G_s(L_1/a)^n\, L_3$. As the strength of the flow increases, the ratio of viscous to elastic forces grows as $\mathrm{Ca}(L_1/a)^{1-n}$. Numerical results show that $L_1/a \propto \mathrm{Ca}$ (Lac *et al.* 2004, Dodson & Dimitrakopoulos 2008). Consequently, the ratio of viscous to elastic forces initially assumed to be Ca is given by Ca^{2-n}. For $n < 2$, this ratio grows faster than Ca, the deforming viscous force dominates the elastic tension, and a steady state cannot be reached. Conversely, for a strain hardening membrane $n > 2$, the elastic forces can always compensate the viscous forces and a steady state can be established.

Effect of the numerical method

The values of Ca_L predicted with the B-spline and finite-element method are similar,

$$\mathrm{Ca}_L = 0.14 \quad \text{for NH law}\,, \quad \mathrm{Ca}_L = 0.15 \quad \text{for SK law } (C=1). \tag{2.6.6}$$

However, different values for Ca_∞ are found,

$$\begin{aligned} &\mathrm{BS}: \quad \mathrm{Ca}_\infty = 0.22 \quad \text{for NH law}\,, \quad \mathrm{Ca}_\infty = 0.7 \quad \text{for SK law } (C=1), \\ &\mathrm{FE}: \quad \mathrm{Ca}_\infty = 0.35 \quad \text{for NH law}\,, \quad \text{no } \mathrm{Ca}_\infty \quad \text{for SK law } (C=1). \end{aligned} \tag{2.6.7}$$

As the critical value $\mathrm{Ca}_\infty = 0.35$ is close to that found by Dodson & Dimitrakopoulos (2008), we may assume that it is more accurate than the lower value obtained with the B-spline method. An explanation for this discrepancy is provided by the results of Dodson & Dimitrakopoulos (2008), showing that capsules in hyperbolic flow develop high tip curvature during the transient phase before settling to a lower curvature at steady state. This behavior is captured with the finite-element method but not with a B-spline method. It is possible that the B-spline method with 30×60 nodes becomes numerically unstable because of high transient curvatures and a more refined mesh would be necessary.

2.7 Discussion

We have presented two numerical techniques for solving a fluid-structure interaction problem concerning the deformation of a capsule in flow. The methods differ by the membrane discretization technique. However identical results are obtained when stable deformed capsule shapes develop. In particular, the same deformed profiles and tank-treading frequencies, if any, are found. The two methods predict the same interval of values of critical capillary number for which there are no compressive tensions in the wall and a simple membrane description with no bending resistance is appropriate.

Local equilibrium requires that the membrane load, $\mathbf{q}$, be equal to the surface divergence of the elastic tensions, $\mathbf{T}$. For the surface derivatives to be finite, the tensions must be continuous, and the position field $\mathbf{x}(\xi^1, \xi^2)$ must be at least C^1 continuous over the surface. Unfortunately, the Poincaré-Brouwer "hairy ball" theorem states that a C^1 continuous map of position is not possible on a sphere (Milnor 1978). This topological problem is unavoidable when using a local equilibrium equation. For example, a sphere mapped along spherical coordinates has two singular points at the poles. Conversely, when the sphere is meshed with finite elements, there is no pole singularity but the spatial derivatives are discontinuous between elements.

The B-spline method allows for a C^2 description of position over the entire capsule but introduces two singular points at the poles where a local covariant basis cannot be defined and the tensions cannot be evaluated. The problem is alleviated by introducing a second mesh whose poles are orthogonal to those of the first mesh. While giving very precise results, the B-spline method is computationally costly as it requires solving the membrane mechanics problem twice and transferring the position and load from one grid onto the other. Numerical instability arises when high curvatures develop.

Instead of enforcing local equilibrium, the finite-element method relies on a variational formulation that is free of topological constraints. Because of the integral formulation, equation (2.2.20) permits the position and displacement fields to reside in the Sobolev space H^1. The solid mechanics problem can be discretized using a finite-element space with discontinuous first spatial derivatives of the position without resorting to ad-hoc averaging or smoothing.

A major difference between the B-spline and finite-element method concerns their performance when the membrane undergoes in-plane compression. Whereas the B-spline method fails when negative tensions appear, the finite-element method remains stable and is able to describe steady equilibrium states. The most probable reason is numerical stiffness due to the finite elements as compared to the bicubic B-spline functions. The stiffness introduced by the numerical method enhances the stability of the motion by endowing the membrane with an effective bending rigidity. The numerical simulation then remains stable during transient phases where in-plane compression and high curvatures may render other methods unstable. As a result, steady states with negative tensions can be captured. The observed folds depend on the mesh with a wavelength equal to the element size and are thus purely numerical. Consequently, the numerical stiffness cannot be used to model the physical bending stiffness.

A real capsule membrane has a finite thickness and therefore a finite bending stiffness. However, as long as the membrane is locally taut and the local curvature remains small, the bending effects can be neglected. For a capsule that is initially spherical and must increase its surface area to deform, these hypotheses are frequently valid. This explains why the finite-element method with a small non-physical stiffness generally produces results identical to those obtained with the B-spline method, which has a vanishing bending stiffness.

When negative tensions appear, flexural stiffness determines the behavior of the capsule membrane. A membrane model is sufficient for determining the onset and location of negative tension. A more thorough study of wrinkling requires modeling the capsule wall as a thin shell with a physical bending stiffness. The finite-element framework seems appropriate for implementing this extended shell model.

References

BARTHÈS-BIESEL, D., DIAZ, A. & DHENIN, E. (2002) Effect of constitutive laws for two dimensional membranes on flow-induced capsule deformation. *J. Fluid Mech.* **460**, 211–222.

CARIN, M., BARTHÈS-BIESEL, D., EDWARDS-LEVY, F., POSTEL, C. & ANDREI, C. D. (2002) Compression of biocompatible liquid filled HSA-alginate capsules: determination of the membrane mechanical properties. *Biotech. Bioeng.* **82**, 207.

CERDA, E. & MAHADEVAN, L. (2003) Geometry and physics of wrinkling. *Phys. Rev. Lett.* **90**, 074302–1–074302–5.

CHANG, K. S. & OLBRICHT, W. L. (1993) Experimental studies of the deformation and breakup of a synthetic capsule in steady and unsteady simple shear flow. *J. Fluid Mech.* **250**, 609–633.

CHARRIER, J. M., SHRIVASTAVA, S. & WU, R. (1989) Free and constrained inflation of elastic membranes in relation to thermoforming – non-axisymmetric problems. *J. Strain Anal.* **24**, 55–74.

CRISFIELD, M. A. (1997) *Non-Linear Finite Element Analysis of Solids and Structures, Volume 2, Advanced Topics.* John Wiley & Sons, Chichester.

DIAZ, A. & BARTHÈS-BIESEL, D. (2001) Effect of membrane viscosity on the dynamic response of an axisymmetric capsule. *Phys. Fluids* **13**, 3835–3838.

DIAZ, A. & BARTHÈS-BIESEL, D. (2002) Entrance of a bioartificial capsule in a pore. *CMES* **3**, 321–337.

DIAZ, A., PELEKASIS, N. A. & BARTHÈS-BIESEL, D. (2000) Transient response of a capsule subjected to varying flow conditions: effect of internal fluid viscosity and membrane elasticity. *Phys. Fluids* **12**, 948–957.

DODDI, S. & BAGCHI, P. (2008*a*) Effect of inertia on the hydrodynamic interaction between two liquid capsules in simple shear flow. *Int. J. Multiph. Flow* **34**, 375–392.

DODDI, S. & BAGCHI, P. (2008*b*) Lateral migration of a capsule in a plane Poiseuille flow in a channel. *Int. J. Multiph. Flow* **34**, 966–986.

DODSON, W. R. & DIMITRAKOPOULOS, P. (2008) Spindles, cusps, and bifurcation for capsules in Stokes flow. *Phys. Rev. Lett.* **101**, 208102.

EGGLETON, C. D. & POPEL, A. S. (1998) Large deformation of red blood cell ghosts in a simple shear flow. *Phys. Fluids* **10**, 1834–1845.

GREEN, A. E. & ADKINS, J. E. (1960) *Large Elastic Deformations.* Oxford University Press, Oxford.

HAMMER, P., MARLOWE, O. & STROUD, A. (1956) Numerical integration over simplexes and cones. *Mathematical Tables and Other Aids to Computation* **10**, 130–137.

KÜHTREIBER, W. M., LANZA, R. P. & CHICK, W. L. (1999) *Cell Encapsulation Technology and Therapeutics.* Birkhäuser, Boston.

LAC, E. & BARTHÈS-BIESEL, D. (2005) Deformation of a capsule in simple shear flow: effect of membrane prestress. *Phys. Fluids* **17**, 0721051–0721058.

LAC, E. & BARTHÈS-BIESEL, D. (2008) Pair-wise interaction of capsules in simple shear flow: three-dimensional effects. *Phys. Fluids* **20**, 040801–1– 040801–6.

LAC, E., BARTHÈS-BIESEL, D., PELEKASIS, N. A. & TSAMOPOULOS, J. (2004) Spherical capsules in three-dimensional unbounded Stokes flow: effect of the membrane constitutive law and onset of buckling. *J. Fluid Mech.* **516**, 303–334.

LAC, E., MOREL, A. & BARTHÈS-BIESEL, D. (2007) Hydrodynamic interaction between two identical capsules in a simple shear flow. *J. Fluid Mech.* **573**, 149–169.

LEYRAT-MAURIN, A. & BARTHÈS-BIESEL, D. (1994) Motion of a deformable capsule through a hyperbolic constriction. *J. Fluid Mech.* **279**, 135–163.

LI, X. & SARKAR, K. (2008) Front tracking simulation of deformation and buckling instability of a liquid capsule enclosed by an elastic membrane. *J. Comp. Phys.* **227**, 4998 – 5018.

LI, X. Z., BARTHÈS-BIESEL, D. & HELMY, A. (1988) Large deformations and burst of a capsule freely suspended in an elongational flow. *J. Fluid Mech.* **187**, 179–196.

LUO, H. & POZRIKIDIS, C. (2007) Buckling of a pre-compressed or pre-stretched membrane. *Int. J. Solids Struct.* **44**, 8074–8085.

MILNOR, J. (1978) Analytic proofs of the "Hairy ball theorem" and the Brouwer fixed point theorem. *Amer. Math. Month.* **85**(7), 521–524.

ODEN, J. T. (1972) *Finite Elements of Nonlinear Continua.* McGraw–Hill, New York.

POZRIKIDIS, C. (1990) The axisymmetric deformation of a red blood cell in uniaxial straining Stokes flow. *J. Fluid Mech.* **216**, 231–254.

POZRIKIDIS, C. (1992) *Boundary Integral and Singularity methods for Linearized Viscous Flow.* Cambridge University Press, New York.

POZRIKIDIS, C. (1995) Finite deformation of liquid capsules enclosed by elastic membranes in simple shear flow. *J. Fluid Mech.* **297**, 123–152.

POZRIKIDIS, C. & LUO H. (2010) A note on the buckling of an elastic plate under the influence of simple shear flow. *J. Appl. Mech.* **77**, No 021007.

QUÉGUINER, C. & BARTHÈS-BIESEL, D. (1997) Axisymmetric motion of capsules through cylindrical channels. *J. Fluid Mech.* **348**, 349–376.

RAMANUJAN, S. & POZRIKIDIS, C. (1998) Deformation of liquid capsules enclosed by elastic membranes in simple shear flow: Large deformations and the effect of capsule viscosity. *J. Fluid Mech.* **361**, 117–143.

REHAGE, H., HUSMANN, M. & WALTER, A. (2002) From two-dimensional model networks to microcapsules. *Rheol. Acta* **41**, 292–306.

SCHENK, O. & GÄRTNER, K. (2004) Solving unsymmetric sparse systems of linear equations with PARDISO. *Future Generation Computer Systems*, **20**, 475–487.

SCHENK, O. & GÄRTNER, K. (2006) On fast factorization pivoting methods for sparse symmetric indefinite systems. *Electr. Trans. Num. Anal.* **23**, 158–179.

SHERWOOD, J. D., RISSO, F., COLLÉ-PAILLOT, F., EDWARDS-LÉVY, F. & LÉVY, M. C. (2003) Transport rates through a capsule membrane to attain Donnan equilibrium. *J. Coll. Interf. Sci.* **263**, 202–212.

SKALAK, R., TOZEREN, A., ZARDA, R. P. & CHIEN, S. (1973) Strain energy function of red blood cell membranes. *Biophys. J.* **13**, 245–264.

WALTER, A., REHAGE, H. & LEONHARD, H. (2000) Shear-induced deformation of polyamid microcapsules. *Coll. Polymer Sci.* **278**, 169–175.

WALTER, A., REHAGE, H. & LEONHARD, H. (2001) Shear induced deformation of microcapsules: shape oscillations and membrane folding. *Coll. Surf. A: Physicochem. Eng. Aspects* **183–185**, 123–132.

WALTER, J., SALSAC, A.-V., BARTHÈS-BIESEL, D. & LE TALLEC, P. (2010) Coupling of finite element and boundary integral methods for a capsule in a Stokes flow. *Int. J. Numer. Meth. Eng.*, DOI: 10.1002 / nme.2859.

3 A high-resolution fast boundary-integral method for multiple interacting blood cells

J. B. Freund

Department of Mechanical Science & Engineering and
Department of Aerospace Engineering
University of Illinois at Urbana-Champaign

H. Zhao

Department of Mechanical Engineering
Center for Turbulence Research
Stanford University

A high-resolution fast boundary-integral method is developed for simulating red blood cell motion in complex geometries. The algorithm employs a particle-mesh-Ewald method (PME) to achieve computational efficiency superior to that of standard boundary-element implementations. The computational expense scales with $O(N \log N)$, where N is the number of collocation points distributed over the cell membranes. The no-slip boundary condition on vessel walls with complex geometry is enforced implicitly in terms of a linear system for unknown forces required to stop the flow at the walls. In the numerical implementation, cell shapes are represented by spherical harmonic functions interpolated through collocation points. Because the resolution of the global spectral basis functions is perfect in some sense, accurate solutions can be obtained efficiently and error-free interpolation can be achieved for switching resolutions. The procedure facilitates the control of aliasing error and circumvents explicit filtering and implicit numerical dissipation; both would degrade the numerical accuracy. The resolution of the scheme can be set based on desired accuracy, unconstrained by stability considerations. The overall approach, motivation, advantages, and drawbacks of the numerical scheme are discussed in detail. The solver is demonstrated for case studies involving the motion of a red blood cell through a constriction, the computation of the effective blood viscosity in tube flow, the transport of a leukocyte in a microvessel, and flow in a model network. Considering the simplicity of the finite-deformation elastic constitutive model used to describe the cell membrane, the calculated effective viscosity reproduces remarkably well the well-known non-monotonic dependence on vessel diameter.

3.1 Introduction

As cells go, red blood cells are mechanically simple. The cell membrane has a nearly constant area due to strong resistance to dilation, but exhibits a relatively small resistance to shearing and bending deformation (e.g, Hochmuth & Waugh 1987, Evans 1983). The membrane lipid bilayer is sufficiently mobile to flow as though it were a two-dimensional fluid (Evans & Hochmuth 1976). The significance of the membrane viscosity is not entirely clear. Red blood cells lack the constitutional complexity of other cells. Early in the cell formation, the nucleus and other densely packed organelles that generally clutter the interior of eukaryotic cells are ejected, leaving behind a relatively homogeneous cytoplasm, consisting predominantly of a hemoglobin solution (Alberts *et al.* 2009, Goodsell 1998). The viscosity of the cytoplasma is higher by a factor of not more than five than the viscosity of the plasma surrounding the cells (Dintenfass 1968).

Despite the relative mechanical simplicity of its cellular constituents, the phenomenology of blood flow is rich. Blood is knwon to exhibit shear-thinning and some elastic behavior in large-scale flow (Whitmore 1968, Merrill 1969). In small conduits, the particulate constitution plays an important role in determining the microstructural and associated rheological properties. It is well known that the apparent viscosity of blood decreases significantly in blood vessels whose diameter is comparable to the size of red blood cells, apparently due to flow-induced reorganization (e.g., Fåhræhus & Lindqvist 1931, Pries *et al.* 1992, Pries & Secomb 2003). By a related mechanism, plasma skimming leads to a widely varying local red-cell volume fraction (tube hematocrit) in capillary networks (Krogh 1992, Pries & Secomb 2003). The motion of red blood cells marginates white blood cells (leukocytes) toward the wall of blood vessels with significant implications in inflammation response (Firrell & Lipowsky 1989, Abbitt & Nash 2003).

Red blood cell lysis (hemolysis) due to high shear stress is an important consideration in the design of devices processing or manipulating blood. Examples include heart-lung machines, cardiac mechanical assist devices, and artificial valves. The quantitative prediction of hemolysis remains a challenge (Garon & Farinas 2004).

The character of red blood cells presents challenges in numerical simulation. Strong resistance to surface area introduces surface compression modes with time scales that challenge numerical stability. Circumventing this difficulty and imposing the surface incompressibility constraint, as in standard formulations for incompressible fluid mechanics, is attractive but requires further investigation. Under physiological conditions, red blood cells undergo large deformation. In fact, the cells are so deformable that their well-known biconcave resting shape is not observed in most flows. Numerical schemes must have sufficient resolution to accurately represent significantly deformed membranes.

The spectral methods discussed in this chapter have perfect resolution in the sense that each mode of the discretized cell shape is an exact representation. Most discretizations do not share this property. Although flow at the cellular scale occurs at Reynolds numbers that are low enough for the governing equations to be effectively linear, the nonlinearity of the membrane mechanics provides an entry for numerical instability. This is true even if linear membrane constitutive laws are used, as finite deformation introduces geometric nonlinearity. The numerical scheme discussed in this chapter is specifically designed to address nonlinear instability mechanisms by facilitating dealiasing. Since cell interactions are of interest, numerical efficiency is imperative.

Our simulation scheme is based on a standard boundary-integral formulation. The velocity at a point $\mathbf{x}$ in the plasma is given by

$$u_j(\mathbf{x}) = U_j - \frac{1}{8\pi\mu} \iint_{W \cup D} f_i(\mathbf{y})\, \mathcal{G}_{ij}(\mathbf{y}, \mathbf{x})\, \mathrm{d}S(\mathbf{y}) + \frac{1-\lambda}{8\pi} \iint_D u_i(\mathbf{y})\, \mathcal{T}_{ijk}(\mathbf{y}, \mathbf{x})\, n_k(\mathbf{y})\, \mathrm{d}S(\mathbf{y}), \tag{3.1.1}$$

where D to be the union of all cell membrane surfaces, W is the blood vessel wall, u_j are Cartesian components of the fluid velocity, U_j is the velocity of a specified driving flow, μ is the plasma viscosity, $\lambda\mu$ is the interior fluid viscosity, f_i is the traction exerted on the membranes or vessels walls, $\mathbf{n}$ is the cell surface normal pointing into the plasma, and $\mathcal{G}_{ij}$ and $\mathcal{T}_{ijk}$ are velocity and stress Green's functions for the Stokes-flow equation of motion.

As the evaluation point $\mathbf{x}$ approaches a point $\mathbf{y}$ on the cell surface from the plasma side, a jump occurs,

$$\lim_{\mathbf{x} \to \mathbf{x}_0} \iint_D u_i(\mathbf{y})\, \mathcal{T}_{ijk}(\mathbf{y}, \mathbf{x})\, n_k(\mathbf{y})\, \mathrm{d}S(\mathbf{y}) = 4\pi u_i(\mathbf{x}_0) + \iint_D^{PV} u_i(\mathbf{y})\, \mathcal{T}_{ijk}(\mathbf{y}, \mathbf{x}_0)\, n_k(\mathbf{y})\, \mathrm{d}S(\mathbf{y}), \tag{3.1.2}$$

where PV indicates the principle-value integral. A boundary-integral equation for the cell surface velocity then arises,

$$u_j(\mathbf{x}) = \frac{2}{1+\lambda} U_j - \frac{1}{4(1+\lambda)\pi\mu} \iint_{W \cup D} f_i(\mathbf{y})\, \mathcal{G}_{ij}(\mathbf{y}, \mathbf{x})\, \mathrm{d}S(\mathbf{y}) + \frac{1-\lambda}{4(1+\lambda)\pi} \iint_{D}^{PV} u_i(\mathbf{y})\, \mathcal{T}_{ijk}(\mathbf{y}, \mathbf{x})\, n_k(\mathbf{y})\, \mathrm{d}S(\mathbf{y}), \tag{3.1.3}$$

where $\mathbf{x} \in D$ is any cell surface point (e.g., Rallison & Acrivos 1978, Pozrikidis 1992).

In discretizing (3.1.3), it is convenient to represent the membrane of each cell by a set of collocation points, $\vec{\mathbf{x}}$. The vector above $\mathbf{x}$ indicates a list of position vectors. Approximating the first integral in (3.1.3) using a quadrature with weights w^β for each collocation point β, we obtain

$$\iint f_j(\mathbf{y})\, \mathcal{G}_{ji}(\mathbf{y}, \mathbf{x}^\alpha)\, \mathrm{d}S(\mathbf{y}) \simeq \sum_{\beta} \mathcal{G}_{ij}(\mathbf{y}^\beta, \mathbf{x}^\alpha)\, f_j^\beta w^\beta, \tag{3.1.4}$$

where $\vec{\mathbf{f}}$ is a list of all components of all the surface tractions. The right-hand side is a matrix-vector multiplication involving all collocations points, $\mathbf{A}\vec{\mathbf{f}}$, where the matrix $\mathbf{A}$ includes the Green's function values and quadrature weights. A similar discretization of the second integral in (3.1.3) yields a corresponding matrix-vector multiplication, $\mathbf{B}\vec{\mathbf{u}}$. Equation (3.1.3) then takes the form

$$\vec{\mathbf{u}} = c\,\mathbf{U} + \mathbf{A}\vec{\mathbf{f}} + \mathbf{B}\vec{\mathbf{u}}, \tag{3.1.5}$$

where c is a constant.

A practical difficulty in simulating large systems becomes immediately apparent. The slow decay of the Green's function kernels prevents the truncation of interactions at any reasonably short distance (Pozrikidis 1992). Consequently, the system (3.1.5) involves dense matrices. Even when the viscosity ratio λ is unity, in which case $\mathbf{B} = \mathbf{0}$, $O(N^2)$ operations are required to evaluate $\mathbf{u}$ directly. For large systems, the computational cost is significant to the extent that it is not clear that the boundary-integral formulation can compete with mesh-only formulations, such as the immersed-boundary formulation (e.g., Doddi & Bagchi 2009). When $\lambda \neq 1$, a direct dense-matrix solver requires $O(N^3)$ operations. Periodic boundaries introduce additional interactions between collocation points and all periodic images of the union of all collocation points.

3.1.1 Fast summation methods

Despite advances in available computational resources, the direct application of (3.1.3) to large systems without algorithmic improvement for the evaluation of the matrix-vector products in (3.1.5) is prohibitively expensive. Difficulties associated with periodic boundaries can be resolved by utilizing the Ewald (1921) approach, first formulated for Stokes flow by Hasimoto (1959). By carefully balancing error and computational expense, the formulation leads to an $O(N^{3/2})$ method (Frenkel & Smit 1996, Karasawa & Goddard 1989). Advanced algorithms provide even better $O(N)$ or $O(N \log N)$ methods for evaluating matrix-vector products in (3.1.5). With this capability, the $\lambda = 1$ evaluation of (3.1.5) is relatively efficient. When $\lambda \neq 1$, iterative methods can be used to solve for $\vec{\mathbf{u}}$ with nearly linear scaling.

Fast multipole methods

In the fast multipole method, expansions of the Green's function kernels in (3.1.3) are used to facilitate the grouping of long-range interactions (Greengard 1988). Accurate expressions for the velocity are obtained without explicitly calculating each pairwise interaction. Multipole methods are typically implemented with a free-space Green's function. Formulations based on a macroscale Ewald-like decomposition compatible with periodic boundary conditions are available (Lambert *et al.* 1996).

The standard formulation yields an $O(N \log N)$ operation count. Additional manipulations yield an $O(N)$ scaling for Stokes flow (Sangani & Mo 1996). Because long-range interactions are grouped together, multipole methods can be advantageous on distributed memory parallel computers. However, the bookkeeping is tedious and the actual operation count can be huge because the operation count coefficient multiplying N or $N \log N$ is large (Greengard 1988). Estimates from electrostatics where a slowly decaying and singular interaction potential is also encountered suggest that calculations for systems with less than 10^{60} points are more efficient if mesh-based methods are employed (Pollock & Glosli 1996).

Fast mesh-based methods

An alternative is to use a fast mesh-based method, splitting the Green's functions into short-range components that are singular but decays exponentially in space, and smooth components that decay slowly but are infinitely differentiable and thus admits an exponentially convergent Fourier representation. This decomposition is main idea behind the Ewald decomposition (Hasimoto 1959). When an appropriate length scale is selected for the decomposition, only a small number of close force–velocity interactions is necessary and the fast Fourier transform (FFT) can be used to evaluate the smooth

(Fourier) component with $O(N \log N)$ operations. Although the algorithm fails to match the $O(N)$ scaling of fast multipole methods, the operation count for an M-mesh-point FFT is only approximately $\frac{5}{2} M \log M$ and the overall operation count remains smaller for large values of N. These methods are employed in our formulation discussed in Section 3.3.

3.1.2 Boundary conditions

Fast summation methods cannot directly accommodate the no-slip boundary condition. The reason is that mathematical expressions are naturally formulated with the free-space or periodic Green's function. No-slip boundary conditions can be implemented by modifying the Green's function, but this is possible only for simple geometries (Stabel *et al.* 2003, Weinbaum *et al.* 1990). Taking advantage of the linearity of the equations governing Stokes flow, the boundary-integral solver can sometimes be used in conjunction with a mesh-based flow solver to enforce the no-slip boundary condition independent of the underlying algorithm (Hernández-Ortiz *et al.* 2007). In this approach, the wall-bounded geometry is discretized into a mesh where a fluid flow is computed with boundary conditions that negate the velocity on the wall of the boundary-integral portion of the algorithm. Unfortunately, the procedure is incompatible with equation (3.1.3) when $\lambda \neq 1$.

The proposed algorithm imposes an implicit constraint on the velocity field. Unknown force distributions are then computed using an efficient iterative process, as discussed in Section 3.2.

3.1.3 Membrane constitutive equations

The specific constitutive model used in the simulations was formulated by Pozrikidis (2005). However, the underlying numerical scheme is general and not restricted to any particular model. Advanced models can be developed to describe experimental observations (Li *et al.* 2005, Lim *et al.* 2004). The spectral basis functions presently employed are well suited to accurately compute high-order derivatives appearing in constitutive models.

Even with the simple finite-deformation elasticity models presently employed, the term $\mathbf{f}$ in (3.1.3) is strongly nonlinear at finite deformation. This nonlinearity transfers energy between different surface deformation modes and promotes the appearance of short waves that are hard to resolve. The inter-mode deformation energy transport is physical and thus unavoidable whether or not spectral basis functions are employed.

Unfortunately, the nonlinearity promotes numerical instability. Even when a certain resolution is sufficient for representing a cell shape, nonlinear interactions deposit some energy into deformations with a smaller length scale.

Energy that is not resolved by the available discretization must be removed to be consistent with the selected resolution. Instead, the numerical method causes the energy to be aliased back into resolved modes. This concept has been discussed at length in the context of turbulence simulations (Canuto *et al.* 1987, Kravchenko & Moin 1997). The principal motivation for utilizing spectral basis functions is that a rigorous way of suppressing aliasing feedback without degrading the fidelity of the resolved solution becomes available. A dealiasing procedure is discussed in Section 3.4.

3.1.4 Preamble

We will discuss the numerical method based on the boundary-integral formulation and present illustrative examples. Transport of leukocytes in a two-dimensional model microvessel has been reported elsewhere (Freund 2007). In that case, the algorithm was shown to be efficient enough to simulate cells passing thousands of times through this streamwise vessel model, which is sufficient to converge statistics relevant to the leukocyte margination process. In three dimensions, even the simple two-parameter elastic model employed reproduces the nonmonotonic effective viscosity versus diameter for blood in small tubes. Examples of leukocyte transport will be shown in three dimensions, and a simulation of blood flow through a complex geometry network of vessels will be described.

Simulations of blood flow are particularly well suited to our algorithm. However, we cannot expect that any particular method will be the best performer in all conceivable situations. We conclude in section 3.7 with a concise recapitulation of the benefits of our approach, a frank discussion of the drawbacks, and suggestions for further improvement.

3.2 Mathematical framework

We consider an arbitrary number of red blood cells denoted by Ω_i, where $i = 1, \ldots, N_c$, flowing inside boundaries with complex geometry where the no-slip boundary condition is required, as shown in figure 3.2. The interior cell viscosity is $\lambda\mu$, where μ is the viscosity of the suspending fluid, denoted by Ω_0. A membrane, denoted by $\partial\Omega_i$, encloses the ith cell. Because the membrane mass density is negligible, the cell moves at the local fluid velocity, and its membrane residual force due to deformation is exactly balanced by a fluid traction force $\mathbf{f}$.

The velocity is required to be zero on the closed surface W where the no-slip boundary condition is required, as in the case of a vessel wall. The fluid traction force $\mathbf{f}$ exerted on W necessary to enforce this condition is *a priori* unknown. The domain Ω_0 includes fluid outside the closed surface W.

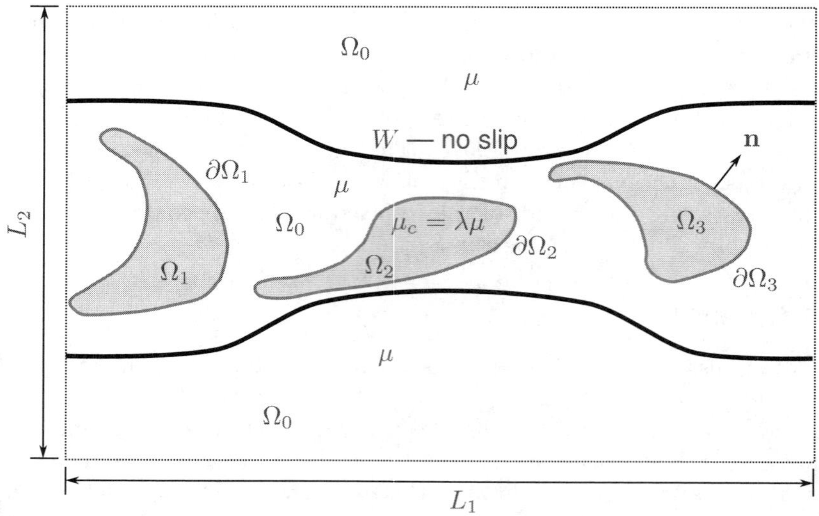

Figure 3.2.1 Schematic illustration of a simulation domain involving multiple cells.

The reason is that the numerical method requires that the composite domain $\Omega = \Omega_0 \cup \Omega_1 \cup \cdots \cup \Omega_{N_c}$ is a rectangular cuboid,

$$\Omega = [0, L_1] \times [0, L_2] \times [0, L_3]. \tag{3.2.1}$$

Periodic boundary conditions are employed, as discussed in Section 3.3. Because of the no-slip condition on W, the flow outside W does not influence the flow inside W.

3.2.1 Integral formulation

The Reynolds number on the scale of a red cell for velocities observed in the microcirculation is so small that the flow is governed by the continuity equation and the Stokes equation,

$$\nabla \cdot \mathbf{u} = 0, \qquad \nabla p = \mu \nabla^2 \mathbf{u}, \tag{3.2.2}$$

where $p(\mathbf{x}, t)$ is the pressure and $\mathbf{u}(\mathbf{x}, t)$ is the velocity. The viscosity inside the cells is $\mu = \mu_c = \lambda\mu$. The velocity field is triply periodic in Ω. The pressure is triply periodic only in the absence of a mean pressure gradient driving a mean flow (Hasimoto 1959). The nonperiodic part of the pressure can be expressed as $\mathbf{x} \cdot \langle \nabla p \rangle$, where $\langle \nabla p \rangle$ is a mean pressure gradient balancing the wall friction.

Ewald decomposition

The boundary-integral solution (3.1.3) of (3.2.2) is well established for fluids with the same or different viscosity (e.g., Pozrikidis 1992). The Green's functions $\mathcal{G}_{ij}$ and $\mathcal{T}_{ijk}$ in (3.1.3) must be compatible with the periodicity of the domain Ω. The sum of all periodic images can be decomposed into two parts

$$\mathcal{G}_{ij} = \sum_{\mathbf{a}} \left(\frac{\delta_{ij}}{r} + \frac{r_i r_j}{r^3} \right) = \mathcal{G}_{ij}^{\mathrm{sr}} + \mathcal{G}_{ij}^{\mathrm{sm}}, \tag{3.2.3}$$

where the sum over $\mathbf{a}$ is over all periodic images (Metsi 2000, Saintillan *et al.* 2005). The periodic offsets are

$$\mathbf{a} = (m_1 L_1, m_2 L_2, m_3 L_3) \tag{3.2.4}$$

for integer triplets, $m_{1,2,3}$, so that the distance between the point force and a velocity point in (3.1.3) is $\mathbf{r} = \mathbf{y} - \mathbf{x} + \mathbf{a}$.

The first component of the Green's function is

$$\mathcal{G}_{ij}^{sr} = \sum_{\mathbf{a}} \operatorname{erfc}(\tilde{r}) \left(\frac{\delta_{ij}}{r} + \frac{r_i r_j}{r^3} \right) + \frac{2}{\sqrt{\alpha}} \sum_{\mathbf{a}} e^{-\tilde{r}^2} \left(\frac{r_i r_j}{r^2} - \delta_{ij} \right), \tag{3.2.5}$$

where $\tilde{r}$ is the nondimensional modulus of $\tilde{\mathbf{r}} = (\pi/\alpha)^{1/2}\,\mathbf{r}$.

The second component of the Green's function is expressed as a sum in wave number space,

$$\mathcal{G}_{ij}^{\mathrm{sm}} = \frac{2\alpha}{V} \sum_{\mathbf{k}\neq 0} \Phi_1(\tilde{k}^2)(\tilde{k}^2 \delta_{ij} - \tilde{k}_i \tilde{k}_j) \exp[2\pi \mathrm{i}\, \mathbf{k} \cdot (\mathbf{x} - \mathbf{y})], \tag{3.2.6}$$

where $\mathbf{k} \neq 0$ denotes sum over

$$\mathbf{k} = (\frac{m_1}{L_1}, \ \frac{m_2}{L_2}, \ \frac{m_3}{L_3}), \tag{3.2.7}$$

m_i are integers and $\tilde{\mathbf{k}} = \sqrt{\pi\alpha}\,\mathbf{k}$.

The parameter $\sqrt{\alpha}$ is an arbitrary length scale determining the balance of the decomposition. In practice, α is chosen to reduce the number of operations in the numerical evaluation of this expression. The first component (3.2.5) decays super-exponentially in r, meaning that, for small α, high accuracy can be achieved when truncating interactions at short distances, much less than $L_{1,2,3}$. The function Φ_1 in (3.2.6) is defined via

$$\Phi_\gamma(z) = \int_1^\infty \mathrm{e}^{-zt}\, t^\gamma \, \mathrm{d}t. \tag{3.2.8}$$

Thus, the summed terms in (3.2.6) also exhibit a super-exponential decay for large $\tilde{k}$. However, in contrast to the decay of the first component, the decay of (3.2.6) is slower for smaller α.

Green's function for the stress

Like the pressure, the stress Green's function $\mathcal{T}_{ijk}$ in (3.1.3) has a non-periodic and a periodic component,

$$\mathcal{T}_{ijk} = -\frac{8\pi}{V} x_j \delta_{ik} + \mathcal{T}^{p}_{ijk}. \tag{3.2.9}$$

As for $\boldsymbol{\mathcal{G}}$, the periodic component can be decomposed as

$$\mathcal{T}^{p}_{ijk} = -6 \sum_{\mathbf{a}} \frac{r_i r_j r_k}{r^5} = \mathcal{T}^{\mathrm{sr}}_{ijl} + \mathcal{T}^{\mathrm{sm}}_{ijl}, \tag{3.2.10}$$

where

$$\mathcal{T}^{\mathrm{sr}}_{ijl} = -\frac{8\sqrt{\pi}}{\alpha} \sum_{\mathbf{a}} \Phi_{\frac{3}{2}}(\tilde{r}^2) \tilde{r}_i \tilde{r}_j \tilde{r}_l \tag{3.2.11}$$

and

$$\begin{aligned}\mathcal{T}^{\mathrm{sm}}_{ijl} = \frac{\alpha}{V} \sum_{\mathbf{k}\neq 0} \Big[& 2(2\pi\mathrm{i})(k_i\delta_{jl} + k_j\delta_{il} + k_l\delta_{ij})\Phi_0(\tilde{k}^2) \\ & + \frac{\alpha}{\pi}(2\pi\mathrm{i})^3 k_i k_j k_l \Phi_1(\tilde{k}^2)\Big] \exp[2\pi\mathrm{i}\,\mathbf{k}\cdot(\mathbf{x}-\mathbf{y})]. \end{aligned} \tag{3.2.12}$$

Numerical integration

Because the Green's function is singular as $\mathbf{x} \to \mathbf{y}$ when $\mathbf{a} = \mathbf{0}$, special treatment is required to ensure the convergence of the numerical integration quadrature. Nearly singular points also degrade the accuracy of integration quadratures and require special treatment. Near a collocation point, $\mathbf{x}_0$, a generic singular integral can be decomposed into two parts,

$$\begin{aligned}\iint_D \mathcal{K}(\mathbf{y}-\mathbf{x}_0) f(\mathbf{y})\, \mathrm{d}S(\mathbf{y}) = & \iint_D \mathcal{K}(\mathbf{y}-\mathbf{x}_0)\, \eta[\rho(\mathbf{y},\mathbf{x}_0))]\, f(\mathbf{y})\, \mathrm{d}S(\mathbf{y}) \\ & + \iint_D \mathcal{K}(\mathbf{y}-\mathbf{x}_0)\, [1-\eta(\rho(\mathbf{x},\mathbf{x}_0))]\, f(\mathbf{x})\, \mathrm{d}S(\mathbf{x}), \end{aligned} \tag{3.2.13}$$

where $\rho(\mathbf{y}, \mathbf{x}_0)$ is the distance between the points $\mathbf{y}$ and $\mathbf{x}_0$ along the surface of a reference sphere where the cell shape is mapped, and

$$\eta(\rho) = \begin{cases} \exp\left(2\, \frac{\exp(-1/t)}{t-1}\right) & \text{for } t = \rho/\rho_1 < 1, \\ 0 & \text{for } \rho \geq \rho_1. \end{cases} \tag{3.2.14}$$

Because the second integrand in (3.2.13) is smooth, a standard quadrature corresponding to the discretized membrane shape converges rapidly with increasing number of base points. The first integral is evaluated in plane polar coordinates over a patch centered at $\mathbf{x}_0$. In the case of a nearly-singular integral, the polar coordinates are centered at a surface point that is closest to the evaluation point. Bicubic splines are used to interpolate the integrand on this patch. The necessary derivatives are computed via fast Fourier transforms (FFT). The integrand on this polar coordinate patch is nonsingular but discontinuous at the origin. Since the integrand is smooth elsewhere, applying a Gaussian quadrature in the radial direction on a sinh-compressed mesh and a simple sum in the azimuthal direction provides us with an accurate approximation. The integration converges as $O(h^3)$, where h is the local mesh spacing (Zhao *et al.* 2010).

System assembly

The first integral in (3.1.3) is computed over all cell surfaces, $D = \{\partial\Omega_i \text{ for } i > 0\}$, and over all wall boundaries, W. The second integral is computed only over cell surfaces, and then only in the case of discontinuous viscosity. Treating D and W as separated surfaces in (3.1.3) and forming expressions analogous to (3.1.5) separately for points on cells and walls, we obtain

$$\vec{\mathbf{u}}_c = \mathbf{A}_{cc}\vec{\mathbf{f}}_c + \mathbf{A}_{cw}\vec{\mathbf{f}}_w + \mathbf{B}_{cc}\vec{\mathbf{u}}_c \tag{3.2.15}$$

for the velocity at all collocation points on the cells, $\vec{\mathbf{u}}_c$, and

$$\vec{\mathbf{u}}_w = \mathbf{A}_{wc}\vec{\mathbf{f}}_c + \mathbf{A}_{ww}\vec{\mathbf{f}}_w + \mathbf{B}_{wc}\vec{\mathbf{u}}_c \tag{3.2.16}$$

for the velocity at collocation points on the wall, $\vec{\mathbf{u}}_w$, where $\mathbf{A}_{pq}$ and $\mathbf{B}_{pq}$ are influence matrices. The wall velocities are set to zero due to the no-slip boundary condition. For a given configuration of collocation points, $\vec{\mathbf{x}}$, all matrices $\mathbf{A}_{pq}$ and $\mathbf{B}_{pq}$ in (3.2.15) and (3.2.16) are known, and the discrete cell forces $\vec{\mathbf{f}}_c$ are available. Combining (3.2.15) with (3.2.16) we formulate a linear system,

$$\begin{bmatrix} \mathbf{I} - \mathbf{B}_{cc} & -\mathbf{A}_{cw} \\ -\mathbf{B}_{wc} & -\mathbf{A}_{ww} \end{bmatrix} \begin{bmatrix} \vec{\mathbf{u}}_c \\ \vec{\mathbf{f}}_w \end{bmatrix} = \begin{bmatrix} \mathbf{A}_{cc}\vec{\mathbf{f}}_c \\ \mathbf{A}_{wc}\vec{\mathbf{f}}_c \end{bmatrix}. \tag{3.2.17}$$

Given the position vectors of all collocation points, $\vec{\mathbf{x}}$, this system can be assembled and solved for all cell velocities and wall forces. However, explicitly formulating the system is both inefficient and unnecessary. Instead, the fast methods discussed in Section 3.3 are used to carry out the matrix-vector multiplication in (3.2.17), in conjunction with an iterative solver. Rather than doing this directly for the full system, a simple split-stepping time advancement scheme is employed.

3.2.2 Time advancement

The split-time algorithm is based on the first-order Euler scheme. More accurate schemes can be implemented by extending this basic approach (Zhao *et al.* 2010). Given the collocation point position, $\mathbf{x}^n$, and wall forces, $\mathbf{f}^n$, at time level n, the nodal velocities are calculated directly from (3.2.15),

$$(\mathbf{I} - \mathbf{B}_{cc}^n)\, \vec{\mathbf{u}}_c = \mathbf{A}_{cc}^n\, \vec{\mathbf{f}}_c(\vec{\mathbf{x}}_c^n) + \mathbf{A}_{cw}\, \vec{\mathbf{f}}_w. \tag{3.2.18}$$

For $\lambda = 1$, we obtain an explicit expression for $\vec{\mathbf{u}}_c$. For $\lambda \neq 1$, the GMRES algorithm is used to obtain a solution after approximately 20 iterations for the configurations discussed in Section 3.6 (Saad & Schultz 1986). Each iteration requires a fast matrix-vector multiplication.

Collocation points on the cells move with the local fluid velocity,

$$\frac{\mathrm{d}\vec{\mathbf{x}}_c}{\mathrm{d}t} = \vec{\mathbf{u}}_c. \tag{3.2.19}$$

For first-order explicit time advancement,

$$\vec{\mathbf{x}}^{n+1} = \vec{\mathbf{x}}^n + \Delta t \vec{\mathbf{u}}^n, \tag{3.2.20}$$

where Δt is the time step.

An estimate for $\mathbf{f}^{n+1}$ to be used with (3.2.16) is made based on the available solution at the current and previous time levels,

$$-\mathbf{A}_{ww}\vec{\mathbf{f}}_w^{n+1} = \mathbf{B}_{wc}^{n+1}\vec{\mathbf{u}}_c^n + \mathbf{A}_{wc}^{n+1}\vec{\mathbf{f}}_c^{n+1}. \tag{3.2.21}$$

Approximating $\mathbf{u}_c^{n+1}$ with $\mathbf{u}_c^n$ introduces an error that is consistent with the first-order time integration error in computing $\mathbf{u}_c^{n+1}$ by (3.2.20), while avoiding the solution of (3.2.17) as a complete system. In the examples presented in section 3.6, 20 GMRES iterations are sufficient to reduce the relative residual to less than 10^{-3}.

3.2.3 Flow specification

Equation (3.1.3) introduces subtleties in setting up parameters of interest for studying cellular flow in confined geometries. The mean velocity of the flow in Ω determined by $\mathbf{U}$ in (3.1.3) is not convenient to be used as a control parameter. The reason is that, for geometries like that shown in figure 3.2, the mean velocity takes into account the flow outside the vessel. In the case of a straight circular tube, accurate series solutions for the exterior flow can be used to relate the mean velocity along the vessel to the mean flow rate inside the vessel (Zhao *et al.* 2010). For more complex geometries, the velocity for any particular velocity $\mathbf{U}$ needs to be explicitly integrated over a volume

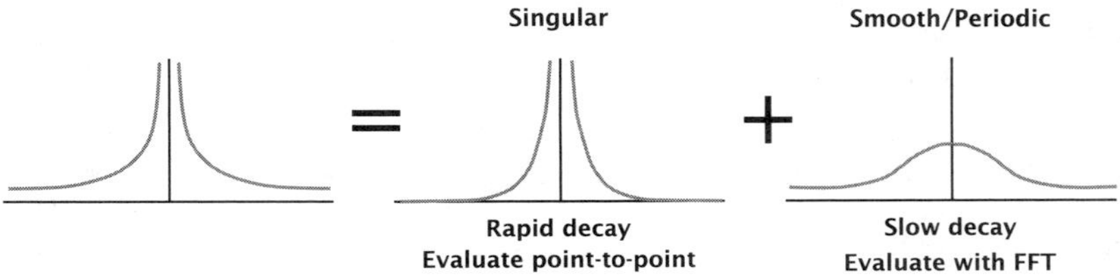

Figure 3.3.1 Schematic illustration of the periodic Green's function decomposition (3.2.3) and (3.2.9) for fast numerical approximation of the convolution integrals (3.1.3).

of interest in order to calculate a relevant mean flow rate inside a confined geometry. Similar difficulties are encountered in specifying the mean pressure gradient corresponding to the velocity $\mathbf{U}$ driving the flow. Because the applied pressure gradient is exactly balanced by the net forces on a wall, it could be included in (3.2.17) as an additional constraint.

These concerns make the present boundary-integral formulation less direct in terms of setting up desired flow properties compared to mesh-based Eulerian solutions, at least in terms of specifying the mean velocity or pressure gradient. However, in parametric studies of flow rate against effective viscosity, a range of $\langle \nabla p \rangle$ can be studied by varying $\mathbf{U}$ directly without solving the full $\langle \nabla p \rangle$-constrained system. Setting up iterative methods that adjust $\mathbf{U}$ in order to set up the desired $\langle \nabla p \rangle$ are straightforward.

3.3 Fast summation in boundary-integral computations

The Ewald decomposition introduced in (3.2.3) and (3.2.10) facilitates the implementation of periodic boundary conditions and provides us with a basis for a fast algorithm. The full Green's function tensor $\mathcal{G}$ is both singular and slowly decaying, as depicted schematically in figure 3.3.1. Slow decay requires accounting for long-range influences and a large number interactions for accurate velocity evaluations. Direct discretization of (3.2.2) on a mesh avoids explicitly accounting of all interaction. However, the singular forces corresponding to the singular Green's functions hinder accurate direct discretization.

The two challenging aspects of the Green's function are separated by the Ewald decomposition. The first component, $\mathcal{G}^{\mathrm{sr}}$, is singular but decays rapidly with distance from the point force. The second component, $\mathcal{G}^{\mathrm{sm}}$, decays slowly but is smooth and periodic and thus well approximated on a mesh. Although other decompositions share this property (Pozrikidis 1996), it appears that only the Ewald decomposition has been incorporated into fast solvers as presently described.

3.3.1 Short-range component evaluation

Apart from handling singular and nearly singular points discussed in section 3.2, evaluating the short-range component, $\mathcal{G}_{ij}^{sr} f_j$ as in (3.1.3) is straightforward. Because $\mathcal{G}_{ij}^{sr}$ decays rapidly in space when α is sufficiently small, only nearby interactions need to be considered. This is done most efficiently if, for each collocation point, a list of nearly interacting collocation points is maintained. If interactions are deemed negligible beyond a certain cutoff distance, r_c, the list is easily constructed with $O(N)$ operations by dividing the entire periodic computational domain into approximately cubic cells with dimension greater than r_c. Standard methods from molecular dynamics are available (Frenkel & Smit 1996).

Looping over all nearby points is an $O(N)$ process, and the short-range contributions to the discrete approximations scale with $O(N)$ multiplied by a small coefficient. Thus, the short-range portions of the $\mathbf{A}\vec{\mathbf{f}}$ and $\mathbf{B}\vec{\mathbf{u}}$ products in (3.1.5) and (3.2.17) are computed with high accuracy with this scaling. Since increasing r_c decreases the error in the truncation as $\exp(-r_c^2)$, higher accuracy can be achieved by including only a small number of additional interactions.

3.3.2 Smooth component evaluation

The efficient evaluation of the smooth part of the sum is more involved. Unlike the short-range part, no interactions can be neglected outright. Although the sums in (3.2.6) and (3.2.12) are rapidly convergent, a key to efficiency is the fast Fourier transform (FFT).

Since the FFT is naturally formulated on regular uniform meshes, the term $\mathbf{F}^\beta = \mathbf{f}^\beta w^\beta$ in (3.1.4) must be interpolated onto a regular mesh. The FFT of the approximation on this mesh, $\vec{\mathbf{F}}$, is $\hat{\mathbf{F}}(\mathbf{k})$. In compact notation,

$$\hat{\mathbf{F}} = \mathcal{F}\{\mathcal{W}\vec{\mathbf{F}}\}, \tag{3.3.1}$$

where $\mathcal{W}$ is a distribution operation discussed by Hockney & Eastwood (1988) for the B-splines presently employed.

The smooth part of the Green's function reformulated as a sum on $\mathbf{k}$ is essentially an inverse Fourier transform,

$$\mathcal{G}_{ij}^{\text{sm}} = \sum_{\mathbf{k}} \hat{\mathcal{G}}_{ij}^{\text{sm}} \exp(2\pi \mathrm{i}\, \mathbf{k}\cdot\mathbf{x}), \tag{3.3.2}$$

and likewise for $\mathcal{T}_{ijk}^{\mathrm{sm}}$. Necessary convolution sums, such as those seen in (3.1.4), are constructed as products in the $\mathbf{k}$ space and then inversely transformed to physical coordinates as

$$\mathcal{F}^{-1}\left\{\hat{\mathcal{G}}^{\mathrm{sm}}\cdot\hat{\mathbf{F}}\right\}. \tag{3.3.3}$$

The procedure provides us with an estimate for the smooth Green's function convolutions in (3.1.3) as a field represented by discrete points on the FFT mesh. The result needs to be transferred back to collocation points, which amounts to undoing the previous distribution onto the mesh. Combining these steps, we obtain the approximation

$$A\vec{\mathbf{f}} \simeq \mathcal{W}^{-1}\left[\mathcal{F}^{-1}\left\{\hat{\mathcal{G}}^{\mathrm{sm}}\cdot\mathcal{F}\{\mathcal{W}\vec{\mathbf{F}}\}\right\}\right]. \tag{3.3.4}$$

The procedure is abstractly presented here. Operations for each component in this expression are implemented in separate subroutines. For compact distribution operators, $\mathcal{W}$, the distribution and interpolation operations scale with N. Assuming that the mesh scales with the system size, N, the FFT operations are $O(N\log N)$, far better than those required for direct evaluation involving matrix-vector multiplication. A corresponding approximation can be made for $B\vec{\mathbf{u}}$ in (3.1.5).

3.3.3 Particle–particle/particle–mesh method (PPPM)

Any sensible choice of distribution and interpolation operations would provide a readily computed estimate of the matrix–vector multiplication in (3.1.5). However, significant benefits can be gained by tailoring this overall particle–particle/particle–mesh (PPPM) approach to minimize errors. A number of methods have been developed for electrostatic interactions. Differences pertain to the means of interpolating velocities, distribution of forces between points (the collocation points on the cells) and the FFT mesh, and management of the error.

The basic approach was introduced by Hockney & Eastwood (1988) for B-spline interpolants. A continuum-based reasoning was employed to develop a modified Green's function, called the influence function, that minimizes errors. These authors developed and extensively analyzed detailed expressions similar to those appearing on the right-hand side of (3.3.4). A similar method focusing on the Ewald decomposition for electrostatics using the exact Green's function and Lagrange interpolants, known as the particle-mesh Ewald (PME) method, was proposed by Darden *et al.* (1993). Their approach was generalized by Essemann *et al.* (1995) using cardinal B-splines into the smooth-particle-mesh Ewald (SPME) method, making it close to the original PPPM. The smoothness of the B-spline interpolation is advantageous.

In electrostatics, the gradient of the electrostatic potential energy is required to calculate the force. In the original implementation, the computation of derivatives violated conservation of momentum to some extent (Deserno & Holm 1998a). This concern does not arise in the case of Stokes flow where a corresponding derivative operation does not appear and the original SPME formulation does not violate the momentum balance. It appears that particle–mesh methods for Stokes flow were applied first by Higdon and coworkers (Guckel 1997, 1999; Metsi 2000). A similar approach built on the standard Lagrange interpolation PME was implemented in an accelerated Stokesian dynamics algorithm by Sierou & Brady (2001). The procedure was generalized to smooth B-spline interpolants and used to simulate sedimenting filaments by Saintillan *et al.* (2005). Their formulation is adopted in the present implementation (Zhao *et al.* 2010).

The accuracy of particle–mesh schemes is hard to estimate and optimize when simultaneously attempting to minimize the computational expense. Choices must be made for α, the B-spline order, p, the FFT mesh size, N_f, and the cutoff radius, r_c. Although some guidance is available in the literature, precise rules are not available (Hockney & Eastwood 1988, Deserno & Holm 1998*ab*). Different problems are best solved with different adjustments of these parameters. Given the important role of α in setting the decay rates of the sums in (3.2.5) and (3.2.6), it is not surprising that, for fixed p, M and r_c, the accuracy strongly depends on this parameter. Graphs of the log of the error versus α are V-shaped, as seen in figure 3.3.2 (Deserno & Holm 1998*ab*). The results presented in this figure pertain to the calculation of the velocity induced by 100 point forces placed randomly in a $5\times5\times5$ computational box. In this example, the PME error is compared with an accurate and expensive direct evaluation of the Ewald sum (3.2.3).

In our simulations with dimensionless red blood cell volume $V = 4\pi/3$, we take $\alpha = 0.04$ and use eighth-order ($p = 8$) B-splines. The cutoff radius for the short-range interactions is set such that

$$\frac{\mathcal{T}^{\rm sr}_{ijk}}{\mathcal{T}^{p}_{ijk}} \lesssim \epsilon, \tag{3.3.5}$$

evaluated at r_c, where $\epsilon = 10^{-3}$. Similarly, the mesh spacing for the FFT is set based on the decay of the smooth part of the potential, such that

$$\Phi_1(\tilde{k}^2_{\max})\tilde{k}^2_{\max} = \epsilon. \tag{3.3.6}$$

This corresponds to a Fourier mesh size

$$N_f \gtrsim L\Big(\frac{-\log\epsilon}{\pi\alpha}\Big)^{1/2} \tag{3.3.7}$$

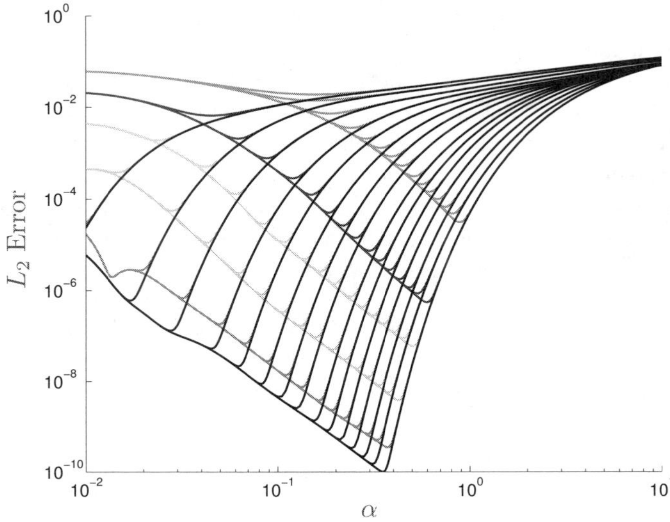

Figure 3.3.2 Particle–mesh Ewald error for various parameters for a case discussed in the text with $N_f = 200$ (lowest downward sloping diagonal curve), 160, 100, 60, 40 and 20, with cutoff radius varying from $r_c = 0.15$ to 1.55. (*Color in the electronic file.*)

in each direction, where L is the period of Ω_0. The lowest value of N_f satisfying this inequality is chosen. The same value $\epsilon = 10^{-3}$ is used in all simulations discussed in this chapter.

3.4 Membrane mechanics

The force balance equations governing the membrane mechanics provide us with a force field $\mathbf{f}$ to be used with the boundary-integral equation (3.1.3).

3.4.1 Spectral basis functions

To calculate the membrane load $\mathbf{f}$, we interpolate through membrane collocation points using a finite sum of orthogonal global basis functions. Spherical harmonics are a natural choice for sphere-like cells. The cell surface is represented as a function of colatitude angle, $\theta = \theta_1 \in [0, \pi]$, and longitudinal angle, $\phi = \theta_2 \in [0, 2\pi)$, as

$$\mathbf{x}(\theta, \phi) = \sum_{n=0}^{N_s-1} \sum_{m=0}^{n} \bar{P}_n^m(\sin\theta)(\mathbf{a}_{nm}\cos m\phi + \mathbf{b}_{nm}\sin m\phi), \tag{3.4.1}$$

where $\mathbf{a}_{nm}$ and $\mathbf{b}_{nm}$ are expansion coefficients and $\bar{P}_n^m(x)$ are normalized associated Legendre polynomials (Swarztrauber & Spotz 2000). With these basis functions, singularities at the poles $\theta = 0$ and π do not arise and the resolution is uniform over the unit sphere. The modes are distributed in a way that avoids over-resolution in ϕ near the poles that would arise if this same latitude–longitude surface mesh were discretized in finite difference or finite element implementations. Such over-resolution would significantly suppress the time step permissible for stable explicit time advancement. These properties are discussed by Zhao *et al.* (2010) in detail.

Since all derivatives of (3.4.1) are well defined, it is straightforward to analytically differentiate and compute quantities necessary for calculating membrane tensions. The surface tangent vectors, $\mathbf{a}_1$ and $\mathbf{a}_2$, and the normal vector, $\mathbf{a}_3 = \mathbf{n}$, are

$$\mathbf{a}_1 = \frac{\partial \mathbf{x}}{\partial \theta}, \qquad \mathbf{a}_2 = \frac{\partial \mathbf{x}}{\partial \phi}, \qquad \mathbf{a}_3 = \mathbf{n} = \frac{\mathbf{a}_1 \times \mathbf{a}_2}{|\mathbf{a}_1 \times \mathbf{a}_2|}. \tag{3.4.2}$$

The surface deformation gradient can be constructed from the tangent vectors of the deformed and reference surface,

$$\hat{\mathbb{F}} = \mathbf{a}_1 \mathbf{A}^1 + \mathbf{a}_2 \mathbf{A}^2, \tag{3.4.3}$$

where the vectors $\mathbf{A}^1$ and $\mathbf{A}^2$ are the reciprocal tangent vectors of the reference surface satisfying the relation

$$\mathbf{A}^\alpha \cdot \mathbf{A}_\beta = \delta_\beta^\alpha \qquad \alpha, \beta = 1, 2 \tag{3.4.4}$$

with $\mathbf{A}_{1,2}$ being the tangent vectors of the reference surface.

3.4.2 Constitutive equations

An in-plane neo-Hookean elastic strain energy can be defined in terms of the invariants of left Cauchy–Green tensor, $\hat{\mathbb{V}}^2 = \hat{\mathbb{F}} \cdot \hat{\mathbb{F}}^T$,

$$I_1 = \lambda_1^2 + \lambda_2^2 - 2, \qquad I_2 = \lambda_1^2 \lambda_2^2 - 1, \tag{3.4.5}$$

where λ_1 and λ_2 are the eigenvalues of the in-plane components of $\hat{\mathbb{V}}^2$. The elastic strain-energy function is given by

$$W_s = \frac{E_S}{4}\left(\frac{1}{2} I_1^2 + I_1 - I_2\right) + \frac{E_D}{8} I_2^2, \tag{3.4.6}$$

and the corresponding Cauchy stress is

$$\tau = \frac{E_S}{2 J_S}(I_1 + 1)\hat{\mathbb{V}}^2 + \frac{J_S}{2}(E_D I_2 - E_S)\hat{\mathbb{P}}, \tag{3.4.7}$$

where $J_S = \lambda_1\lambda_2$ is the dilatation, $\hat{\mathbb{P}} = \hat{\mathbb{I}} - \mathbf{nn}$ is a projection matrix, $\hat{\mathbb{I}}$ is the identity matrix, E_S is a shear elastic modulus, and E_D is a large surface dilation modulus resisting surface area changes.

A linear isotropic model for the bending moment $\mathbf{M}$ is used,

$$M^{\alpha}_{\beta} = -E_B(b^{\alpha}_{\beta} - b^{\alpha\,R}_{\beta}) \tag{3.4.8}$$

for $\alpha, \beta = 1, 2$, where E_B is the bending modulus and $b_{\alpha\beta} = \mathbf{a}_{\alpha,\beta} \cdot \mathbf{n}$ is the second fundamental form of the surface (covariant component of the curvature tensor) of the deformed surface, and $b^R_{\alpha\beta}$ is the corresponding second fundamental form of the reference surface.

3.4.3 Equilibrium equations

A torque balance requires

$$M^{\alpha\beta}_{|\alpha} - Q^{\beta} = 0, \qquad e_{\alpha\beta}(N_{\alpha\beta} - b^{\alpha}_{\gamma}M^{\gamma\beta}) = 0, \tag{3.4.9}$$

where $\mathbf{Q}$ is the surface transverse tensor, $\mathbf{N}$ is the in-plane tension, and the subscript '$|\alpha$' denotes a covariant derivative. A force balance provides us with the hydrodynamic surface traction $\mathbf{f}$ to be used in (3.1.3),

$$N^{\alpha\beta}_{|\alpha} - b^{\beta}_{\alpha}Q^{\alpha} + f^{\beta} = 0 \tag{3.4.10}$$

for $\beta = 1, 2$, and

$$Q^{\alpha}_{|\alpha} + N^{\alpha\beta}b_{\alpha\beta} + f^3 = 0, \tag{3.4.11}$$

where f^3 denotes the normal component of $\mathbf{f}$ (Pozrikidis 2003*b*).

To verify the implementation, the following elastic moduli were chosen: $E_S = 4.2 \times 10^{-6}$ N/m, $E_B = 1.8 \times 10^{-19}$ N m, and $E_D = 6.8 \times 10^{-5}$ N / m. Our three-dimensional simulations closely match Pozrikidis's (2005) results (Zhao *et al.* 2010). These parameter values provide us with a reasonable quantitative model for computing the effective viscosity of blood in small tubes, as discussed in Section 3.6.

We have neglected viscous dissipation due to membrane deformation. Membrane viscosity is expected to affect relaxation times in some cases (Evans & Hochmuth 1976, Pozrikidis 2005). However, the significance of the membrane viscosity on normal blood flow is unclear.

3.5 Numerical fidelity

All derivatives were computed in terms of the spherical harmonic basis functions (3.4.1) which are insensitive to accumulation of error by repeated differentiation. Extension to more general and complex strain energy models require straightforward modifications.

Boundary element discretizations have been used by previous authors in simulations of capsule and cell deformation (Pozrikidis 2001, 2003*a*; Doddi & Bagchi 2009, Yechun & Dimitrakopoulos 2007, Muldowney & Higdon 1995). These methods are advantageous for representing complex geometries that are not well suited to spherical harmonic expansions. In the present work, a standard boundary-element discretization is used for the vessel walls.

Spectral element methods can provide super-geometric convergence with respect to the order of the element spectral basis functions. However, there are key advantages to using a global spectral basis stemming from practical aspects of the simulation with a finite number of collocation points, as opposed to theoretical estimates of convergence as the number of collocation points is increased. The essentially perfect finite-mesh resolution of the scheme is discussed in the following section. The ability to suppress nonlinear instability by explicit dealiasing is discussed in the section after that and demonstrated by numerical experimentation.

3.5.1 Truncation errors, convergence and resolution

The rate of convergence of a numerical approximation is determined by the truncation error of the numerical scheme expressed, for example, by the first neglected term in an asymptotic series. A three-point centered finite difference approximation with mesh spacing h converges as $O(h^2)$, yielding a second-order method. Global spectral bases allow for an exponential convergence in the case of infinitely differentiable smooth functions, where the error scales as $e^{-1/h}$. For a spectral-element method, refinement of the basis functions inside elements shows the same order of convergence.

Convergence in the limit $h \to 0$ is not a clear descriptor of the accuracy of the scheme at finite h. Although it is essential to obtain h-independent results for observable quantities of interest, different observables demand different levels of accuracy. For apparently chaotic systems where the observables are often statistical, as in turbulence flow or molecular dynamics (Mittal & Moin 1997, Frenkel & Smit 1996), low-order schemes with favorable conservation properties can significantly outperform high-order schemes in terms of accuracy. This may also be the case in the statistical investigation of cellular flows (Freund 2007).

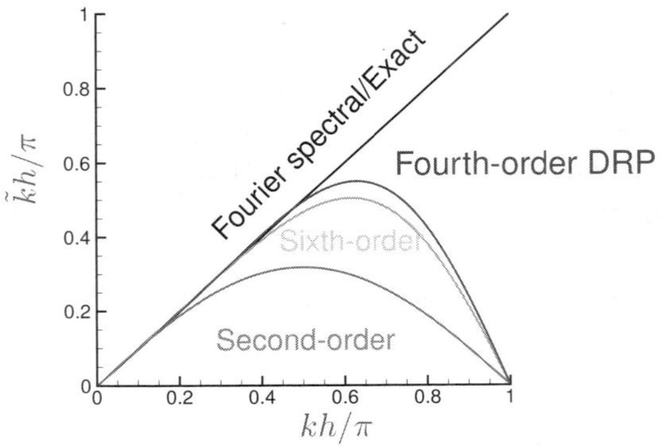

Figure 3.5.1 Modified wave numbers for standard order-optimized centered second- and sixth-order finite-difference schemes, and illustration of a common coefficient optimized seven-point fourth-order scheme (Tam & Webb 1993) and the exact behavior of a Fourier scheme. (*Color in the electronic file.*)

The finite-h behavior of a scheme is tied to the resolution quantified by the ability of the method to represent different wavelengths. Resolution can be measured by testing a scheme with the function $g(x) = e^{\mathrm{i}kx}$, where i is the imaginary unit. For the second-order central finite-difference scheme, we find that

$$g' \simeq \frac{g(x+h) - g(x-h)}{2h} = \mathrm{i}\tilde{k}g, \tag{3.5.1}$$

where

$$\tilde{k} \equiv k\frac{\sin(kh)}{kh} \tag{3.5.2}$$

is the modified wave number. Plotting $\tilde{k}$ against the exact wave number k provides us with a visual representation of the resolution, as shown in figure 3.5.1, The order of the scheme is deduced by examining how fast $\tilde{k}$ approaches k as $h \to 0$. The rest of the plot reveals how well the scheme performs for finite h. Finite-difference schemes can be designed to sacrifice formal order for superior resolution (Lele 1992), as seen in figure 3.5.1. For the global spectral method, it is clear that the collocation points used are used with optimal efficiency: $\tilde{k}$ is exact $\tilde{k} = k$.

Spectral-element methods share this advantage of global spectral methods on an element basis, exhibiting rapid convergence for $h \to 0$. However, because a low degree continuity is enforced between elements, the exact resolution of the global spectral methods is no longer possible. In fact, the modified wave number behavior of finite-element schemes can be alarming, especially in the case of higher derivatives (Kwok *et al.* 2001), and large numbers of points are needed before rapid convergence is achieved. This undesirable property is manifested indirectly in Stokes flow spectral-element discretizations (Dimitrakopoulos & Jingtao 2007, Yechun & Dimitrakopoulos 2006). The solution can display kinks incurring significant errors in $\tilde{k}$ even on virtually uniform meshes. Although advantageous in representing complex geometries, mesh nonuniformity can further degrade the accuracy.

The perfect resolution of the global spectral basis functions makes the scheme particularly efficient at finite resolutions. Every collocation point constitutes an exact representation of the shape for the represented corresponding mode. Exponential convergence is attractive when extremely accurate solutions are desired. However, when simulating a population of interacting cells, there is little hope of closely tracking the solution trajectory for a long time. It is the finite-h resolution of the scheme that facilitates efficient computation in this case.

3.5.2 Aliasing errors, nonlinear instability and dealiasing

Aliasing errors are different from truncation errors. Whereas truncation errors originate from the numerical approximation of continuous functions, aliasing errors originate from nonlinear terms due to finite resolution. This is most easily seen in one dimension where the Fourier transform is a natural way to describe spectra. The present algorithm was first developed for two-dimensional cells where Fourier series expansions of the position of collocation points were used (Freund 2007). The spherical harmonics employed for three-dimensional cells are generalizations of these.

Assume that a two-dimensional cell is parameterized by a discrete set $\theta_j = 2\pi j/N$, where $j = 0, 1, \ldots, N-1$. The spectrum of one component of the position vector $\mathbf{x}$ is

$$\hat{x}_n = \frac{1}{N} \sum_{j=0}^{N-1} x_j \, \mathrm{e}^{-in\theta_j}, \tag{3.5.3}$$

and the inverse is

$$x_j = \sum_{n=-N/2}^{N/2-1} \hat{x}_n \, \mathrm{e}^{in\theta_j}. \tag{3.5.4}$$

An evolution governed by a system of linear differential equations with constant coefficients, $\partial_t q = L(q)$, will not mix Fourier modes. Consequently, the spectrum will retain the support prescribed in the initial state provided that the discretization does not introduce nonlinear interactions or dispersion. Mesh metrics for stretching effectively introduce non-constant coefficients that may spread the spectrum.

A nonlinear right-hand side $\partial_t q = N(q)$ will broaden the spectrum. For example, a quadratic nonlinearity doubles the highest possible wave number. Consider the spectrum of x_j^2,

$$\begin{aligned}(\hat{x}^2)_n &= \frac{1}{N}\sum_{j=0}^{N-1} x_j x_j \mathrm{e}^{-in\theta_j} \\ &= \frac{1}{N}\sum_{j=0}^{N-1}\Big(\sum_{m=-N/2}^{N/2-1} \hat{x}_m \mathrm{e}^{im\theta_j}\Big)\Big(\sum_{l=-N/2}^{N/2-1} \hat{x}_l \mathrm{e}^{il\theta_j}\Big)\mathrm{e}^{-in\theta_j} \\ &= \sum_{m=-N/2}^{N/2-1}\sum_{l=-N/2}^{N/2-1} \hat{x}_m \hat{x}_l \Big(\frac{1}{N}\sum_{j=0}^{N-1} \mathrm{e}^{\mathrm{i}(m+l-n)\theta_j}\Big). \end{aligned} \tag{3.5.5}$$

The last factor has a discrete orthogonality property

$$\frac{1}{N}\sum_{j=0}^{N-1} \mathrm{e}^{\mathrm{i}(m+l-n)\theta_j} = \begin{cases} 1 & m+l-n = pN \text{ for } p = 0, \pm 1, \pm 2, \ldots, \\ 0 & \text{otherwise.} \end{cases} \tag{3.5.6}$$

As $N \to \infty$, providing us with infinite resolution, only the $p = 0$ condition of (3.5.6) is relevant and (3.5.5) becomes

$$(\hat{x}^2)_n = \sum_{m+l=n} \hat{x}_m \hat{x}_l = \sum_{m=-N/2=-\infty}^{N/2-1=\infty} \hat{x}_m \hat{x}_{n-m}, \tag{3.5.7}$$

which is a statement of the discrete convolution theorem: products in physical coordinates are exact convolutions in Fourier space.

For finite N, the terms $p = \pm 1$ in (3.5.6) also play a role,

$$(\hat{x}^2)_n = \underbrace{\sum_{m+l=n} \hat{x}_m \hat{x}_l}_{p=0} + \underbrace{\sum_{m+l=n\pm N} \hat{x}_m \hat{x}_l}_{p=\pm 1}, \tag{3.5.8}$$

where $p = \pm 1$ represent spurious terms. Energy that would have been deposited by the x_j^2 product into modes with $|n| > N/2$ is aliased into modes with $|n| < N/2$ as consistent with the maximum resolution of a mesh of size N (figure 3.5.1). If the product had been computed as a convolution, only

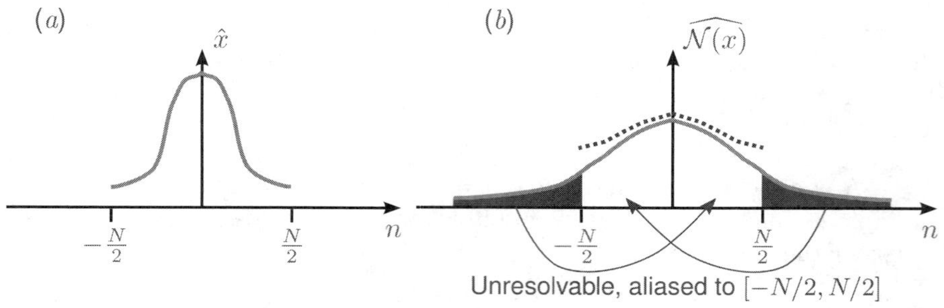

Figure 3.5.1 (a) Schematic illustration of the spectrum of a cell shape component, x. (b) Aliasing error due to a physical coordinate nonlinear operation $\mathcal{N}(x)$ on x. The dotted blue curve shows an increased resolved energy over the exact red curve due to aliasing. The effect is exaggerated for clarity in this illustration. (*Color in the electronic file.*)

the physical $p = 0$ terms would be retained. However, the procedure would be unreasonably expensive and only limited to simple product nonlinearities.

Aliasing errors are commonly discussed in the context of the Navier–Stokes equations which have a quadratic nonlinearity (Canuto *et al.* 1987, Patterson & Orszag 1971, Kravchenko & Moin 1997). Although the governing equations are linear in the case of Stokes flow, the membrane forces $\mathbf{f}$ are nonlinear functions of the point particle position. This nonlinearity will broaden spectra, as shown schematically in figure 3.5.1. This affects $\mathbf{x}$ as it evolves in time according to the local velocity.

The underlying spectral broadening is physical and thus independent of the discretization (Orszag 1971). Because of the finite numerical resolution, energy that cannot be described on a numerical mesh is aliased into resolved modes, as shown in figure 3.5.1(b). Retention of this energy in the solution is inconsistent with the specified resolution. If the resolution is N, energy in modes higher than N should be removed.

Dealiasing

Aliasing errors decrease with increasing resolution and are not necessarily any more of a problem than truncation errors. Aliasing errors are typically most noticeable in high-resolution nondissipative implementations where numerical dissipation does not provide a suppressing mechanism. In the case of spectral methods, a linear problem is, in a sense, solved exactly up to its selected resolution, so that all errors can be interpreted as originating from an aliasing mechanism.

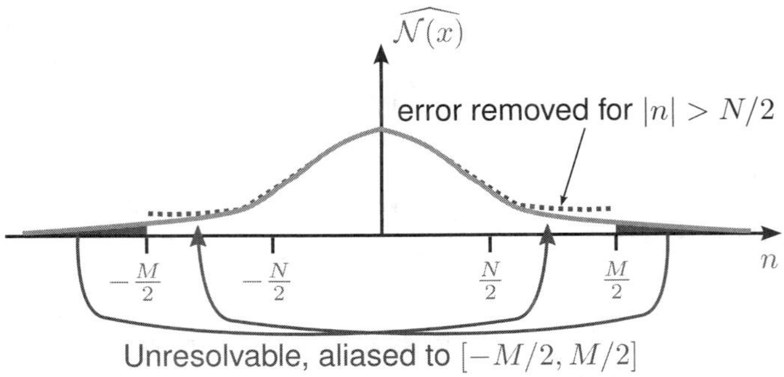

Figure 3.5.2 Schematic illustration of dealiasing. Nonlinear terms are computed at M rather than N collocation points. Upon filtering back to N, the aliasing error is suppressed or even removed for certain nonlinearities.

An important concern is that aliasing enables nonlinear instability. Energy that would have been removed in a way that is consistent with the selected resolution, is instead retained and can introduce further aliasing, leading to the growth of disturbances. Increasing the resolution suppresses the amplification by reducing the amount of energy aliased. However, if increasing the mesh size is done only to maintain stability, the additional effort is potentially wasteful.

A better and significantly less expensive alternative for suppressing aliasing is to dealias the nonlinear terms, as shown schematically in figure 3.5.2. For quadratic nonlinearity with N mesh points, aliasing is suppressed when the product operation is computed on $M = 3N/2$ points and then filtered back to N points (Canuto *et al.* 1987).

The nonlinearity of the membrane model is far more intricate, involving products, roots, and inverses associated with geometric terms in the finite deformation constitutive model. A more nonlinear constitutive model would introduce additional nonlinearity. These terms preclude exact dealiasing of the kind commonly used for the Navier–Stokes equation.

However, approximate dealiasing is straightforward. For N collocation points, performing the nonlinear operations on $M > N$ points and filtering the result back to N points suppresses aliasing errors. This is significantly less expensive than simply performing a calculation with M points at the outset in order to ensure stability. The main reason is that the maximum time in explicit integration scales approximately as N^{-2} for the shear and dilatational terms, and approximately as N^{-4} for the bending term (Zhao *et al.* 2010). Significantly longer stable time steps are permitted by this approach. In ad-

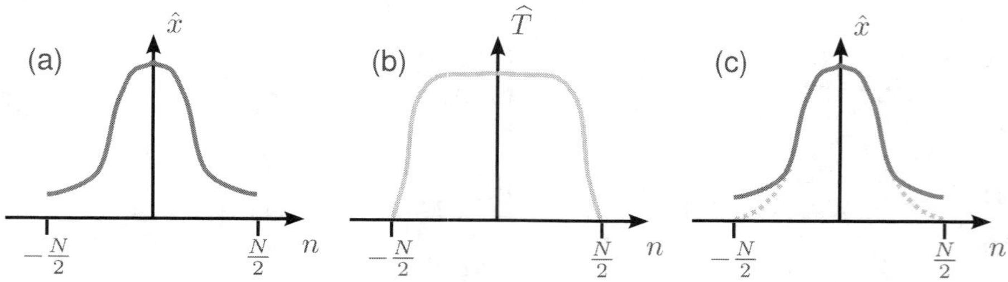

Figure 3.5.3 Schematic illustration of (*a*) solution spectrum, and (*b*) a filter or numerical dissipation transfer function, and (*c*) the result of applying filtering or implementing the dissipation. The procedure suppresses instability due to aliasing errors by damping the solution where aliasing is most important (see figure 3.5.2), but does not necessarily exactly balance the errors and potentially degrades the solution.

dition, the iterative solver for the components of (3.2.17) generally converges faster for smaller systems. If N resolution is acceptable, these savings far outweigh the extra cost incurred by using more points to ensure numerical stability.

Explicit filtering or smoothing can also suppress aliasing errors by damping the aliasing feedback mechanism. However, the resolved solution is damped as well, as shown in figure 3.5.3. Filtering, in essence a convolution operation, is tantamount to multiplying the spectrum by a transfer function. For compact spatial filters, the transfer function affects all wave numbers and dampens the entire solution, unphysically smoothing the fine features of the deformation. The broader the spectrum of the high wave number damped, the weaker the restriction on the time step. The rigorous approach is dealiasing with a sharp cutoff, which leaves the resolved part of the solution unaffected. Numerical dissipation inherent in many discretization schemes has a similar effect.

3.6 Simulations

Simulations will be presented to demonstrate and validate various aspects of the numerical method. The results of the three-dimensional simulations discussed in this section were confirmed by Zhao *et al.*(2010) to accurately reproduce the results of axisymmetric simulations presented by Pozrikidis (2005).

3.6.1 Resolution and dealiasing

In the first example, we demonstrate the high resolution of the numerical scheme and the ability to independently control resolution and stability. The flow is visualized in figure 3.6.1 at high resolution. Lengths have been reduced by the equivalent radius so that the dimensionless cell volume is equal to $V = 4\pi/3$. A single cell is placed inside a tube with length $L = 10$ and diameter $D = 3.5$, necking down to a local diameter $d(z) = 3.5 - 2.2\exp[-1.2(z-5)^2]$. The transverse dimensions of the periodic domain are $L_x = L_y = 4$. A dimensionless scaled velocity $U_z = 2$ drives the flow. A single cell is introduced at the axial position $z = 3.25$ and then shifted by 0.25 off the center of the tube along the x axis. We take $\lambda = 1$ for computational expedience and also because the cell more rapidly develops small length-scale corrugations.

The cell is significantly distorted as it moves through the narrow neck and the membrane is drawn out forming rib-like corrugations, as shown in figure 3.6.1(*f*). A crown-like shape develops when the cell exits the neck region and begins relaxing, as shown in figure 3.6.1(*h*). The small dimensions of the rib-like features are expected to challenge the resolution of the numerical discretization. With such obvious distortion, there is a clear opportunity for geometric nonlinearity to cause aliasing errors.

Without dealiasing, cases with $N_s = 32$ and $N_s = 48$ run stably up to dimensionless time $t \approx 0.9$. At that point, ribbing corrugations appear and the simulation becomes unstable. Increasing the spatial resolution by setting $N_s = 64$ is necessary for a stable calculation. In all cases without dealiasing, $M_s = N_s + 2$, so that the highest mode number where numerical derivatives are ambiguous is removed. All cases shown in figure 3.6.1 with $M_s = 2N_s$ are stable for $N_s \geq 12$.

For $N_s = 8$, dealiasing at least by a factor of three with $M_s = 24$ is necessary for stability. The illustration shown in figure 3.6.2(a) reveals that the prominent rib-like corrugation of the cell membrane artificially disappears when $N_s = 8$. When $N_s = 12$, the corrugations are visible but clearly thicker and oriented somewhat differently than those obtained with larger values of N_s. For $N_s = 16$, the corrugations take the shape observed at high N_s. For $N_s \geq 24$, the cell shapes are indistinguishable even upon close inspection.

Capturing with high accuracy every feature of the cell membrane may be unnecessary and thus wasteful in some applications. A quantitative assessment of the resolution is made in figure 3.6.3(a) showing the evolution of the streamwise pressure drop. Even a low-order expansion with $N_s = 8$ produces accurate results in the early stage of the motion. The curves for $N_s \geq 12$ are indistinguishable in the early stage of the motion but become apparent at later times.

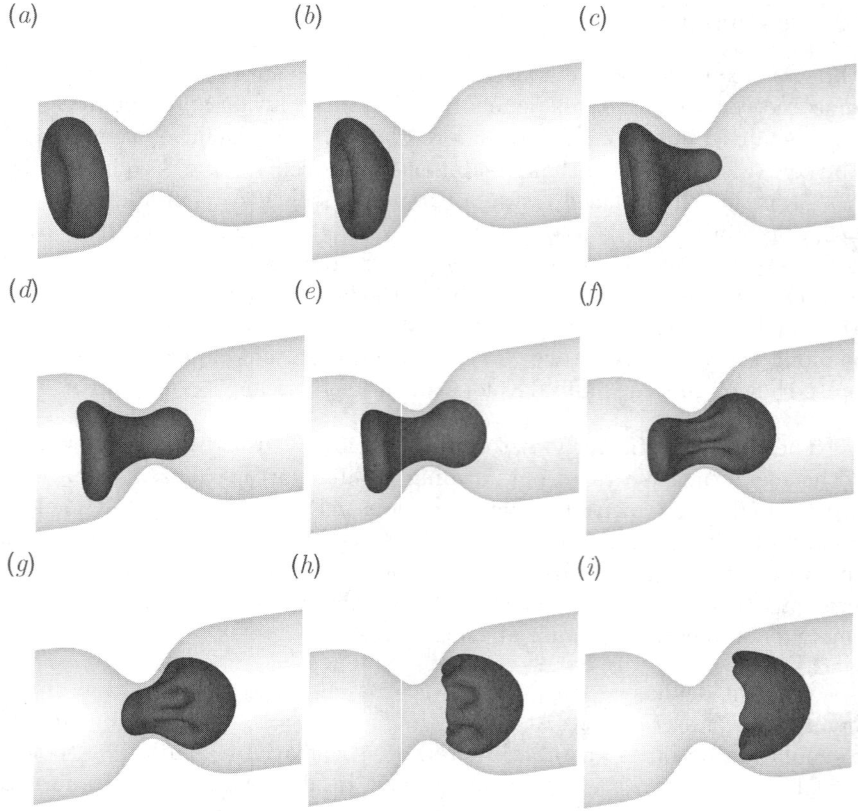

Figure 3.6.1 Evolution of a well-resolved cell ($N = 32$, $M = 64$) passing through a narrow neck at dimensionless time (*a*) 0, (*b*) 0.2, (*c*) 0.4, (*d*) 0.6, (*e*) 0.8, (*f*) 1.0, (*g*) 1.2, (*h*) 1.4, and (*i*) 1.6,

Figure 3.6.3(*b*) shows that the surface spectra are insensitive to N_s and virtually independent of N_s when $N_s \geq 12$. It is remarkable that the spectrum in the marginally resolved case, $N_s = 12$, falls nearly on top of the spectra of the better resolved cases. The modes in low resolution simulations are good approximations of those in high-resolution simulations.

Large savings are possible by judiciously selecting N_s, while M_s is adjusted to ensure numerical stability. Zhao *et al.* (2010) showed that, for relatively small N, $\Delta t_{\max} \sim N_s^{-2}$. Accordingly, the time step for $N_s = 12$ can be 32 times that for $N_s = 64$. These savings compound reduced work associated with a smaller N_s, including rapid convergence of integrative linear solvers.

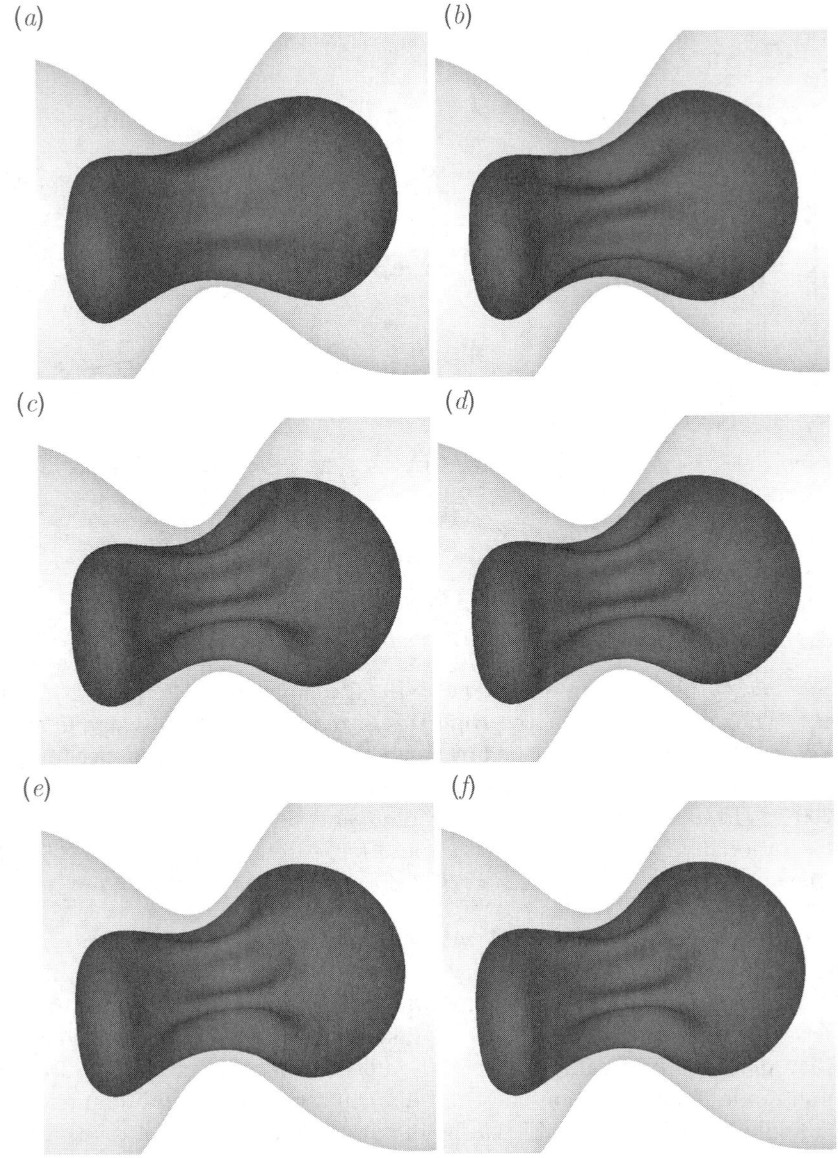

Figure 3.6.2 Cell shape visualized in figure 3.6.1(*f*) at dimensionless time $t = 1$ for increasing (N, M) (*a*) $(8, 48)$, (*b*) $(12, 48)$, (*c*) $(16, 48)$, (*d*) $(24, 48)$, (*e*) $(32, 64)$, and (*f*) $(64, 66)$.

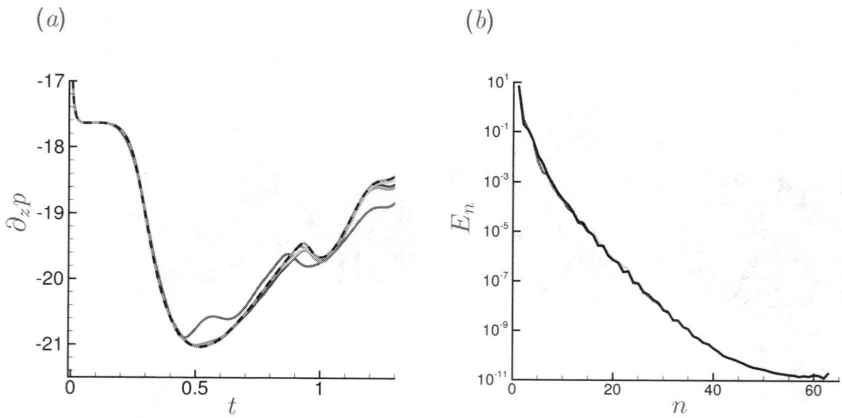

Figure 3.6.3 (a) Net domain pressure drop and (b) surface spectra for the cases visualized in figure 3.6.1 for $N = 8, 12, 16, 24, 32$, and 64. (*Color in the electronic file.*)

3.6.2 Effective viscosity

An elementary physical model of the red cell membrane is employed in our studies. The actual constitutive properties of red blood cells are significantly more complex than those described by a neo-Hookean model. Evidence for this is provided by the recent experimental data of Puig-de-Morales-Marinkovic *et al.* (2007). However, because the choice of geometric and constitutive parameters is rationally based on observation and measurement (Pozrikidis 2003a), some level of agreement with experimental observations is expected.

Our results were compared with an empirical fit of extensive experimental data for the effective viscosity of blood in small cylindrical capillaries (Pries *et al.* 1992). Although the experimental data are widely scattered and the dependence on the shear rate is not taken into consideration, the effective viscosity deduced by assuming the Poiseuille flow profile clearly exhibits a nonmonotonic dependence on the tube diameter, reaching a minimum when the tube diameter is comparable to the maximum dimension of a relaxed cell. Results of simulations presented in figure 3.6.4 reproduce this behavior and provide a reasonable match to the fitted data within an apparent scatter window. Given the uncertainty of the experimental data, we have not attempted to modify the cell properties in order to improve the match.

Some cells in a concentrated suspension are pressed together for a long period of time. However, a thin lubrication layer always separates the membranes and physical contact should occur in a mathematical idealization of

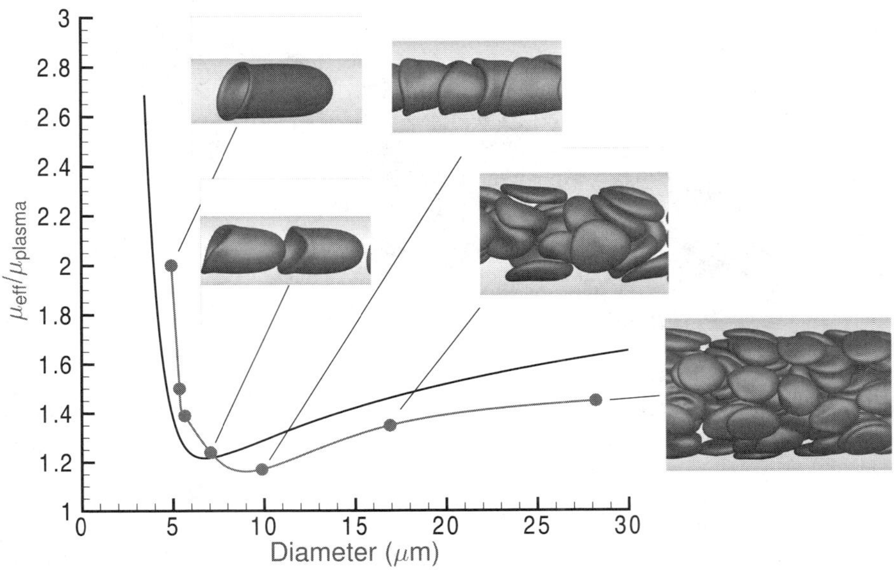

Figure 3.6.4 Dependence of the effective Newtonian-fluid Poiseuille-flow viscosity on the tube diameter for hematocrit $H_c = 0.30$. The solid black line is an empirical fit through a wide range of experimental data (Pries *et al.* 1992). (*Color in the electronic file.*)

this configuration only after an infinite period of time. Numerical error can lead to the lubrication layers erroneously disappearing. To prohibit contact and crossing in these simulations targeted at computing the effective viscosity, the membrane separation is constrained to be larger than one percent of the sphere-equivalent diameter, $0.02\,(\frac{3}{4}\,V)^{1/3}$, where V is the cell volume.

3.6.3 Leukocyte transport

Figure 3.6.5 illustrates a round vessel with diameter 16.9 μm containing 23 red blood cells and one white blood cell. The leukocyte is modeled as an elastic shell with a reference shape smaller than that of the red blood cells, resembling a stiff balloon. The periodic vessel length is $L = 35.8\mu$m and the mean hematocrit is 0.30. This physical arrangement is the three-dimensional version of a two-dimensional configuration recently considered (Freund 2007). The two-dimensional simulations showed that the probability that a leukocyte lies near the vessel wall decreases rapidly with increasing the flow rate, in agreement with experimental observations (Firrell & Lipowsky 1989, Abbitt & Nash 2003). The effect is attributed to changes in the shape and distribution

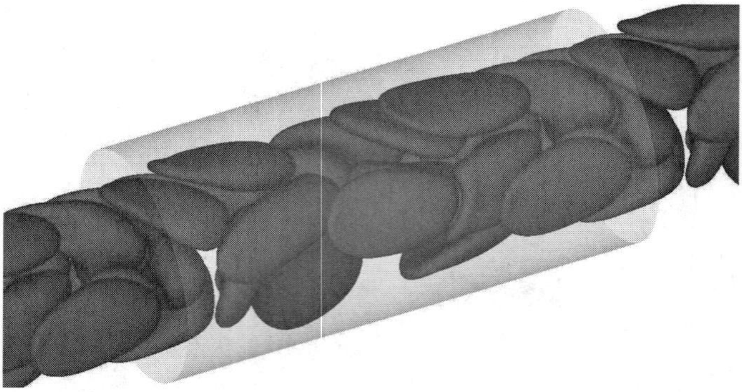

Figure 3.6.5 Illustration of a leukocyte (blue) surrounded by 23 red blood cells in a 16.9μm diameter vessel. (*Color in the electronic file.*)

of red blood cells in the neighborhood of the leukocyte. The thickness of the cell-free layer near the wall increases with increasing the flow rate.

Figure 3.6.6 illustrates the average red blood cell density in the neighborhood of the leukocyte near the wall for a low and a high flow rate. In the case of low flow rate described in figure 3.6.6(a), the red cell distribution is concentrated on the upstream side of the leukocyte and a wake-like structure is observed on the downstream side. A significant back up of red cells is not observed in the case of high flow rate described in figure 3.6.6(b). Consistent with experimental observations, the leukocyte is positioned closer to the wall, and it is thus more likely to bind to the wall in the case of low flow rate. As seen in previous simulations of two-dimensional flow (Freund 2007), the cell-free layer far from the leukocyte is thicker when the flow rate is high. Documenting the effect of the flow rate on the cell-free layer thickness is important for understanding the effective rheology of blood flow in small vessels.

3.6.4 Complex geometries

Since the vessel walls are described with standard boundary-element discretization, the geometry of the flow domain is not restricted to simple shapes. As an example, figure 3.6.7 shows a visualization of blood flow through a model vessel network involving 30 periodically repeated cells. The branches diameters range from 10 to 20 μm, and the length of the periodic network in the flow direction is 85.8 μm. Each cell is represented with $N_s = 30$, $M_s = 60$ collocation points, and the vessel walls are described by 22,503 nodes defining 44,864 triangular elements.

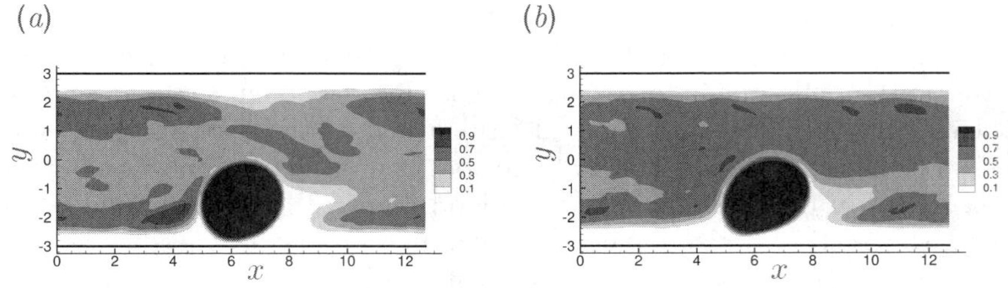

Figure 3.6.6 Contours of cell density relative to leukocyte centroid in the plane passing through the centerline of the microvessel and the leukocyte centroid. The mean velocity is (a) $\bar{u}_z = 0.9$ mm/s, and (b) $\bar{u}_z = 3.6$ mm/s.

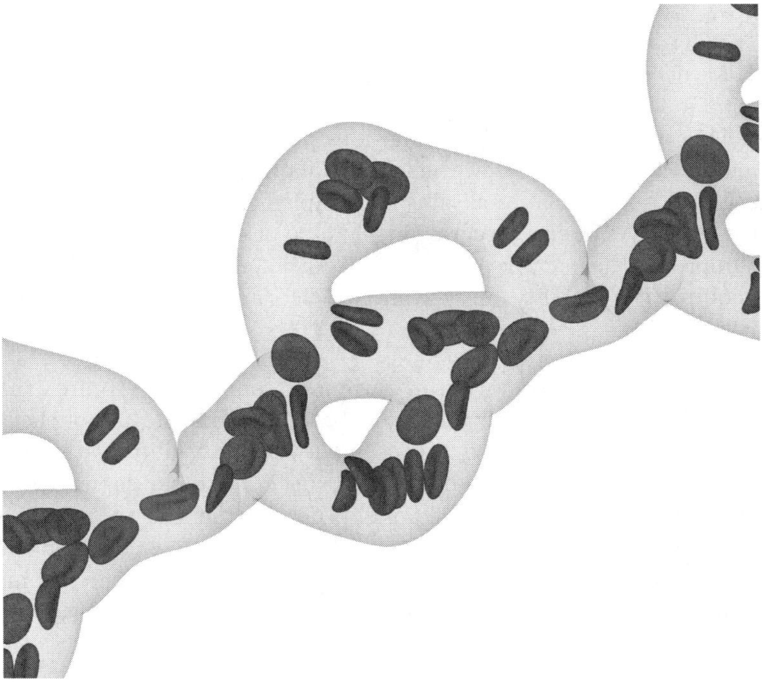

Figure 3.6.7 Simulated blood flow through a model vessel network with complex geometry.

3.7 Summary and outlook

We have discussed an algorithm for simulating cellular blood flow with emphasis on the philosophy motivating the specifics of the numerical design and implementation. While the proposed formulation enjoys favorable properties, any particular algorithm is generally not best suited for all conceivable applications even in the limited context of blood flow in the microcirculation. In closing this section, we summarize its favorable properties and discuss some weaknesses. None of the drawbacks are sufficiently strong to warrant a different approach for the range of problems considered.

3.7.1 Pros

The pros have been discussed at length in this chapter. The small coefficient in the overall $O(N \log N)$ scaling law makes the PME method attractive over multipole methods. Although mesh-based methods, such as the immersed-boundary methods, are $O(N)$, boundary-integral implementations have certain advantages.

High accuracy in any boundary-integral scheme is primarily achieved by refining a surface grid rather than a three-dimensional volume mesh. The close spacing of cells seen in the simulations presented in this chapter imposes demands on the accuracy. However, the availability of analytical expressions for the Green's functions facilitates the implementation of quadratures for computing nearly singular integrals in the case of closely spaced surfaces and greatly improves the accuracy with little additional expense. A similar adaptivity requires significantly more complex algorithms in mesh-based implementations.

The Fourier representation of the mesh component requires that the domain is triply periodic. The constraint-based implementation of the no-slip boundary condition in the time-advancement scheme facilitates the inclusion of general surfaces inside a periodic box and effectively removes the periodicity, as needed. Free-space solutions are possible despite the underlying periodicity of the Fourier representation (Hockney & Eastwood 1988, Pollock & Glosli 1996). Since we are principally interested in confined or periodic geometries, we have not pursued these extensions.

The spectral representation of the cell surfaces facilitates dealiasing to counteract nonlinear instability. Applying filtering or dissipative smoothing would degrade the quality of the solution. A consequence is that every mesh point makes an exact contribution to the accuracy of the solution in that no points restrict the time step while only marginally contributing to accuracy. Accuracy is thus the only criterion in setting the resolution.

3.7.2 Cons

Three-dimensional FFTs do not scale particularly well on distributed memory parallel computer systems. The main reason is that nonlocality requires a high degree of communication between different processors. Although we have not attempted to optimize these algorithms, we anticipate that communication cost will be a factor in selecting α, N_f and r_c as discussed in Section 3.3. Fast multipole methods are advantageous for large systems in that a minimal amount of information is exchanged in parallel implementations even when the floating-point operation count exceeds that of the PME method.

Multipole methods are naturally adaptive to local demands in resolution and thus advantageous when the cells are widely distributed throughout the flow domain. The PME method is only partially adaptive: although the short-range sum (3.2.5) is local and active only in the presence of cells, the FFTs in (3.2.6) depend on the volume. In our discussion of computational cost with regard to number of collocation points, we have assumed a somewhat uniform distribution throughout the domain. Recent multisummation techniques resembling multigrid methods provide us with a venue for evaluating the smooth part of the potential with a small operation count that may scale well with the number of processors (Skeel *et al.* 2002).

Because the boundary-integral formulation hinges on the linearity of the governing equations, a generalization of the present approach to cases of finite fluid inertia is not possible. Although inertia is negligible in the low-Reynolds-number flow through the microcirculation, this restriction reduces the applicability of the method to a broader range of finite-Reynolds-number flows that can be simulated with mesh-based methods.

Another subtle drawback of the boundary-integral method concerns the specification of flow parameters. A direct method of setting the pressure gradient or the flow velocity in a blood vessel is not available (Zhao *et al.* 2010). However, this is only a mild inconvenience in parametric studies where a one-to-one correspondence exists between $\mathbf{U}$ in (3.1.3) and the resulting mean flow and mean pressure. Iterating to any desired conditions is straightforward. It might be possible to modify the formulation in a way that includes the flow conditions as additional constraints in the full formulation (Zhao *et al.* 2010). The additional required expense does not appear to be warranted in general applications.

The computational scheme involves two distinct spectral representations, interpolation between collocation points and FFT-mesh cells, special methods for evaluating singular and nearly singular integrals, and iterative solvers for implementing boundary constraints and handling the implicit system arising in the case of unequal viscosities. The complexity of the numerical procedure is a drawback in a new implementation. Additional effort is also required for analyzing data bases. For example, the solver itself produces the velocity at collocation points over the cells. Additional quadratures are needed to generate the velocity field. Plans are under way to publicly distribute the computer code. Information on the distribution venue can be obtained from the first author.

Outlook

The algorithm can be improved in generality and efficiency. The time step currently used in the explicit time advancement scheme is restricted by the usual constraints. Implicit schemes permit larger time steps but are useful only if the rapid time scales in the numerical solution are spurious in some sense. For the time scales of importance to the physical mechanisms being studied, accuracy degrades when the time step is not too much longer than the stability limit for corresponding explicit schemes. Unimportant time scales are associated with the near incompressibility of the membrane.

The high membrane dilatation modulus imposes a restriction on the time step. However, the dilatation modulus can be set to a high value to significantly suppress dilatation without restricting the time step much beyond that imposed by the bending moments. For significantly stiffer cells, such as leukocytes or platelets, implicit time advancement might allow significantly more efficient solutions. Unfortunately, implementing implicit schemes is cumbersome (Dimitrakopoulos 2007). Further investigation is required to improve the efficiency of the algorithms.

Another potential direction for investigation is the implementation of multiple time stepping. In molecular dynamics simulations, it is recognized that slowly varying long-range components of the potentials do not need to be evaluated frequently. Taking advantage of this can significantly reduce the overall expense of the algorithm. The Green's function decomposition used in our implementation appears to share this property. Multistep simplectic integrators typically used in molecular dynamics are not necessary in the present non-Hamiltonian formulation.

Acknowledgment

We gratefully acknowledge support from the National Science Foundation and the University of Illinois. Amir H. G. Isfahani conducted the simulation of the network presented in Section 3.6 and provided the image, for which we are also grateful. The first author is grateful for discussions with Robert Moser early in the development of this approach.

References

ABBITT, K. B. & NASH, G. B. (2003) Rheological properties of the blood influencing selectin-mediated adhesion of flowing leukocytes. *Am. J. Physiol. Heart Circ. Physiol.* **285**, H229–H240.

ALBERTS, B., JOHNSON, A., LEWIS, J., RAFF, M., ROBERTS, K. & WALTER, P. (2008) *Molecular Biology of the Cell.* Fifth Edition, Garland Science, New York.

CANUTO, C., HUSSAINI, M. Y., QUARTERONI, A. & ZANG, T. A. (1987) *Spectral Methods in Fluid Dynamics.* Springer–Verlag, Berlin.

DARDEN, T., YORK, D. & PEDERSEN, L. (1993) Particle mesh Ewald: An $n\log(n)$ method for Ewald sums in large systems. *J. Chem. Phys.* **98**, 10089–10092.

DESERNO, M. & HOLM, C. (1998*a*) How to mesh up Ewald sums. I. A theoretical and numerical comparison of various particle mesh routines. *J. Chem. Phys.* **109**, 7678–7693.

DESERNO, M. & HOLM, C. (1998*b*) How to mesh up Ewald sums. II. An accurate error estimate for the particle-particle-particle-mesh algorithm. *J. Chem. Phys.* **109**, 7694–7701.

DIMITRAKOPOULOS, P. (2007) Interfacial dynamics in Stokes flow via a three-dimensional fully-implicit interfacial spectral boundary element algorithm. *J. Comp. Phys.* **225**, 408–426.

DIMITRAKOPOULOS, P. & JINGTAO, W. (2007) A spectral boundary element algorithm for interfacial dynamics in two-dimensional Stokes flow based on Hermitian interfacial smoothing. *Eng. Anal. Bound. Elem.* **31**, 646–656.

DINTENFASS, L. (1968) Internal viscosity of the red cell and a blood viscosity equation. *Nature* **219**, 956–958.

DODDI, S. K. & BAGCHI, P. (2009) Three-dimensional computational modeling of multiple deformable cells flowing in microvessels. *Phys. Rev. E* **79**, 046318.

ESSEMANN, U., PERERA, L., BERKOWITZ, M. L., DARDEN, T., LEE, H. & PEDERSEN, L. G. (1995) A smooth particle mesh Ewald method. *J. Chem. Phys.* **103**, 8577–8593.

EVANS, E. A. (1983) Bending elastic modulus of red blood cell membrane derived from buckling instability in micropipette aspiration tests. *Biophys. J.* **43**, 27–30.

EVANS, E. A. & HOCHMUTH, R. M. (1976) Membrane viscoelasticity. *Biophys. J.* **16**, 1–11.

EWALD, P. (1921) Die berechnung optischer und elektrostatischer gitterpotentiale. *Ann. Phys.* **369**, 253–287.

FÅHRÆUS, R. & LINDQVIST, T. (1931) The viscosity of the blood in narrow capillary tubes. *Am. J. Physiol.* **96**, 562–568.

FIRRELL, J. C. & LIPOWSKY, H. H. (1989) Leukocyte margination and deformation in mesenteric venules of rat. *Am. J. Physiol., Heart Circ. Physiol.* **256**, H1667–H1674.

FRENKEL, D. & SMIT, B. (1996) *Understanding Molecular Simulation.* Academic Press, San Diego.

FREUND, J. B. (2007) Leukocyte margination in a model microvessel. *Phys. Fluids* **19**, 023301.

GARON, A. & FARINAS, M.-I. (2004) Fast three-dimensional numerical hemolysis approximation. *Artificial Organs* **28**, 1016–1025.

GOODSELL, D. S. (1998) *The Machinery of Life.* Copernicus, New York.

GREENGARD, L. (1988) *The Rapid Evaluation of Potential Fields in Particle Systems.* The MIT Press, Boston.

GUCKEL, E. K. (1997) *A P^3M Method for Calculation of Stokes Interactions.* Master's thesis, University of Illinois, Urbana-Champaign.

GUCKEL, E. K. (1999) *Large Scale Simulations of Particulate Systems using the PME Method.* Doctoral Dissertation, University of Illinois, Urbana-Champaign.

HASIMOTO, H. (1959) On the periodic fundamental solutions of the Stokes equations and their application to viscous flow past a cubic array of cylinders. *J. Fluid Mech.* **5**, 317–328.

HERNÁNDEZ-ORTIZ, J. P., DE PABLO, J. J. & GRAHAM, M. D. (2007) Fast computation of many-particle hydrodynamic and electrostatic interactions in a confined geometry. *Phys. Rev. Let.* **98**, 140602.

HOCHMUTH, R. M. & WAUGH, R. E. (1987) Erythrocyte membrane elasticity and viscosity. *Ann. Rev. Physiol.* **49**, 209–219.

HOCKNEY, R. W. & EASTOOD, J. W. (1988) *Computer Simulation Using Particles.* Institute of Physics Publishing, Bristol.

KARASAWA, N. & GODDARD III, W. A. (1989) Acceleration of convergence for lattice sums. *J. Phys. Chem.* **93**, 7320–7327.

KRAVCHENKO, A. G. & MOIN, P. (1997) On the effect of numerical errors in large eddy simulations of turbulent flows. *J. Comp. Phys.* **131**, 310–322.

KROGH, A. (1922) *The Anatomy and Physiology of Capillaries.* Yale University Press, New Haven.

KWOK, W. Y., MOSER, R. D. & JIMÉNEZ, J. (2001) A critical evaluation of the resolution properties of B-spline and compact finite difference methods. *J. Comp. Phys.* **174**, 510–551.

LAMBERT, C. G., DARDEN, T. A. & BOARD JR., J. A. (1996) A multipole-based algorithm for efficient calculation of forces and potentials in macroscopic periodic assemblies of particles. *J. Comp. Phys.* **126**, 274–285.

LELE, S. K. (1992) Compact finite difference schemes with spectral-like resolution. *J. Comp. Phys.* **103**, 16–42.

LI, J., DAO, M., LIM, C. T. & SURESH, S. (2005) Spectrin-level modeling of the cytoskeleton and optical tweezers stretching of the erythrocyte. *Biophys. J.* **88**, 3707–3719.

LIM, C. T., DAO, M., SURESH, S., SOW, C. H. & CHEW, K. T. (2004) Large deformation of living cells using laser traps. *Acta Materialia* **52**, 1837–1845.

MERRILL, E. W. (1969) Rheology of blood. *Physiol. Rev.* **49**, 863–888.

METSI, E. (2000) *Large Scale Simulations of Bidisperse Emulsions and Foams.* PhD thesis, University of Illinois, Urbana-Champaign.

MITTAL, R. & MOIN, P. (1997) Suitability of upwind-biased finite-difference schemes for large-eddy simulation of turbulent flows. *AIAA J.* **35**, 1415.

DE MORALES-MARINKOVIC, M. P., TURNER, K. T., BUTLER, J. P., FREDBERG, J. J. & SURESH, S. (2007) Viscoelasticity of the human red blood cell. *Am. J. Physiol., Cell Physiol.* **293**, C597–C605.

MULDOWNEY, G. P. & HIGDON, J. J. L. (1995) A spectral boundary element approach to three-dimensional Stokes flow. *J. Fluid Mech.* **298**, 167–192.

ORSZAG, S. A. (1971) On the elimination of aliasing in finite-difference schemes by filtering high-wavenumber components. *J. Atmosph. Sci.* **28**, 1074–1074.

PATTERSON, G. S. & ORSZAG, S. A. (1971) Spectral calculations of isotropic performance of a subgrid scale model can be improved by turbulence: Efficient removal of aliasing interactions. *Phys. Fluids A* **14**, 2538–2541.

POLLOCK, E. & GLOSLI, J. (1996) Comments on PPPM, FMM, and the Ewald method for large periodic Coulombic systems. *Comp. Phys. Comm.* **95**, 93–110.

POZRIKIDIS, C. (1992) *Boundary Integral and Singularity Methods for Linearized Viscous Flow.* Cambridge University Press, New York.

POZRIKIDIS, C. (1996) Computation of periodic Green's functions of Stokes flow. *J. Eng. Math.* **30**, 79–96.

POZRIKIDIS, C. (2001) Effect of membrane bending stiffness on the deformation of capsules in simple shear flow. *J. Fluid Mech.* **440**, 269–291.

POZRIKIDIS, C. (2003*a*) Numerical simulation of the flow-induced deformation of red blood cells. *Ann. Biomed. Eng.* **31**, 1194–1205.

POZRIKIDIS, C. (2003*b*) Shell theory for capsules and shells. In *Modeling and Simulation of Capsules and Biological Cells,* Pozrikidis, C. (Ed.), Chapman & Hall/CRC, Boca Raton.

POZRIKIDIS, C. (2005) Axisymmetric motion of a file of red blood cells through capillaries. *Phys. Fluids* **17**, 031503.

PRIES, A. R., NEUHAUS, D. & GAEHTGENS, P. (1992) Blood viscosity in tube flow: Dependence on diameter and hematocrit. *Am. J. Physiol., Heart Circ. Physiol.* **263**, H1770–H1778.

PRIES, A. R. & SECOMB, T. W. (2003) Rheology of the microcirculation. *Clin. Hemorh. Microcirc.* **29**, 143–148.

RALLISON, J. M. & ACRIVOS, A. (1978) A numerical study of the deformation and burst of a viscous drop in an extensional flow. *J. Fluid Mech.* **89**, 191–200.

SAAD, Y. & SCHULTZ, M. H. (1986) GMRES: A generalized minimal residual algorithm for solving nonsymmetric linear systems. *SIAM J. Sci. Stat. Comp.* **7**, 856–869.

SAINTILLAN, D., DARVE, E. & SHAQFEH, E. S. G. (2005) A smooth particle-mesh Ewald algorithm for Stokes suspension simulations: The sedimentation of fibers. *Phys. Fluids* **17**, 033301.

SANGANI, A. S. & MO, G. (1996) An $O(N)$ algorithm for Stokes and Laplace interactions of particles. *Phys. Fluids* **8**, 1990–2010.

SIEROU, A. & BRADY, J. F. (2001) Accelerated Stokesian dynamics simulations. *J. Fluid Mech.* **448**, 115–146.

SKEEL, R. D., TEZCAN, I. & HARDY, D. J. (2002) Multiple grid methods for classical molecular dynamics. *J. Comp. Chem.* **23**, 673–684.

STABEN, M. E., ZINCHENKO, A. Z. & DAVIS, R. H. (2003) Motion of a particle between two parallel plane walls in low-Reynolds-number Poiseuille flow. *Phys. Fluids* **15**, 1711–1733.

SWARZTRAUBER P. N., SPOTZ F. (2000) Generalized discrete spherical harmonic transforms. *J. Comp. Phys.* **159**, 213–230.

TAM, C. K. W. & WEBB, J. C. (1993) Dispersion-relation-preserving finite difference schemes for computational acoustics. *J. Comp. Phys.* **107**, 262–281.

WEINBAUM, S., GANATOS, P. & YAN, Z.-Y. (1990) Numerical multipole and boundary integral equation techniques in Stokes flow. *Ann. Rev. Fluid Mech.* **22**, 275–316.

WHITMORE, R. L. (1968) *Rheology of the Circulation.* Pergamon Press, New York.

YECHUN, W. & DIMITRAKOPOULOS, P. (2006) A three-dimensional spectral boundary element algorithm for interfacial dynamics in Stokes flow. *Phys. Fluids* **18**, 1–16.

ZHAO, H., ISFAHANI, A. H. G., OLSON, L. N. & FREUND, J. B. (2010) A spectral boundary integral method for flowing blood cells. *J. Comp. Phys.*, doi:10.1016/j.physletb.2003.10.071

4 Simulating microscopic hemodynamics and hemorheology with the immersed-boundary lattice-Boltzmann method

J. Zhang

School of Engineering
Laurentian University
Sudbury, Canada

P. C. Johnson

Department of Bioengineering
University of California, San Diego
La Jolla

A. S. Popel

Department of Biomedical Engineering
School of Medicine
Johns Hopkins University
Baltimore

Red blood cells are the most important constituents of blood due to their physiological importance and hemodynamic significance. To study the motion of red blood cells in the microcirculation, an immersed-boundary/lattice-Boltzmann method is developed integrating fluid flow and membrane mechanics, and also accounting for cell aggregation. A detailed description of the computational modules is provided, including a lattice-Boltzmann method for fluid mechanics, an immersed-boundary method for fluid-membrane interaction, an explicit fluid property updating algorithm, and a Morse potential for modeling intercellular aggregation. Simulations are presented for several flow configurations to demonstrate the potentiality and usefulness of the computational approach.

4.1 Introduction

Blood performs a number of crucial physiological functions by supplying oxygen and nutrients to tissue, removing waste from tissue, transporting white blood cells, antibodies, and platelets for immunization and self-repair, and regulating the body pH and temperature. Abnormal blood flow is often associated with a broad range of disorders including heart disease, hypertension, diabetes, malaria, anemia, ischemia, atherosclerosis, and thrombosis (e.g., Stoltz *et al.* 1999). Understanding the rheological properties of blood is important in a variety of biomedical and bioengineering applications. Examples include the development of blood substitutes, the design of blood contacting devices, biomedical imaging and drug delivery.

At a microscopic level, blood is a concentrated suspension of several cellular components including erythrocytes or red blood cells (RBCs), leukocytes or white blood cells (WBCs), and platelets. The cells are suspended in an aqueous solution containing an assortment of molecules. Among these constituents, red blood cells play a prominent role because of their large number density ($\sim 5 \times 10^6/\text{mm}^3$) and distinctive mechanical properties. In large-scale flow occurring in the heart and large blood vessels, blood behaves like a shear-thinning fluid whose viscosity decreases as the shear rate becomes higher (e.g., Stoltz *et al.* 1999). As the vessel diameter drops below 100 microns (μm), the particulate constitution becomes apparent and the cell motion plays an important role in regulating flow at the outer branches of the circulatory system, commonly referred to as the microcirculation.

In vitro and *in vivo* studies have shown that flexible red blood cells migrate to the central region of a channel or tube. A cell-rich core with a high apparent viscosity appears, accompanied by a cell-free layer (CFL) of pure plasma acting as a lubrication layer near the vessel walls. This phase separation is responsible for a blunting of the velocity profile compared to the parabolic profile of a pure liquid, for the Fåhraeus effect describing a reduced discharge hematocrit, and for the Fåhraeus-Lindqvist effect describing a reduced effective viscosity in small channels and tubes. The flow-induced blood restructuring plays an important role in the microcirculation by determining flow resistance and biological transport (e.g., Popel & Johnson 2005).

Quantitative studies of the microcirculation have undergone rapid advances in the 1960s, benefiting from the advent of intravital and video microscopy and the development of methods for measuring cell motion and transvascular exchange of water and macromolecules. Electron microscopy has allowed us to extract critical information on the structure of microcirculatory vessels and develop insight into the mechanism of biological exchange. These developments, combined with a growing supply of information, require an in-depth analysis to identify the fundamental underlying physical processes and appreciate the significance of new discoveries.

Important physical and flow properties, such as pressure, velocity, shear stress, membrane stress, and molecular concentration, are difficult to measure on the micron scale in the laboratory. Because *in vivo* processes are highly interconnected, the interpretation and analysis of experimental data is not straightforward. With progress in computational methods, certain aspects of the microcirculatory function, specifically blood rheological properties, can now be addressed by numerical simulation to extract important information inaccessible by direct measurement.

The boundary-integral method and the immersed-boundary method provide us with two frameworks for computing the motion of deformable capsules and cells in flow (Pozrikidis 1992, 2001; Peskin 1977). In the boundary-integral method, discretization is required only over the cell membrane surface. Because microscopic blood flow occurs at Reynolds numbers much less than unity, the restriction of creeping flow is not a serious concern. Successful applications of the boundary-integral method have been reported for red blood cell motion in shear and channel flow (e.g., Pozrikidis 1995, 2001, 2003, 2005), and white blood cell migration in a suspension in channel flow (Freund 2007).

In the immersed-boundary method (IBM), fluid-membrane interaction is implemented by projecting membrane forces onto the fluid and then updating the membrane configuration according to the flow field. The advantage of this approach is that the fluid flow problem can be solved by standard numerical schemes over a fixed Eulerian mesh. The IBM can be readily combined with a variety of computational fluid dynamics formulations based on finite-element, finite-volume, and lattice-Boltzmann methods (LBM). Other numerical approaches include multi-particle collision dynamics (Noguchi & Gompper 2005), dissipative particle dynamics (Pivkin & Karniadakis 2008, Dzwinel *et al.* 2003; see Chapter 6), and lubrication approximations (Secomb *et al.* 2001, Pries & Secomb 2003; see Chapter 7).

The immersed-boundary method has been used extensively to simulate suspensions of red blood cells, vesicles, and other biological cells (e.g., Sui *et al.* 2008, Dupin *et al.* 2007; see Chapter 5). Popel and coworkers conducted investigations of red blood cell motion in shear flow (Eggleton & Popel 1998, Bagchi *et al.* 2005, Zhang *et al.* 2008). Bagchi (2007) simulated the flow of suspensions in small vessels with diameter 20–300 μm. Zhang *et al.* (2007, 2009) investigated the effect of membrane deformability and cell aggregation. White cell and platelet dynamics in microvessels was investigated by other authors (AlMomani *et al.* 2008, N'Dri *et al.* 2003).

In this chapter, an immersed-boundary/lattice-Boltzmann method (IB-LBM) is described, and results of recent studies on microscopic blood flow are discussed (Zhang *et al.* 2007, 2008, 2009). Important features, such as the membrane deformability, viscosity difference across the cell membrane, and intercellular aggregation are incorporated in the formulation. Well-known hemodynamic and hemorheological phenomena are reproduced, including tumbling, swinging, and tank-treading of single cells in shear flow, the formation of cell-free layers near walls, the blunting of the velocity profile in tube flow, and the Fåhraeus and Fåhraeus-Lindqvist effects in channel flow. The results demonstrate the potential usefulness of the numerical model for conducting further studies of blood flow.

4.2 The lattice-Boltzmann method

The lattice-Boltzmann method (LBM) is a mesoscopic simulation technique developed in the past two decades for computing fluid flow. In traditional computational fluid dynamics (CFD), systems of differential equations governing macroscopic fluid properties are solved. In the LBM, the fluid is represented by a collection of fictitious particles moving and colliding on a lattice mesh arising from the discretization of the fluid domain. LBM has several advantages over CFD, especially for multiphase flow and complex boundary geometries (Succi 2001, Sukop & Thorne 2006). Extra effort is not required to obtain the fluid pressure, and code development is straightforward and amenable to parallelization.

4.2.1 General algorithm

The basic principles of the LBM will be summarized in this section. A more detailed discussion can be found in monographs by Succi (2001) and Sukop & Thorne (2006).

Lattice dynamics of fluid particle

The method will be implemented on a square two-dimensional (2D) lattice with nine lattice velocities (D2Q9), as illustrated in figure 4.2.1. Other lattice structures are possible (Succi 2001, Sukop & Thorne 2006). A central concept of the LBM is the density distribution, $f_i(\mathbf{x}, t)$, representing the population of particles moving in the ith lattice direction at position $\mathbf{x}$ and time t. The time evolution of the density distribution is governed by the lattice-Boltzmann equation, which is a discrete representation of the Boltzmann equation in classical statistical physics,

$$f_i(\mathbf{x} + \mathbf{c}_i \triangle t, t + \triangle t) - f_i(\mathbf{x}, t) = \Omega_i, \qquad (4.2.1)$$

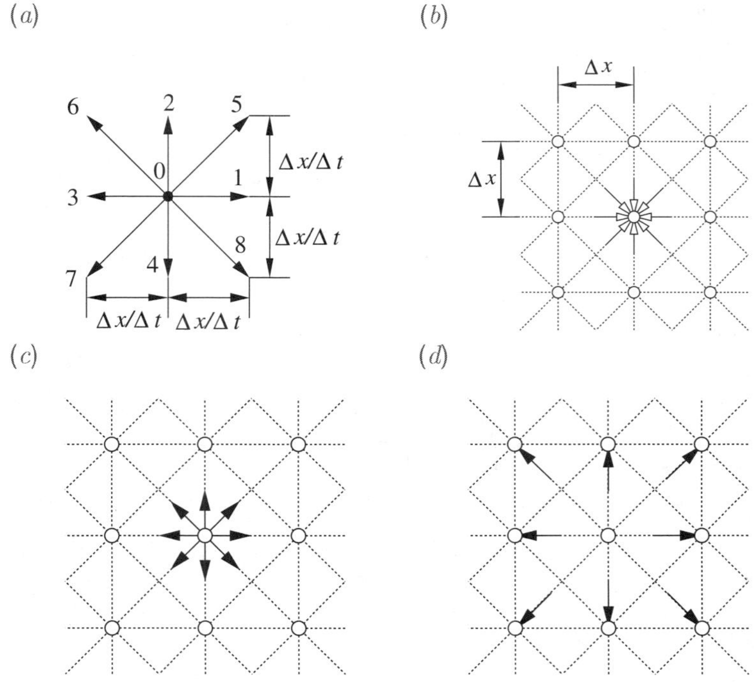

Figure 4.2.1 Illustration of (*a*) the D2Q9 lattice structure, (*b*, *c*) particle collision, and (*c*, *d*) particle propagation over the lattice grid. In (*b*–*d*), circles represent fluid nodes and dashed lines indicate lattice directions. The open arrows in (*b*) are particle populations arriving at the same node. The particles then collide and are redistributed in different directions–filled arrows in (*c*). After a time step, particles move to neighboring nodes along their individual lattice directions–filled arrows in (*d*).

where $\mathbf{c}_i$ is the ith lattice velocity and $\triangle t$ is the time step. The left-hand side of (4.2.1) represents the migration of particles from lattice node $\mathbf{x}$ at time t to a neighboring lattice node $\mathbf{x} + \mathbf{c}_i \triangle t$ at time $t + \triangle t$. The factor Ω_i on the right-hand side is a collision operator expressing the change in f_i due to particle collision.

The collision operator is typically simplified using the Bhatnagar-Gross-Krook (BGK) single-relaxation-time approximation

$$\Omega_i = -\frac{f_i(\mathbf{x}, t) - f_i^{eq}(\mathbf{x}, t)}{\tau}, \tag{4.2.2}$$

where τ is a relaxation parameter (Bhatnagar *et al.* 1954). The equilibrium distribution f_i^{eq} can be expressed as

$$f_i^{eq} = \rho t_i \left[1 + \frac{\mathbf{u} \cdot \mathbf{c}_i}{c_s^2} + \frac{1}{2} \left(\frac{\mathbf{u} \cdot \mathbf{c}_i}{c_s^2} \right)^2 - \frac{\mathbf{u} \cdot \mathbf{u}}{2c_s^2} \right], \tag{4.2.3}$$

where

$$\rho = \sum_i f_i \tag{4.2.4}$$

is the fluid density and

$$\mathbf{u} = \frac{\sum_i f_i \mathbf{c}_i}{\rho} \tag{4.2.5}$$

is the fluid velocity (Succi 2001). Other parameters, including the lattice sound speed, c_s, and weight factors, t_i, depend on the lattice structure.

In the D2Q9 model, the nine lattice velocities $\mathbf{c}_i$ are

$$\begin{aligned} \mathbf{c}_0 &= [0, \quad 0], \\ \mathbf{c}_i &= \left[\cos \frac{i-1}{2} \pi, \quad \sin \frac{i-1}{2} \pi \right] c, \quad \text{for } i = 1-4, \\ \mathbf{c}_i &= \sqrt{2} \left[\cos \frac{2i-9}{4} \pi, \quad \sin \frac{2i-9}{4} \pi \right] c, \quad \text{for } i = 5-8, \end{aligned} \tag{4.2.6}$$

where $c = \triangle x / \triangle t$, and $\triangle x$ is the lattice spacing. The weighting factors are $t_0 = 4/9$, $t_{1-4} = 1/9$, and $t_{5-8} = 1/36$. The speed of sound is $c_s = c/\sqrt{3}$.

From lattice to continuum

The macroscopic continuity equation,

$$\frac{\partial \rho}{\partial t} + \nabla \cdot (\rho \mathbf{u}) = 0, \tag{4.2.7}$$

and Navier-Stokes equation,

$$\frac{\partial \mathbf{u}}{\partial t} + \mathbf{u} \cdot \nabla \mathbf{u} = -\frac{1}{\rho} \nabla p + \nu \nabla^2 \mathbf{u}, \tag{4.2.8}$$

can be derived by applying the Chapman-Enskog expansion, where

$$\nu = \frac{2\tau - 1}{2} c_s^2 \triangle t \tag{4.2.9}$$

is the kinematic viscosity (Succi 2001). The pressure can also be obtained as

$$p = c_s^2 \rho \quad . \tag{4.2.10}$$

Body force

To account for the presence of an external body force, $\mathbf{F}$, the lattice-Boltzmann equation is modified by adding an extra term to the collision operator, yielding

$$f_i(\mathbf{x} + \mathbf{c}_i \triangle t, t + \triangle t) - f_i(\mathbf{x}, t) = -\frac{f_i(\mathbf{x}, t) - f_i^{eq}(\mathbf{x}, t)}{\tau} + \frac{t_i \triangle t}{c_s^2} \mathbf{F} \cdot \mathbf{c}_i, \tag{4.2.11}$$

Alternatively, for the purpose of calculating equilibrium distributions, f_i^{eq}, the velocity $\mathbf{u}$ in equation (4.2.3) can be replaced by the equilibrium velocity

$$\mathbf{u}^{eq} = \frac{1}{\rho} \left(\sum_i f_i \mathbf{c}_i + \triangle t \mathbf{F} \right). \tag{4.2.12}$$

These modifications increase the fluid momentum at a lattice node by $\triangle t \mathbf{F}$ without affecting the fluid density in each time step. The Navier-Stokes equation resulting from the Chapman-Enskog analysis has the correct body-force term on the right-hand side.

4.2.2 Boundary conditions

Appropriate boundary conditions are crucial for a meaningful simulation. A principal variable in LBM is the density distribution, f_i, representing the particle population moving from one lattice node to another along a certain lattice direction. For a lattice node near a boundary, density distributions entering the fluid domain after the propagation step are not available. Boundary conditions must then be enforced by specifying the unknown f_i entering the simulation domain across boundaries. These stipulations pose some difficulties in the LBM implementation.

A number of schemes have been proposed for different boundary conditions (Succi 2001, Ladd 1994, Zhang & Kwok 2006, Zou & He 1997, Mei *et al.* 1999, Guo & Zheng 2002). Periodic boundary conditions and the bounce-back method for stationary and moving boundaries employed in our simulations are discussed in this section.

Periodic boundary conditions

Periodic boundary conditions are easily implemented in LBM. All particles leaving the domain across a periodic boundary reenter the domain from the opposite side, as illustrated in figure 4.2.2. In our simulations, periodic boundary conditions are employed at open flow boundaries. The actual system simulated is an infinitely long horizontal domain with identical repeated

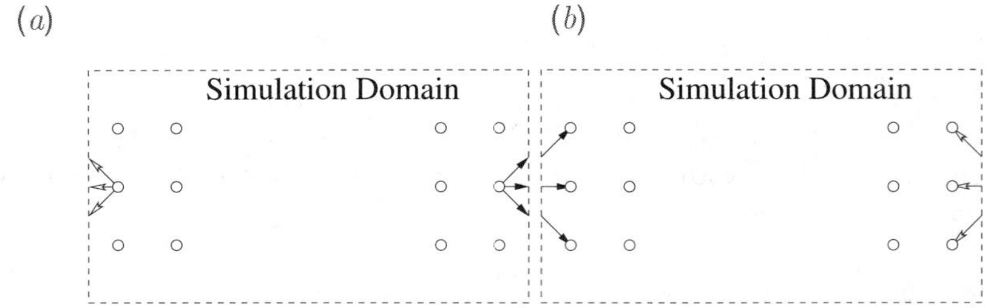

Figure 4.2.2 Illustration of a periodic boundary condition in the horizontal direction. After a collision step, particle distributions leaving the domain across the left boundary–open arrows in (*a*)–reenter the domain from the right side–open arrows in (*b*). The same treatment is applied to distributions crossing the right boundary (filled arrows).

units. To impose a pressure gradient along the channel length, the periodic boundary condition is somewhat modified, as discussed by Zhang & Kwok (2006).

Stationary boundaries

In fluid dynamics, boundary conditions are expressed in terms of macroscopic fluid properties, such as velocity and pressure. In LBM, there are usually more unknown distributions than constraints arising from the boundary conditions. To model flow over a stationary no-slip boundary, we use the mid-grid bounce-back scheme (Succi 2001). Particles leaving a boundary fluid node $\mathbf{x}_b$ bounce back from the boundary to the original site in the reversed lattice velocity, as shown in figure 4.2.3,

$$f_{\bar{i}}(\mathbf{x}_b, t + \triangle t) = f_i^*(\mathbf{x}_b, t), \tag{4.2.13}$$

where $\mathbf{c}_{\bar{i}} = -\mathbf{c}_i$ is the reversed lattice velocity and f_i^* represents the distribution leaving $\mathbf{x}_b$ after collision at time t. The lattice velocity $\mathbf{c}_i$ points outward from the fluid domain at the boundary node $\mathbf{x}_b$.

Moving boundaries

Ladd (1994) proposed a modification of the bounce-back scheme for moving boundaries by adding a new term to incorporate momentum injection to the fluid from a moving boundary,

$$f_{\bar{i}}(\mathbf{x}_b, t + \triangle t) = f_i^*(\mathbf{x}_b, t) - \frac{2\rho t_i}{c_s^2}\mathbf{u}_b \cdot \mathbf{c}_i, \tag{4.2.14}$$

where $\mathbf{u}_b$ is the boundary velocity at the bounce-back point. With $\mathbf{u}_b = \mathbf{0}$,

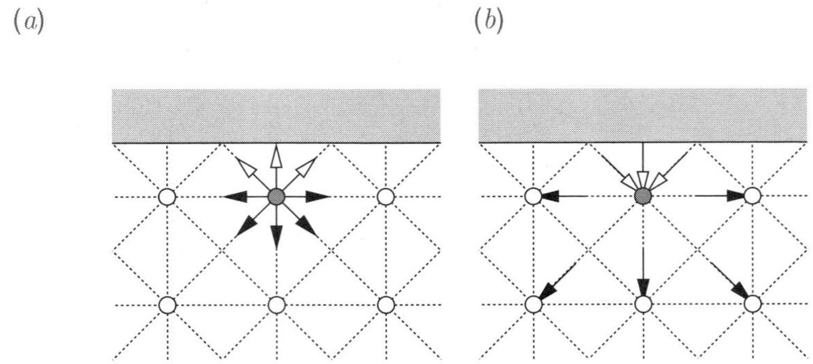

Figure 4.2.3 Illustration of the bounce-back process on a stationary boundary. The grey circle is a boundary node, x_b, and the solid line at the bottom of the grey area represents a boundary. After collision at the gray node, three particle distributions, $f_2^*(x_b, t)$, $f_5^*(x_b, t)$, and $f_6^*(x_b, t)$, bounce back to this same node x_b in reversed direction, respectively, as $f_4(x_b, t + \triangle t)$, $f_7(x_b, t + \triangle t)$, and $f_8(x_b, t + \triangle t)$ (open arrows). Other distributions (filled arrows) follow the general propagation process, moving to corresponding neighboring nodes.

the last term in (4.2.14) disappears and the original bounce-back boundary condition is recovered.

Because both the original and modified bounce-back schemes assume that the boundary is located at the midpoint of the lattice link across the boundary, some loss of accuracy occurs for arbitrary boundary shapes. Aidun *et al.* (1998) pointed out that the additional term in (4.2.14) violates mass conservation and proposed a correction. These concerns are not important in our simulations. Equations (4.2.13) and (4.2.14) are then used for stationary and moving boundaries.

4.3 The immersed-boundary method

Figure 4.3.1 illustrates a membrane segment and adjacent fluid domain. Filled circles represent membrane markers and open circles represent fluid nodes. In the immersed-boundary method (IBM), membrane forces, $\mathbf{f}(\mathbf{x}_m)$, at membrane marker points, $\mathbf{x}_m$, due to the membrane deformation are distributed to neighboring fluid grid points, $\mathbf{x}_f$, using the formula

$$\mathbf{F}(\mathbf{x}_f) = D(\mathbf{x}_f - \mathbf{x}_m)\mathbf{f}(\mathbf{x}_m), \tag{4.3.1}$$

where $D(\mathbf{x})$ is a discrete delta function approximating the Dirac delta function (Peskin 1977, Eggleton & Popel 1998, N'Dri *et al.* 2003).

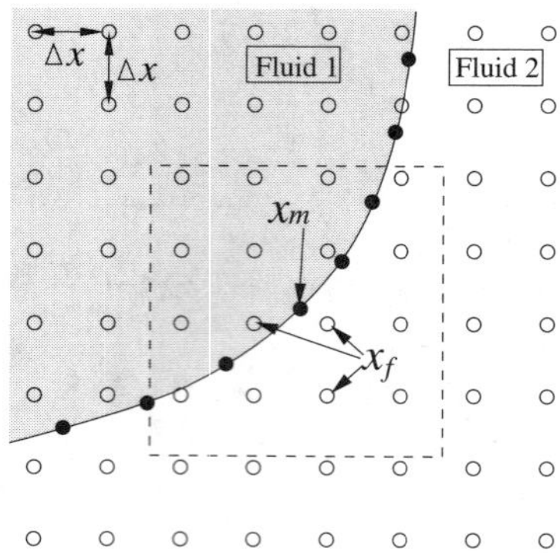

Figure 4.3.1 Schematic illustration of the immersed-boundary method. The open circles are fluid nodes and the filled circles are membrane marker nodes. The membrane force calculated at $\mathbf{x}_m$ is distributed to fluid nodes $\mathbf{x}_f$ inside a $2\triangle x \times 2\triangle x$ square (dashed lines) using equation (4.3.1). The position $\mathbf{x}_m$ is updated according to the velocities at fluid nodes $\mathbf{x}_f$ in the same square using equations (4.3.3) and (4.3.4).

In our two-dimensional system, $D(\mathbf{x})$ is given by

$$D(\mathbf{x}) = \frac{1}{4\triangle x^2}\left(1 + \cos\frac{\pi x}{2\triangle x}\right)\left(1 + \cos\frac{\pi y}{2\triangle x}\right), \tag{4.3.2}$$

if $|x| \le 2\triangle x$ and $|y| \le 2\triangle x$, and $D(\mathbf{x}) = 0$ otherwise. The nodal membrane velocity, $\mathbf{u}(\mathbf{x}_m)$, arises by interpolation from the local flow field,

$$\mathbf{u}(\mathbf{x}_m) = \sum_f D(\mathbf{x}_f - \mathbf{x}_m)\mathbf{u}(\mathbf{x}_f). \tag{4.3.3}$$

The membrane position $\mathbf{x}_m$ is updated using the explicit Euler method,

$$\mathbf{x}_m(t + \triangle t) = \mathbf{x}_m(t) + \mathbf{u}[\mathbf{x}_m(t)]\triangle t \quad . \tag{4.3.4}$$

The force distribution in equation (4.3.1) and velocity interpolation in equation (4.3.3) are computed in a $4\triangle x \times 4\triangle x$ square box indicated by the dashed outline in figure 4.3.1 (Peskin 1977), instead of a circular region of radius $2\triangle x$ employed by other authors (Francois *et al.* 2003, N'Dri *et al.* 2003, Udaykumar *et al.* 1997). It can be shown that the sum of the nonzero values of D in equation (4.3.2) inside the square is equal to unity. Therefore, missing a node inside the square and outside the circle causes the membrane force to be incorrectly projected onto the fluid. Similarly, a membrane marker point will not receive the corresponding velocity contribution from missing fluid nodes.

4.4 Fluid property updating

A membrane can separate fluids with different properties, such as density and viscosity. In our system, the densities of the interior cytoplasm and suspending plasma are identical. However, the viscosity of the cytoplasm is typically five times that of the suspending plasma. As the membrane moves and deforms in the ambient fluid, fluid properties must correspondingly be adjusted.

An index field called a color function is introduced to represent the position of a fluid node relative to the membrane. Tryggvason *et al.* (2001) generated this index field by solving a Poisson equation that arises by taking the divergence of the color function gradient. Other authors suggested updating the fluid index directly from the instantaneous membrane shape (Udaykumar *et al.* 1997, N'Dri *et al.* 2003, Francois *et al.* 2003). Because the Heaviside function employed in the second implementation depends on orientation, the calculated fluid index can be greatly different for different membrane orientations in a chosen coordinate system for the same position of fluid nodes relative to a membrane segment (Zhang *et al.* 2007).

In the present implementation, the approach of Shyy and coworkers is modified by introducing the shortest signed distance of a fluid node from the membrane, d. The sign of d indicates the host fluid phase. A negative value indicates that a node of interest lies outside the membrane in the plasma domain, and a positive value indicates that a node of interest lies inside the membrane in the cytoplasm domain. To smooth the transition of the rapidly varying fluid property α across the membrane, we set

$$\alpha(\mathbf{x}) = \alpha_{out} - (\alpha_{out} - \alpha_{in})\theta[d(\mathbf{x})], \tag{4.4.1}$$

where the subscripts *in* and *out* indicate bulk values inside and outside the membrane, and

$$\theta(d) = \begin{cases} 0\,, & d < -2\triangle x, \\ \frac{1}{2}\left(1 + \frac{d}{2\triangle x} + \frac{1}{\pi}\sin\frac{\pi d}{2\triangle x}\right)\,, & -2\triangle x \leq d \leq 2\triangle x, \\ 1\,, & d > 2\triangle x, \end{cases} \tag{4.4.2}$$

is a smoothed Heaviside function.

Since the membrane displacement in each time step is smaller than the grid size, nodal index values change only at nodes that are sufficiently close to the membrane, $|d| \leq 2\triangle x$. Outside the interfacial zone, $|d| > 2\triangle x$, the fluid properties remain unchanged and do not need to be updated. Since the affected nodes near the membrane constitute only a small portion of the total fluid nodes, performing this update is more efficient than solving a Poisson equation over the entire fluid domain (Tryggvason *et al.* 2001).

4.5 Models of RBC mechanics and aggregation

The unstressed cell geometry, fluid and membrane properties, and membrane interaction forces used in the simulations are discussed in this section.

4.5.1 RBC geometry and fluid viscosity

The main function of red blood cells is to facilitate oxygen exchange in tissue and lung. Unlike other cells, mammalian red blood cells lack a nucleus so that they can accommodate more oxygen and easily deform in order to pass through the capillaries. From the viewpoint of hydrodynamics, an RBC is a capsule filled with a concentrated solution of hemoglobin. The cell absorbs oxygen in the lungs and travels through blood vessels to release oxygen to tissue throughout the body. Hemoglobin serves as a vehicle for removing waste products, such as carbon dioxide.

The cytoplasm of the human RBC is a Newtonian fluid with approximate viscosity 6 cP (centipoise), which is about five times the viscosity of the suspending plasma, $\sim$ 1.2 cP. In our simulations, this viscosity difference is incorporated using the property update scheme described in Section 4.4. Increasing the cytoplasm viscosity reduces the cell deformability and impairs the normal function of blood in the microcirculation, as observed in diabetes, mellitus and sickle cell anemia.

In the absence of flow, human RBCs take a unique biconcave discoidal shape with diameter 6 $\sim$ 8 μm and approximate thickness $\sim$ 2 μm. This geometry provides a surface area 30 $\sim$ 40% in excess of that corresponding to a spherical shape with the same volume, and thereby facilitates gas transport across the membrane. The biconcave shape also allows for large deformation necessary for the cells to negotiate flow in microvessels and small capillaries.

In our calculations, the undeformed biconcave RBC shape is described by an empirical equation proposed by Evans & Fung (1972),

$$\bar{y} = 0.5(1 - \bar{x}^2)^{1/2}(c_0 + c_1\bar{x}^2 + c_2\bar{x}^4) \,, \qquad -1 \le \bar{x} \le 1, \tag{4.5.1}$$

where $c_0 = 0.207$, $c_1 = 2.002$, and $c_2 = 1.122$. The dimensionless coordinates $(\bar{x}, \bar{y})$ are normalized by the radius of a human RBC (3.91 μm).

Membrane mechanics

The cell membrane consists of a phospholipid bilayer and an underlying network of proteins forming the cytoskeleton. The bilayer and cytoskeletal network are stapled together by an assortment of transmembrane proteins. Due to the dual molecular structure, the cell membrane exhibits unique mechanical properties (e.g., Mohandas & Gallagher 2008). In pathological situations such as hypertension, cancer, and malaria, the membrane mechanical properties are altered due to pathological biochemical reconstruction (e.g., Stoltz *et al.* 1999).

Because the membrane thickness ($\sim$ 5 nm) is by orders of magnitude less than the cell size ($\sim$ 8 μm), the membrane can be regarded as a zero-thickness curved sheet. A neo-Hookean (nonlinear) constitutive equation may then be used to describe the developing membrane tensions. The elastic shear modulus lies in the range 2.5 – 10 μN/m, indicating low resistance to shearing deformation. Because of the near incompressibility of the lipid bilayer, the membrane possesses a large modulus of dilation that prevents excessive deformation. Viscous response and flexural stiffness become important when regions of large curvature develop.

Various experimental techniques have been employed to infer the cell membrane properties, including micropipette aspiration, optical tweezer, flow chamber, and microchannel flow experiments. The membrane properties are deduced by fitting experimental observations to theoretical and numerical predictions, subject to an assumed membrane constitutive relation (Evans 1973, Tran-Son-Tay *et al.* 1984).

In our two-dimensional RBC model, we follow Bagchi *et al.* (2005, 2007) and describe the neo-Hookean elastic component of the membrane tension by the equation

$$T_e = \frac{E_s}{\epsilon^{3/2}}(\epsilon^3 - 1), \tag{4.5.2}$$

where E_s is the membrane elastic modulus and ϵ is the stretch ratio (see also Chapter 1). Bending resistance is incorporated in terms of the membrane curvature,

$$T_b = \frac{\mathrm{d}}{\mathrm{d}l}[E_b(\kappa - \kappa_0)], \tag{4.5.3}$$

where E_b is the bending modulus, κ and κ_0 are the instantaneous and initial stress-free membrane curvatures, and l is the arc length along the membrane contour (Pozrikidis 2001, Bagchi *et al.* 2005). The total membrane stress $\mathbf{T}$ is the sum of two terms,

$$\mathbf{T} = T_e\mathbf{t} + T_b\mathbf{n}, \tag{4.5.4}$$

where $\mathbf{t}$ and $\mathbf{n}$ are the tangential and outward normal unit vectors.

Intercellular aggregation

Red blood cells agglomerate to form one-dimensional rouleaux resembling stacks of coins or three-dimensional aggregates (e.g., Popel & Johnson

2005, Stoltz *et al.* 1999, Baumler *et al.* 1999, Baskurt & Meiselman 2007). These structures can be broken up by action of viscous stresses imparted by an ambient flow. Cell aggregation affects *in vivo* hemodynamics, especially in venular regions of slow flow. Pronounced aggregation is observed in clinical states such as diabetes, heart disease, inflammation, acute myocardial infarction, and bacterial sepsis.

The physical mechanism of aggregation is not entirely clear. A bridging mechanism and a depletion mechanism have been proposed (e.g., Baumler *et al.* 1999, Baskurt & Meiselman 2007). The former assumes that surface macromolecules, such as fibrinogen or dextran, adhere and then link on adjacent cell surfaces. The depletion model attributes aggregation to a polymer depletion layer developing near the cell surfaces causing a decrease in osmotic pressure. Both models successfully describe some observed aggregation characteristics.

To model intercellular attraction, Bagchi *et al.* (2005) implemented a ligand–receptor interaction model based on bridging. The formulation involves a number of parameters that are not available from experiments. Chung *et al.* (2006) employed a depletion model proposed by Neu & Meiselman (2002). Unfortunately, a physically unrealistic constant, instead of decaying, attractive force is predicted at large separations (Chung *et al.* 2006).

In the present implementation, we follow Liu *et al.* (2004) and model the intercellular interaction energy ϕ using a Morse potential,

$$\phi(r) = D_e \left(\mathrm{e}^{2\beta(r_0 - r)} - 2\,\mathrm{e}^{\beta(r_0 - r)} \right), \tag{4.5.5}$$

where r is the surface separation, r_0 is the separation at zero force, D_e is an interaction strength, and β is a scaling factor controlling the interaction decay. The interaction force arising from this potential is $f(r) = -\mathrm{d}\phi/\mathrm{d}r$. The Morse potential is selected for its simplicity and tentative physical relevance to cell–cell interaction. An assumption of the underlying physical mechanism is not implied. Other interaction models can be readily incorporated in the IB-LBM algorithm.

4.6 Single cells and groups of cells

In the simulations reported in this chapter, the membrane of each cell is represented by 100 segments. Necessary physical parameters are taken from experimental measurements, as shown in table 4.6.1. Experimental values concerning the parameters in the Morse potential for cell aggregation, described in equation (4.5.5), are not available. In our simulations, to prevent the overlap of immersed boundary layers, the zero force distance r_0 is chosen to be 2.5 lattice units (0.49 μm). A cut-off distance typical in molecular dynamics simulations is introduced to improve computational efficiency (Heyes

Parameter	*Value*	
lattice unit (h)	$0.195\ \mu m$	
time step ($\triangle t$)	1.25×10^{-8} s	
fluid density (ρ)	$1000\ kg/m^3$	SC
plasma viscosity (μ_p)	1.2 cP	SC
cytoplasm viscosity (μ_c)	6.0 cP	WH
membrane elastic modulus (E_s)	$6.0 \times 10^{-6} - 1.2 \times 10^{-4}$ N/m	WH
membrane bending modulus (E_b)	$2.0 \times 10^{-19} - 4.0 \times 10^{-18}$ Nm	WH
intercellular interaction strength (D_e)	$0 - 1.3 \times 10^{-6}\ \mu J/\mu m^2$	Z
scaling factor (β)	$3.84\ \mu m^{-1}$	
zero-force distance (r_0)	$0.49\ \mu m$	

Table 4.6.1 Parameter values employed in the simulations. Some values are listed as ranges where individual effects are investigated. In the third column, SC refers to Skalak & Chien (1987), WH refers to Waugh & Hochmuth (2006), and Z refers to Zhang *et al.* (2008).

1998). The scaling factor β in equation (4.5.5) is adjusted to ensure a negligible attractive force beyond a cut-off distance $r_c = 3.52\ \mu m$. The interaction strength D_e varies from 0 to $1.3 \times 10^{-6}\ \mu J/\mu m^2$.

4.6.1 Deformation of a single cell in shear flow

The motion and deformation of red blood cells suspended in shear flow have been investigated extensively by theoretical, computational, and experimental methods. Simple shear flow is commonly employed to measure macroscopic blood viscosity and deduce cell membrane mechanical properties (e.g., Chien 1970, Tran-Son-Tay *et al.* 1984).

Depending on the magnitude of the hydrodynamic viscous stress exerted on the membrane and on the cell deformability, a cell may exhibit different types of motion in simple shear flow (e.g., Fischer & Schmid-Schönbein 1977, Tran-Son-Tay *et al.* 1984, Abkarian *et al.* 2007). At low shear rates, the cell rotates and flips like a solid particle, exhibiting a so-called tumbling motion (Goldsmith & Marlow 1972). As the shear rate increases, the cell is stretched into a flattened elongated shape, swinging back and forth about a certain orientation angle (Abkarian *et al.* 2007). A further increase in the shear rate reduces the amplitude of the swinging angle and increases the capsule elongation. When the shear rate is sufficiently strong, the cell reaches a steady shape and orientation, while exhibiting a so-called tank-treading motion (Fischer & Schmid-Schönbein 1977). In all cases, the cell membrane rotates around the cell even after the cell has reached a steady state. The details of the tum-

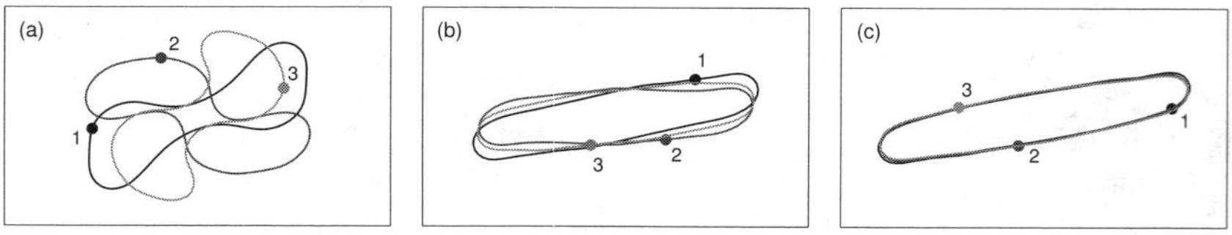

Figure 4.6.1 Deformation of a single cell in simple shear flow for cytoplasm viscosity $\mu_c = 6$ cP and suspending fluid viscosity (*a*) $\mu_p = 1$ cP, (*b*) 2.5 cP, and (*c*) 6 cP. In each case, three consecutive cell shapes indicated by labels (1–3) and colors (black, blue, and red), are displayed during one period: (*a*) flipping, (*b*) swinging, and (*c*) tank-treading motion. The small circles represent a membrane marker initially at the cell rim. (*Color in the electronic file.*)

bling, swinging, and tank-treading motion, including cell rotating/swinging frequency, deformed shape, elongation, and orientation angle, depend on the cell properties and flow conditions.

To demonstrate these modes and further validate the numerical algorithm, we consider the deformation of a single cell in simple shear flow for three different ambient fluid viscosities. In all cases, $E_s = 6 \times 10^{-6}$ N/m and $E_b = 2 \times 10^{-19}$ Nm. Elevated suspending fluid viscosity is relevant to resuscitation where a fluid whose viscosity is higher than that of plasma is employed (Martini *et al.* 2006). The simulations are conducted on a 256×256 lattice grid in a domain with physical size 50 μm $\times$ 50 μm. The general periodic boundary condition is implemented in the horizontal direction. The shear rate is set to 50 s^{-1} using Ladd's modified bounce-back scheme. Representative snapshots are displayed in figure 4.6.1.

When the surrounding medium is less viscous than the cytoplasm ($\mu_p =$ 1 cP and $\mu_c = 6$ cP), the biconcave cell rotates and flips while exhibiting periodic deformation, as shown in figure 4.6.1(*a*). Unlike a solid particle, the cell deforms while the membrane rotates around the interior fluid in response to viscous stresses imparted by the interior and exterior flow. As the exterior fluid viscosity increases ($\mu_p = 2.5$ cP and $\mu_c = 6$ cP), a whole cycle of rotation cannot be completed. Instead, the cell keeps swinging about a certain inclination angle with an elongated shape, as shown in figure 4.6.1(*b*), in agreement with experimental observations (Abkarian *et al.* 2007). A further increase in the suspending fluid viscosity pronounces the cell elongation and decreases the amplitude of the swinging angle. When $\mu_p = \mu_c = 6$ cP, a swinging motion does not appear and the cell reaches a nearly steady shape with a constant

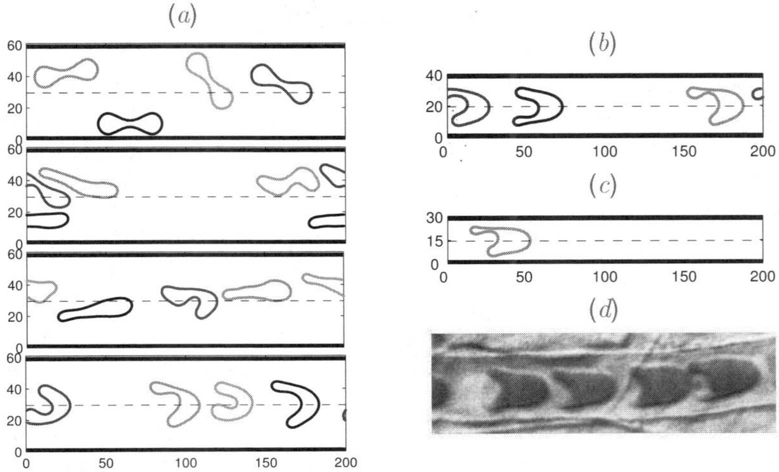

Figure 4.6.2 Motion of RBCs in a channel with plate separation (*a*) 12, (*b*) 8, and (*c*) 6 μm. For comparison, a photograph of rat RBCs flowing in a mesenteric microvessel is shown in (*d*), reproduced with permission from Pries & Secomb (2005). *The simulation figures are reproduced with permission from Zhang et al. (2007).*

inclination angle, as illustrated in figure 4.6.1(*c*). Slight shape variations are observed during the tank-treading motion due to the biconcave unstressed shape. Although the cell retains a nearly steady shape, the cell membrane and cytoplasmic fluid exhibit continuous rotation (tank-treading) around the cell center, indicated by the marker point motion in figure 4.6.1(*c*).

4.6.2 Channel flow

Next, we consider flow in small channels with width 6–12 μm. The general bounce-back scheme for stationary boundaries and periodic boundary conditions are applied to the top/bottom and inlet/outlet boundaries, respectively. A body force or pressure gradient is necessary to drive the flow. Multiple cells are studied to investigate the effect of relative initial position and orientation in the absence of intercellular aggregation.

Figure 4.6.2(*a*) illustrates the motion and deformation of four cells in a 12 μm channel. Despite the different initial orientation and position, all cells migrate toward the centerline to reduce flow resistance. Lateral migration of cells has been observed both *in vitro* (Goldsmith 1971) and *in vivo* (Secomb *et al.* 2007). As the channel width decreases to 8 μm and then to 6 μm, the cell deformation becomes more pronounced, as shown in figure 4.6.2(*b, c*). For comparison, a photograph of rat RBCs flowing in a capillary with diameter 7 μm is shown in figure 4.6.2(*d*). The deformed cell shape depends on the cell size, capillary diameter, membrane properties, and flow conditions. A direct comparison between simulation and experiments is not possible because of the two-dimensional approximation in the numerical model and the lack of specific information on the experimental setup.

4.6.3 Rouleaux formation

Next, we discuss the interaction of four cells under the influence of attractive surface forces parametrized by D_e, in the absence of flow. In the simulations, the cells are initially placed at the central region of the computational domain with a center-to-center separation 3.52 μm, as shown in figure 4.6.3(*a*). The closest gap between two adjacent cells is 0.95 μm. The results show that the cells slowly aggregate due to intercellular attraction until equilibrium has been established where the membrane is deformed and the induced membrane forces counterbalance aggregation forces from adjacent cells. Figure 4.6.3 shows rouleaux shapes developed from different interaction strengths, D_e, in the range 5.2×10^{-8} – 1.3×10^{-6} μJ/μm^2. Consistent with experimental observations by Chien & Jan (1973), the gap is nearly uniform between adjacent cells.

The cells at the end of the rouleaux assume concave shapes in the case of weak interaction, as shown in figure 4.6.3(*a–c*), and convex shapes in the case of strong interaction, as shown in figure 4.6.3(*d–f*). When $D_e = 1.3 \times 10^{-6}$ μJ/μm^2, the rouleau contracts to a nearly spherical clump to maximize the contact area, as shown in figure 4.6.3(*f*). The contact surfaces are flat in the case of weak interaction, and curved in the case of strong interaction. These results are consistent with experimental observations by Skalak *et al.* (1981) shown in figure 4.6.3(*g, h*). The similarity indicates that the Morse potential is acceptable for modeling cell aggregation. Other aggregation mechanisms, such as ligand-receptor dynamics, can be readily implemented in the IB-LBM framework (Bagchi *et al.* 2005, N'Dri *et al.* 2003).

4.6.4 Rouleaux dissociation in shear flow

Experiments have shown that cell aggregates can be broken into smaller pieces or even individual cells at high shear rates (Popel & Johnson 2005, Stoltz *et al.* 1999). To describe this behavior, we have simulated the motion of four cells in shear flow. The physical configuration is analogous to that observed in hemorheological experiments (e.g., Stoltz *et al.* 1999). The simulation strategy is similar to that discussed in Section 4.6.1. A rouleau formed in the absence of flow is placed at the center of the simulation domain, and a shear flow is

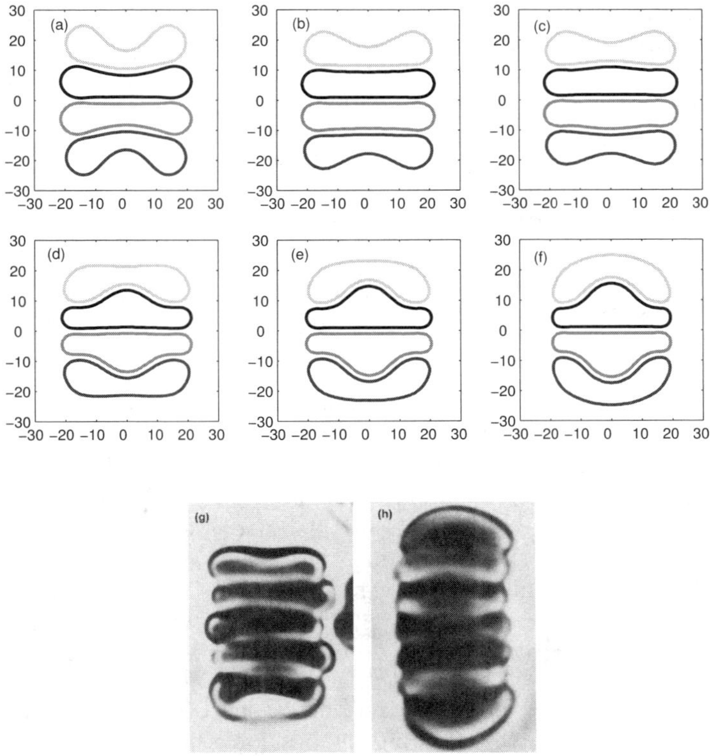

Figure 4.6.3 Rouleaux formed with aggregation strength (a) $D_e = 5.2 \times 10^{-8}$, (b) 2.1×10^{-7}, (c) 2.6×10^{-7}, (d) 3.9×10^{-7}, (e) 6.5×10^{-7}, and (f) 1.3×10^{-6} μJ/μm^2. For comparison, photographs of rouleaux formed in a Dextran Dx 70 (molecular weight 75,000) suspension with concentration (g) 0.01 and (h) 0.04 g/ml are reproduced with permission from Skalak *et al.* (1981). The end-cells have concave shapes in (g) and convex shapes in (h). *The simulation figures are reproduced with permission from Zhang et al. (2008).*

imposed by assigning to the top and bottom boundaries velocities of equal magnitude and opposite sign. Different shear rates are selected by adjusting the boundary velocities. In the parametric investigations, the individual effects of shear rate and aggregation force are investigated. Because of the importance of membrane deformation, it is not appropriate to examine only the relative strength of shear rate and aggregation, as in the case of solid particles. Rather, the effect of the shear rate must be explicitly considered. To underscore this distinction, we note that cell rouleaux formed with different aggregation strengths have different shapes, as shown in figure 4.6.3.

Effect of aggregation force

Figure 4.6.4 shows representative snapshots for shear rate $\gamma = 20$ s^{-1}. The slight asymmetry of the streamlines is an artefact of the plotting software. The aggregation strength is $D_e = 0$ in (a), 5.2×10^{-8} in (b), and 1.3×10^{-6} in (c). In the absence of aggregation, the cell deformation is similar to that discussed in Section 4.6.1 for a single cell in shear flow. During rotation, cell shapes alternate between the biconcave shape and a compressed folded shape resembling a reversed S.

In the presence of a weak intercellular interaction force, $D_e = 5.2 \times 10^{-8}$ μJ/μm^2, the cells adhere to form a rouleau-like aggregate in the absence of flow, as shown in the first panel of figure 4.6.4(b). When a shear flow with shear rate $\gamma = 20$ s^{-1} is applied, the aggregate deforms and rotates by almost 270°. However, the interaction is not strong enough to hold the cells together for an indefinite period of time. At time $\gamma t = 43.75$, the two end-cells separate, and the remnant two-cell aggregate rotates around the center of the computation domain for four periods. Finally, the two center cells separate and the original aggregate becomes completely dismantled.

A stronger intercellular interaction ($D_e = 1.3 \times 10^{-6}$ μJ/μm^2) was introduced to overcome the shearing viscous force. The process of rouleau formation is illustrated in figure 4.6.4(c). Now the rouleau simply rotates in the shear flow with some reconfiguration, but is not broken up into individual cells or smaller units due to strong intercellular attraction.

Effect of the shear rate

To examine the effect of hydrodynamic viscous force, the shear rate was increased to $\gamma = 50$ s^{-1} for a weak interaction, $D_e = 5.2 \times 10^{-8}$ μJ/μm^2, and a strong interaction, $D_e = 1.3 \times 10^{-6}$ μJ/μm^2. In the case of weak interaction, the rouleau rotates by only about 90°, and is then quickly broken up into two two-cell aggregates, as shown in figure 4.6.5(a). The doublets eventually split up into four individual cells. Compared to the case of weak interaction and a lower shear rate $\gamma = 20$ s^{-1} described in figure 4.6.4(b), now the rouleau easily separates into individual cells in a shorter period of time.

A similar trend is observed in the case of strong interaction illustrated in figure 4.6.5(b). Having rotated by almost 270°, the compact rouleau dismantles into two two-cell aggregates at time $\gamma t = 40.625$. The two-cell aggregates rotate attached due to a strong aggregation force even at this high shear rate.

(*a*)

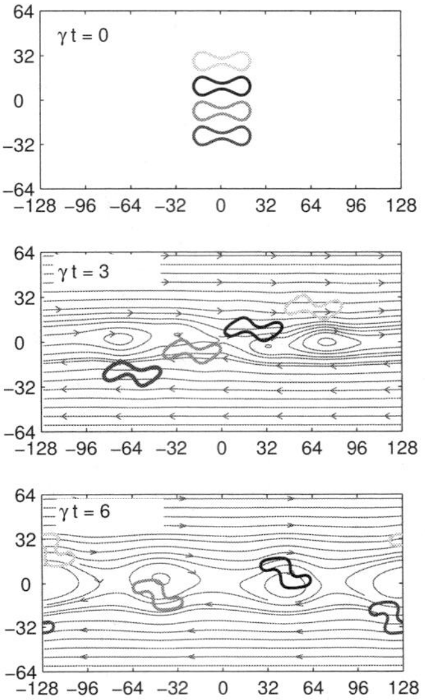

(*b*)

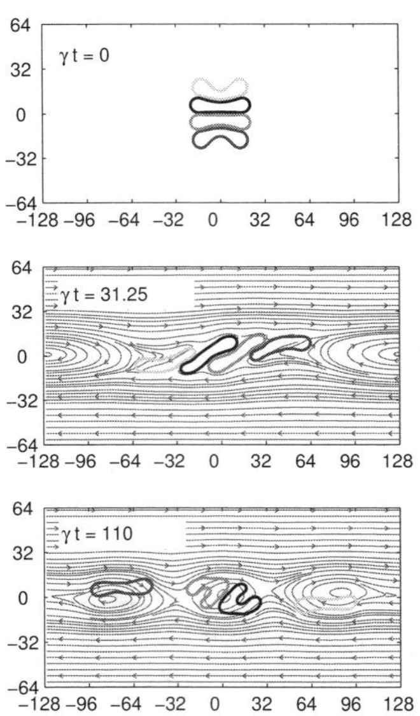

Figure 4.6.4 Continued on the next page.

(c)

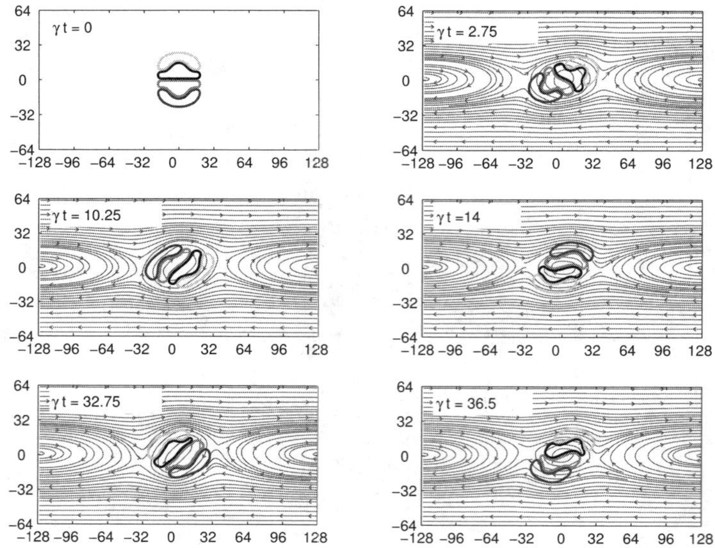

Figure 4.6.4 Representative snapshots of four cells for shear rate $\gamma = 20$ s^{-1} , (*a*) in the absence of intercellular aggregation ($D_e = 0$), (*b*) with weak intercellular aggregation, ($D_e = 5.2 \times 10^{-8}$ μJ/μm^2), and (*c*) with strong intercellular aggregation ($D_e = 1.3 \times 10^{-6}$ μJ/μm^2). *Reproduced with permission from Zhang et al. (2008).*

4.7 Cell suspension flow in microvessels

The numerical method was also applied to study flow in microchannels mimicking blood flow in microvessels. The simulations improve previous studies where cells are treated as rigid particles and aggregation forces were not taken into account (Sun *et al.* 2003, Sun & Munn 2005, Chung *et al.* 2006, Bagchi 2007). Simulations with 27 RBCs were carried out on a 100×300 D2Q9 grid in a 19.5 μm$\times$58.5 μm domain, as shown in figure 4.7.1. No-slip and periodic boundary conditions are imposed, respectively, at the channel walls and inlet/outlet. The tube hematocrit, defined as the volume fraction occupied by the cells, is set to $H_T = 32.2\%$. The flow is initiated by an imposed streamwise pressure gradient 62.5 kPa/m using a modified periodic boundary scheme (Zhang & Kwok 2006). In the case of pure plasma flow devoid of cells, the pressure gradient generates a parabolic flow with mean velocity 1.65 mm/s according to Poiseuille's law. This mean velocity is typical of blood flow in microvessels (Sun & Munn 2005, Schmid-Schönbein *et al.* 1980).

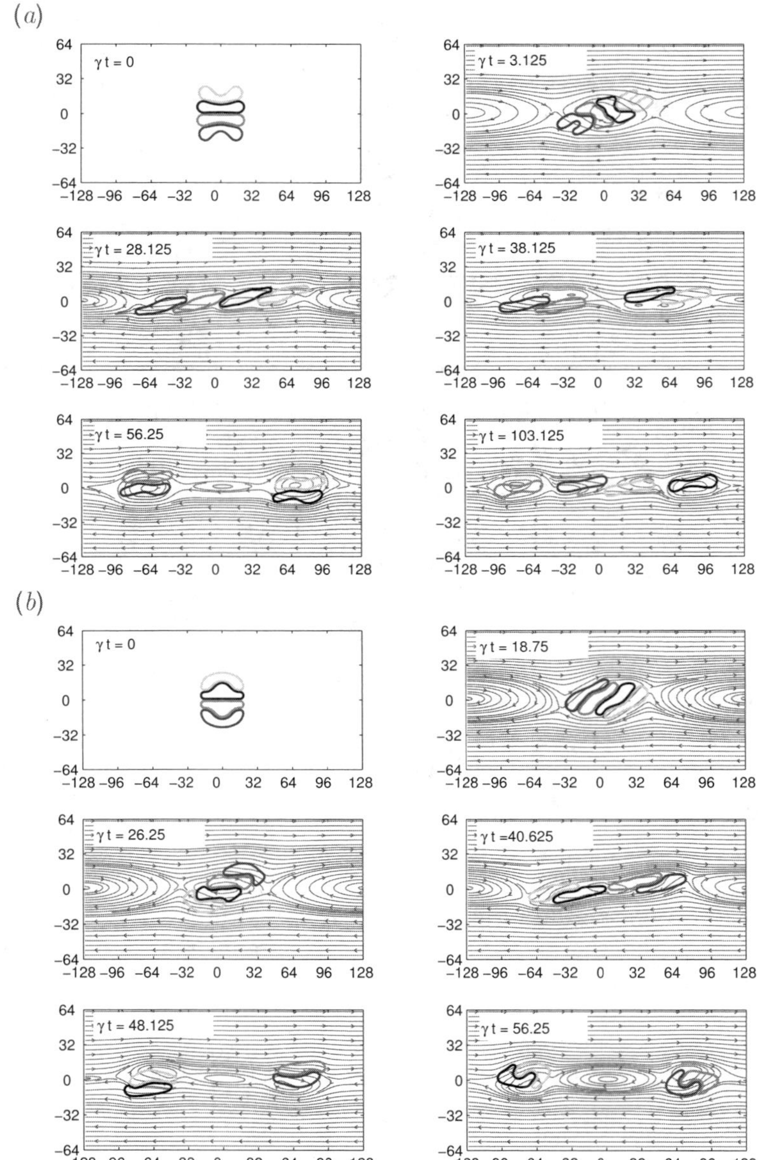

Figure 4.6.5 Representative snapshots of four cells for shear rate $\gamma = 50$ s^{-1} with (a) weak intercellular aggregation ($D_e = 5.2 \times 10^{-8}$ $\mu\mathrm{J}/\mu\mathrm{m}^2$), and ($b$) strong intercellular aggregation ($D_e = 1.3 \times 10^{-6}$ $\mu\mathrm{J}/\mu\mathrm{m}^2$). *Reproduced with permission from Zhang et al. (2008).*

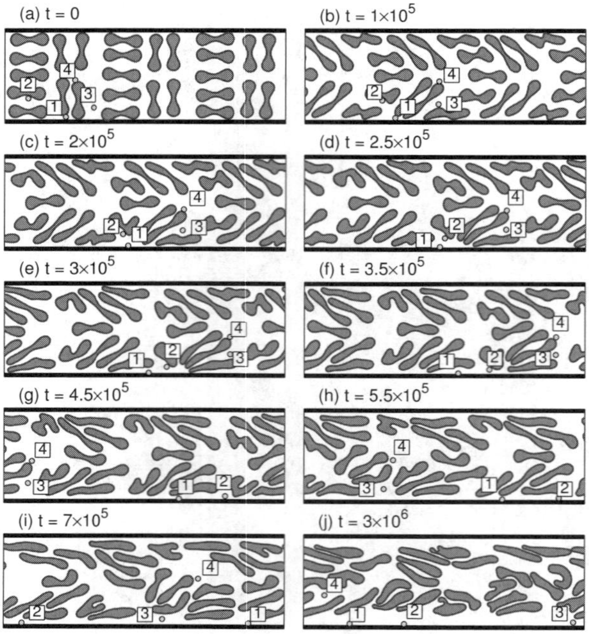

Figure 4.7.1 Snapshots of normal cells with moderate aggregation in a straight channel. Four fluid tracers (small blue circles) are followed in the plasma to illustrate the nature of the flow. *Reproduced with permission from Zhang et al. (2009).* (*Color in the electronic file.*)

The cell properties are the same as those used in the simulations discussed earlier in this chapter, $E_s = 6 \times 10^{-6}$ N/m and $E_b = 2 \times 10^{-19}$ Nm. Since these values are generally acceptable based on a variety of measurements, hereafter RBCs with these properties are designated as normal cells. The aggregation strength is set to $D_e = 5.2 \times 10^{-8}$ μJ/μm^2.

4.7.1 Cell-free layers

Figure 4.7.1 illustrates the evolution of the flow. As soon as a pressure gradient is applied, the suspension begins flowing and large cell deformations gradually develop. Cells near the walls migrate toward the centerline. As a result, cell-free layers appear near the walls and the central region hosts a higher cell concentration. Passive fluid tracers are introduced to visualize the cell migration process. The tracer position is updated using equation (4.3.4). The motion of four representative tracers is displayed in figure 4.7.1. We see that tracer 1, initially near the lower wall, drifts in the flow direction with small transverse fluctuations. Tracers 2 and 3, initially away from walls, are quickly displaced by migrating cells toward the marginal CFL zone, as shown in figure 4.7.1(*b–g*) for tracer 2 and in figure 4.7.1(*e–i*) for tracer 3. Once these tracers have arrived in the CFL zone, they behave similarly to tracer 1, moving mainly in the direction of the flow. Tracer 4 is trapped between cells in the central high-hematocrit region and cannot escape.

These results indicate that, unless a fluid parcel is closely surrounded by cells, it is likely to be displaced to the cell-free layer by migrating cells. Once a small fluid parcel reaches the CFL zone, it stays roughly at the same transverse position. The motion of passive tracers described in the simulations indicates that small particles in blood tend to accumulate in the cell-free zones. This behavior is consistent with experimental observations of platelets (Aarts *et al.* 1988). The numerical method could be adapted to investigate the motion of nanoparticles for drug and gene delivery in blood flow.

4.7.2 RBC distribution and velocity profile

Figure 4.7.2(*a*) illustrates the evolution of the volumetric flow rate of the whole blood suspension, Q_{total}, and the corresponding flow rate of the RBCs, Q_{RBC}. Both are calculated from the streamwise velocity, u_x, and fluid index, θ, using equations

$$Q_{total} = \int u_x \mathrm{d}y, \qquad Q_{RBC} = \int \theta u_x \mathrm{d}y, \tag{4.7.1}$$

where the integration is performed over the channel width. As soon as a pressure gradient is applied, both flow rates gradually increase and reach a quasi-steady fluctuating state, since the RBC configuration is constantly changing. As changes in the RBC configuration affect both the velocity and index distribution, as shown in equation (4.7.1), the fluctuations are more evident in the RBC flow rate Q_{RBC} than in the total flow rate Q_{total}.

A discharge hematocrit, H_D, defined as the RBC volume fraction observed in a discharge reservoir, can be evaluated from Q_{total} and Q_{RBC} averaged over time after the flow has reached a quasi-steady state,

$$H_D = \frac{Q_{RBC}}{Q_{total}}. \tag{4.7.2}$$

The results show that $H_D = 38.9\%$ is higher than the tube hematocrit $H_T = 32.2\%$, confirming the well-known Fåhraeus effect.

Figure 4.7.2(*b*) shows cell distribution profiles during flow development. The fluid index, θ, has been averaged over the channel length. As time progresses, a cell-free layer of increasing thickness develops near each wall.

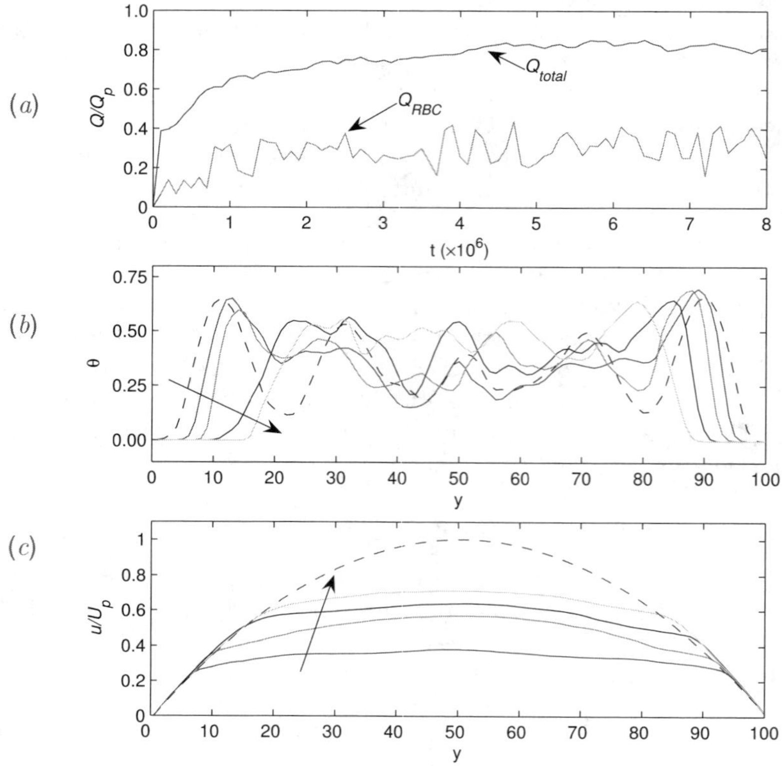

Figure 4.7.2 (a) Total and RBC flow rates, (b) averaged θ distribution, and (c) averaged velocity profiles during flow development. In (b) and (c), profiles are shown at dimensionless times $t = 0.4, 1.0, 3.0$ and 7.0×10^6 increasing in the arrow direction. The dashed lines represent the initial θ distribution in (b) and the parabolic velocity profile of a pure plasma flow for the same pressure gradient in (c). Q_p and U_p are the flow rate and maximum velocity of the parabolic plasma flow. *Reproduced with permission from Zhang et al. (2009).*

Figure 4.7.2(c) describes velocity profiles at several time instants averaged over the channel length. In agreement with *in-vivo* and *in-vitro* experiments, blunt profiles are observed due to the high fluid viscosity in the RBC-rich core. The flow rate of the suspension, Q_{total}, is lower than that of pure plasma flow for the same pressure gradient, described by the dashed line in figure 4.7.2(c).

The relative apparent viscosity of the suspension is defined as

$$\mu_{rel} = \frac{Q_p}{Q_{total}}, \tag{4.7.3}$$

where Q_p is the flow rate of the pure plasma flow. According to Poiseuille's law, $Q_p = \triangle PH^3/12L\mu_p$, where $\triangle P$ is the pressure difference across a channel of length L and width H. In the simulations, $\mu_{rel} = 1.29 > 1$, confirming an increase in the flow resistance due to the presence of the cells.

4.7.3 Effect of cell deformability and aggregation

Additional simulations were carried out to study the effect of membrane deformability and aggregation on flow structure and evolution. Experiments have shown that RBCs exhibit reduced deformability and pronounced aggregation in heart disease, hypertension, diabetes, malaria, and sickle cell anemia (e.g., Popel & Johnson 2005). In computational models, cell rigidity can be adjusted by increasing the membrane elastic modulus and bending resistance. Stiff RBCs were produced by increasing the membrane shear and bending moduli to twenty times the normal values, $E_s = 1.2 \times 10^{-4}$ N/m and $E_b = 4.0 \times 10^{-18}$ Nm. Three different aggregation strengths, $D_e = 0$ (none), 5.2×10^{-8} (moderate), and 1.3×10^{-6} μJ/μm^2 (strong), were selected corresponding to the previous study on rouleaux formation discussed in Section 4.6.3.

Figure 4.7.3 shows instantaneous configurations for different degrees of cell deformability and aggregation strength. Stiff RBCs displayed in the right column nearly retain the original biconcave shape, exhibiting only small deformation compared to normal RBCs displayed in the left column. Cell-free layers are clearly seen near the walls, except in the case of stiff cells and in the absence of aggregation, as shown in the top-right frame. The thickness of the cell-free layer increases as aggregation forces become stronger and the cells become more deformable. Physically, aggregation generates a more concentrated RBC core at the channel centerline. Membrane rigidity prevents large deformation and the onset of a large cell-cell contact area, resulting in a more porous RBC core and thinner cell-free layers than those observed for normal cells. The velocity vector field visualized by the arrows in figure 4.7.3 illustrates the bluntness of the velocity profile compared to the parabolic profile, as discussed in Section 4.7.2.

The cell-free layer thickness is defined as the distance from the channel boundary up to an extrapolated $\theta = 0$ position from the first two points where $\theta \neq 0$ in the θ profile. The computed thickness of the cell-free layer, Δ, is shown in table 4.7.1 for various flow conditions. The results confirm that Δ increases as the cells become more deformable or the aggregation strength

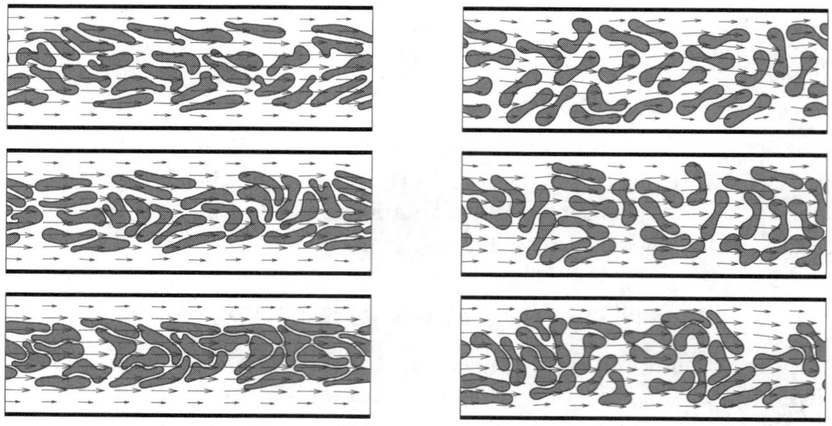

Figure 4.7.3 Quasi-steady states of normal (left) and stiff (right) cells with different aggregation strengths: from top to bottom: none, moderate, and strong. Flow fields are displayed by arrows. *Reproduced with permission from Zhang et al. (2009).*

Deformability	*Aggregation*	Δ (μm)	μ_{rel}	H_D (%)
stiff	none	1.32	1.66	36.8
	moderate	1.52	1.60	37.4
	strong	2.45	1.34	39.6
normal	none	2.44	1.27	38.8
	moderate	2.69	1.29	38.9
	strong	3.18	1.23	40.0

Table 4.7.1 Cell-free layer thickness, Δ, relative apparent viscosity, μ_{rel}, and discharge hematocrit, H_D, for different cell deformabilities and aggregation strength.

becomes stronger. A cell-free layer is hardly observed for stiff RBCs in the absence of aggregation. The computed value of Δ in this case is the smallest of all six combinations considered, 1.32 μm. A similar behavior was observed in simulations of rigid particles by Sun & Munn (2005).

The fourth column in table 4.7.1 shows results for the relative apparent viscosity calculated from equation (4.7.3). In all cases, the relative apparent viscosity is greater than unity because of the presence of the cells. As the cells become stiffer, the relative apparent viscosity becomes higher. However, the effect of cell aggregation is nonmonotonic. With stronger aggregation, the denser cell core has a higher viscosity, while the thicker cell-free layers introduce a significant lubricating effect. The opposing effect of these two mechanisms on the global apparent viscosity of the suspension provides us with a plausible physical explanation for contradictory experimental findings discussed by Baskurt & Meiselman (2007).

For the systems presently studied, the relative apparent viscosity μ_{rel} generally decreases with aggregation due to dominant lubrication effects, in agreement with experimental observations (Alonso *et al.* 1995, Cokelet & Goldsmith 1991, Reinke *et al.* 1987, Murata 1987). However, when a moderate aggregation force is introduced between normal cells, μ_{rel} increases slightly from the reference value in the absence of aggregation. Physically, the high core viscosity counteracts the benefits of the lubricating flow in the cell-free zone.

The discharge hematocrit and the tube to discharge hematocrit ratio are calculated from equation (4.7.2). The tube hematocrit is the same in all cases, 32.3%, while the discharge hematocrit H_D varies from 36.8% to 40.0%. The ratio H_T/H_D varies in the range 0.875–0.805. Values lower than unity are consistent with the well-known Fåhraeus effect. It appears that, unlike the apparent viscosity, H_D increases slightly with both membrane deformability and aggregation strength.

4.7.4 Effect of the channel width

The significance of the channel width was examined by simulating the flow of normal and stiff cells in a smaller 12 μm channel, in the absence of aggregation. The tube hematocrit is set to 35%, which is close to the value specified in the earlier 20 μm channel. Figure 4.7.4 shows an initial arrangement of 12 cells and the emerging flow pattern for normal and stiff cells. As in the case of a wider channel, cell-free layers develop in the case of normal cells but not in the case of stiff cells. The calculated discharge hematocrit is 43.7% for normal cells and 39.8% for stiff cells. Both are larger than the tube hematocrit 35%, consistent with the Fåhraeus effect. A higher discharge hematocrit is observed for normal than stiff cells.

The computed relative apparent viscosity is 1.14 for normal cells and 1.57 for stiff cells. Comparing these to corresponding values in the 20 μm channel shown in table 4.7.3, 1.27 for normal cells and 1.66 for stiff cells, we see that the apparent viscosity decreases as the channel width is reduced from 20 to 12 μm while the tube hematocrit is held approximately constant. This trend is consistent with the Fåhraeus-Lindqvist effect.

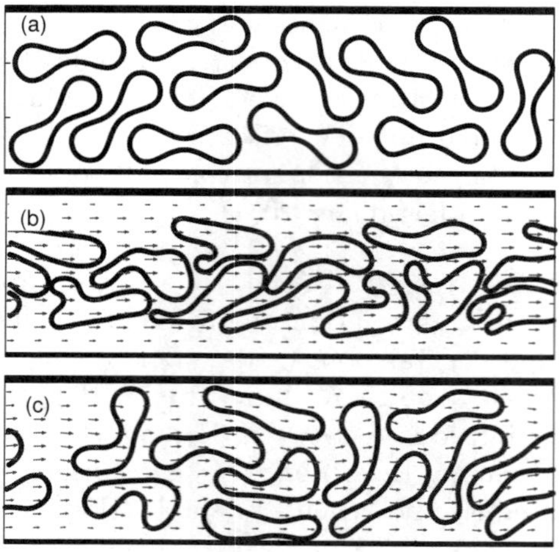

Figure 4.7.4 (*a*) Initial RBC distribution and well-developed flow configurations of (*b*) normal and (*c*) stiff cells in a 12 μm channel. *Reproduced with permission from Zhang et al. (2007).*

4.8 Summary and discussion

We have described an IB-LBM method for simulating RBC deformation and motion in shear and channel flow. Extensive simulations were performed and a number of hemodynamic and hemorheological phenomena have been described. When exposed to a shear flow, a single red blood cell may perform flipping, swinging, or tank-treading, depending on the shear rate. When aggregation is enabled, rouleaux of various shapes arise. Strong aggregation generates compact rouleau structures with large contact areas, curved interfaces, and convex end cells. Applying a shear flow to an existing rouleau may break the cell structure into smaller parts and produce individual cells. The precise action depends on the competition between the dismantling hydrodynamic force and the adherence force.

RBCs migrate to the centerline of microchannels and fold into slipper-like shapes to reduce resistance to flow. A careful examination of the flow development process reveals that small particles suspended in plasma are likely to be displaced toward the boundaries. Cell-free layers, velocity bluntness, and the Fåhraeus and Fåhraeus-Lindqvist effects are observed in the simulations. Cell deformability and aggregation play an important role in determining RBC flow behavior.

Although undoubtedly valuable, two-dimensional models can provide only qualitative information. The ratio of the CFL/RBC-core cross-sectional area is $2\Delta/H$ in two-dimensional flow and $4\Delta(D-\Delta)/D^2$ in three-dimensional flow, where D is the vessel diameter. For $\Delta = 3$ μm and $H = D = 20$ μm, $2\Delta/H = 0.3$ and $4\Delta(D-\Delta)/D^2 = 0.51$, indicating a weaker cell-free-layer lubrication effect and a lower core hematocrit in two-dimensional flow. The apparent viscosity of an RBC suspension may be significantly different in two- and three-dimensional flow, even at the same bulk hematocrit. Caution should be exercised in interpreting the results of simulations for two-dimensional flow. For these reasons, our results have not been compared with empirical formulas of discharge hematocrit and apparent viscosity (Pries *et al.* 1992).

Although the simulations presented in this chapter pertain to RBC flow, the computational framework can be adapted to other microcirculation flows. White blood cells and platelets can easily be included to study particulate transport. The effects of the glycocalyx layer can be incorporated by attaching a thin porous layer to the vessel walls (Pries & Secomb 2005). Several LBM models for porous media flow are available (Ginzburg 2008, Freed 1998). To model complex vessel geometries, implementations with arbitrary boundary shapes can be coded with little additional effort (Guo & Zheng 2002, Mei *et al.* 1999).

Acknowledgment

This work was supported by grants HL18292, HL52684, and HL079087 from the NIH. J. Zhang acknowledges financial support from the Natural Science and Engineering Research Council of Canada (NSERC) and from the Laurentian University via a Research Fund. This work was made possible thanks to the facilities of the Shared Hierarchical Academic Research Computing Network (SHARCNET: www.sharcnet.ca). We thank Professor C. Pozrikidis for constructive comments and editorial improvements.

References

Aarts, P., van der Broek, S., Prins, G., Kuiken G. & Sixma, J. & Heethaar, R. (1988) Blood platelets are concentrated near the wall and red blood cells, in the center in flowing blood. *Arteriosclerosis* **6**, 819–824.

Abkarian, M., Faivre, M. & Viallat, A. (2007) Swinging of red blood cells under shear flow. *Phys. Rev. Lett.* **98**, 188302.

Aidun, C., Lu, Y. & Ding, E. (1998) Direct analysis of particulate suspensions with inertia using the discrete Boltzmann equation. *J. Fluid Mech.* **373**, 287–311.

ALMOMANI, T., UDAYKUMAR, H. S., MARSHALL, J. S. & CHANDRAN, K. B. (2008) Micro-scale dynamic simulation of erythrocyte-platelet interaction in blood flow. *Ann. Biomed. Eng.* **36**, 905–920.

ALONSO, C., PRIES, A. R., KIESSLICH, O., LERCHE, D. & GAEHTGENS P. (1995) Transient rheological behavior of blood in low-shear tube flow–velocity profiles and effective viscosity. *Am. J. Physiol., Heart Circ. Physiol.* **268**, H25–H32.

BAGCHI, P. (2007) Mesoscale simulation of blood flow in small vessels. *Biophys. J.* **92**, 1858–1877.

BAGCHI, P., JOHNSON, P. C. & POPEL, A. S. (2005) Computational fluid dynamic simulation of aggregation of deformable cells in a shear flow. *J. Biomech. Eng.* **127**, 1070–1080.

BASKURT, O. & MEISELMAN, H. (2007) Hemodynamic effects of red blood cell aggregation. *Indian J. Exp. Biol.* **45**, 25–31.

BAUMLER, H., NEU, B., DONATH, E. & KIESEWETTER, H. (1999) Basic phenomena of red blood cell rouleaux formation. *Biorheology* **36**, 439–442.

BHATNAGAR, P., GROSS, E. & KROOK, K. (1954) A model for collisional processes in gases I: Small amplitude processes in charged and neutral one-component system. *Phys. Rev. B* **94**, 511–525.

CHIEN, S. (1970) Shear dependence of effective cell volume as a determinant of blood viscosity. *Science* **168**, 977–979.

CHIEN, S. & JAN, K. (1973) Ultrastructural basis of the mechanism of rouleaux formation. *Microvasc. Res.* **5**, 155–166.

CHUNG, B., JOHNSON, P.C . & POPEL, A. S. (2006) Application of chimera grid to modeling cell motion and aggregation in a narrow tube. *Int. J. Numer. Meth. Fluids* **53**, 105–128.

COKELET, G. & GOLDSMITH, H. (1991) Decreased hydrodynamic resistance in the two-phase flow of blood through small vertical tubes at low flow-rates. *Circ. Res.* **68**, 1–17.

DUPIN, M. M., HALLIDAY, I., CARE, C. M., ALBOUL, L. & MUNN, L. L. (2007) Modeling the flow of dense suspensions of deformable particles in three dimensions. *Phys. Rev. E* **75**, No 066707.

DZWINEL, W., BORYCZKO, K. & YUEN, D. A. (2003) A discrete-particle model of blood dynamics in capillary vessels. *J. Colloid Interf. Sci.* **258**, 163–173.

EGGLETON, C. D. & POPEL, A. S. (1998) Large deformation of red blood cell ghosts in a simple shear flow. *Phys. Fluids* **10**, 1834–1845.

EVANS, E. A. (1973) New membrane concept applied to the analysis of fluid shear- and micropipette-deformed red blood cells. *Biophys. J.* **13**, 941–954.

EVANS, E. A. & FUNG, Y. C. (1972) Improved measurements of the erythrocyte geometry. *Microvasc. Res.* **4**, 335–347.

FISCHER, T. M. & SCHMID-SCHÖNBEIN, H. (1977) Tank tread motion of red cell membranes in viscometric flow: behavior of intracellular and extracellular markers. *Blood Cells* **3**, 351–365.

FRANCOIS, M., UZGOREN, E., JACKSON, J. & SHYY, W. (2003) Multigrid computations with the immersed boundary technique for multiphase flows. *Int. J. Numer. Meth. Heat Fluid Flow* **14**, 98–115.

FREED, D. (1998) Lattice-Boltzmann method for macroscopic porous media modeling. *Int. J. Modern Phys. C* **9**, 1491–1503.

FREUND, J. B. (2007) Leukocyte margination in a model microvessel. *Phys. Fluids* **19**, 023301.

GINZBURG, I. (2008) Consistent lattice Boltzmann schemes for the Brinkman model of porous flow and infinite Chapman-Enskog expansion. *Phys. Rev. E* **77**, No 066704.

GOLDSMITH, H. L. (1971) Red cell motions and wall interactions in tube flows. *Fed. Proc.* **30**, 1578–1590.

GOLDSMITH, H.L. & MARLOW, J. (1972) Flow behavior of erythrocytes. I. Rotation and deformation in dilute suspensions. *Proc. R. Soc. Lond. B* **182**, 351–384.

GUO, Z. & ZHENG, C. (2002) An extrapolation method for boundary conditions in lattice Boltzmann method. *Phys. Fluids* **14**, 2007–2010.

HEYES, D. M. (1998) *The Liquid State: Applications of Molecular Simulations.* Wiley, Chichester.

LADD, A. J. C. (1994) Numerical simulations of particulate suspensions via a discretized Boltzmann equation. Part I. Theoretical foundation. *J. Fluid Mech.* **271**, 285–309.

LIU, Y., ZHANG, L., WANG, X. & LIU, W. K. (2004) Coupling of Navier-Stokes equations with protein molecular dynamics and its application to hemodynamics. *Int. J. Numer. Method. Fluids* **46**, 1237–1252.

MARTINI, J., CARPENTIER, B., NEGRETE, A. C., CABRALES, P., TSAI, A. G. & INTAGLIETTA M. (2006) Beneficial effects due to increasing blood and plasma viscosity. *Clin. Hemorheol. Microcirc.* **35**, 51–57.

MEI, R., LUO, L. S. & SHYY, W. (1999) An accurate curved boundary treatment in the lattice Boltzmann method. *J. Comp. Phys.* **155**, 307–330.

MOHANDAS, N. & GALLAGHER, P. G. (2008) Red cell membrane: past, present, and future. *Blood* **112**, 3939–3948.

MURATA, T. (1987) Effects of sedimentation of small red blood cell aggregates on blood flow in narrow horizontal tubes. *Biorheology* **33**, 267–283.

N'DRI, N. A., SHYY, W. & TRAN-SON-TAY, R. (2003) Computational modeling of cell adhesion and movement using a continuum-kinetics approach. *Biophys. J.* **85**, 2273–2286.

NEU, B. & MEISELMAN, H. J. (2002) Depletion-mediated red blood cell aggregation in polymer solutions. *Biophys. J.* **83**, 2482–2490.

NOGUCHI, H. & GOMPPER, G. (2005) Dynamics of fluid vesicles in shear flow: Effect of membrane viscosity and thermal fluctuations. *Phys. Rev. E* **72**, No 011901.

PESKIN, C. S. (1977) Numerical analysis of blood flow in the heart. *J. Comp. Phys.* **25**, 220–252.

PIVKIN, I.V. & KARNIADAKIS, G.E. (2008) Accurate coarse-grained modeling of red blood cells. *Phys. Rev. Lett.* **101**, 118105.

POPEL, A. S. & JOHNSON, P. C. (2005) Microcirculation and hemorheology. *Annu. Rev. Fluid Mech.* **37**, 43–69.

POZRIKIDIS, C. (1992) *Boundary Integral and Singularity Methods for Linearized Viscous Flow.* Cambridge University Press, New York.

POZRIKIDIS, C. (1995) Finite deformation of liquid capsules enclosed by elastic membranes in simple shear flow. *J. Fluid Mech.* **297**, 123–152.

POZRIKIDIS, C. (2001) Interfacial dynamics for Stokes flow. *J. Comp. Phys.* **169**, 250–301.

POZRIKIDIS, C. (2003) Numerical simulation of the flow-induced deformation of red blood cells. *Ann. Biomed. Eng.* **31**, 1194–1205.

POZRIKIDIS, C. (2005) Axisymmetric motion of a file of red blood cells through capillaries. *Phys. Fluids* **17**, 031503.

PRIES, A. R., NEUHAUS, D. & GAEHTGENS P. (1992) Blood viscosity in tube flow: Dependence on diameter and hematocrit. *Am. J. Physiol., Heart Circ. Physiol.* **263**, H1770–H1778.

PRIES, A. R. & SECOMB, T. W. (2003) Rheology of the microcirculation. *Clin. Hemorheol. Microcirc.* **29**, 143–148.

PRIES, A. R. & SECOMB, T. W. (2005) Microvascular blood viscosity in vivo and the endothelial surface layer. *Am. J. Physiol., Heart Circ. Physiol.* **289**, H2657–H2664.

REINKE, W., GAEHTGENS, P. & JOHNSON, P. C. (1987) Blood viscosity in small tubes–Effect of shear rate, aggregation, and sedimentation. *Am. J. Physiol., Heart Circ. Physiol.* **253**, H540–H547.

SCHMID-SCHÖNBEIN, G. W., USAMI, S., SKALAK, R. & CHIEN, S. (1980) Interaction of leukocytes and erythrocytes in capillary and postcapillary vessels. *Microvasc. Res.* **19**, 45–70.

SECOMB, T. W., HSU, R. & PRIES, A. R. (2001) Motion of red blood cells in a capillary with an endothelial surface layer: effect of flow velocity. *Am. J. Physiol., Heart Circ. Physiol.* **281**, H629–H636.

SECOMB, T. W., STYP-REKOWSKA, B. & PRIES, A. R. (2007) Two-dimensional simulation of red blood cell deformation and lateral migration in microvessels. *Ann. Biomed. Eng.* **35**, 755-765.

SKALAK, R. & CHIEN, S. (1987) *Handbook of Bioengineering.* McGraw–Hill, New York.

SKALAK, R., ZARDA, P., JAN, K. & CHIEN, S. (1981) Mechanics of rouleau formation. *Biophys. J.* **35**, 771–781.

STOLTZ, J., SINGH, M. & RIHA, P. (1999) *Hemorheology in Practice.* IOS Press, Amsterdam.

SUCCI, S. (2001) *The Lattice Boltzmann Equation.* Oxford University Press.

SUI, Y., CHEW, Y. T., ROY, P., CHENG, Y. P. & LOW, H. T. (2008) Dynamic motion of red blood cells in simple shear flow. *Phys. Fluids* **20**, No 112106.

SUKOP, M. C. & THORNE, D. T. (2006) *Lattice Boltzmann Modeling: An Introduction for Geoscientists and Engineers.* Springer, Berlin.

SUN, C., MIGLIORINI, C. & MUNN, L. L. (2003) Red blood cells initiate leukocyte rolling in postcapillary expansions: a lattice Boltzmann analysis. *Biophys. J.* **85**, 208–222.

SUN, C. & MUNN, L. L. (2005) Particulate nature of blood determines macroscopic rheology: a 2-D lattice Boltzmann analysis. *Biophys. J.* **88**, 1635–1645.

TRAN-SON-TAY, R., SUTERA, S. P. & RAO, P. R. (1984) Determination of red-blood-cell membrane viscosity from rheoscopic observations of tank-treading motion. *Biophys. J.* **46**, 65–72.

TRYGGVASON, G., BUNNER, B., ESMAEELI, A., JURIC, D., AL-RAWAHI, N., TAUBER, W., HAN, J., NAS, S. & JAN, Y. J. (2001) A front-tracking method for the computations of multiphase flow. *J. Comp. Phys.* **169** 708–759.

UDAYKUMAR, H. S., KAN, H. C., SHYY, W. & TRAN-SON-TAY, R. (1997) Multiphase dynamics in arbitrary geometries on fixed Cartesian grids. *J. Comp. Phys.* **137**, 366–405.

WAUGH, R. E. & HOCHMUTH, R. M. (2006) Chapter 60: Mechanics and deformability of hematocytes. In *Biomedical Engineering Fundamentals*, Bronzino, J. D. (Ed.), Third Edition, CRC, Boca Raton.

ZHANG, J., JOHNSON, P. C. & POPEL, A. S. (2007) An immersed boundary lattice Boltzmann approach to simulate deformable liquid capsules and its application to microscopic blood flows. *Phys. Biol.* **4**, 285–295.

ZHANG, J., JOHNSON, P. C. & POPEL, A. S. (2008) Red blood cell aggregation and dissociation in shear flows simulated by lattice Boltzmann method. *J. Biomech.* **41**, 47–55.

ZHANG, J., JOHNSON, P. C. & POPEL, A. S. (2009) Effects of erythrocyte deformability and aggregation on the cell free layer and apparent viscosity of microscopic blood flows. *Microvasc. Res.* **77**, 265–272.

ZHANG, J. & KWOK, D. Y. (2006) Pressure boundary condition of the lattice Boltzmann method for fully developed periodic flows. *Phys. Rev. E* **73**, No 047702.

ZOU, Q. & HE, X. (1997) On pressure and velocity boundary conditions for the lattice Boltzmann BGK model. *Phys. Fluids* **9**, 1591–1598.

5 Front-tracking methods for capsules, vesicles and blood cells

P. Bagchi

Department of Mechanical and Aerospace Engineering
Rutgers University
Piscataway

High-accuracy computational modeling and simulation of microhemodynamics is a major challenge, primarily because blood in small vessels must be described as a dense suspension of deformable cells. Flow-induced deformation of erythrocytes combined with inter-cellular and cell–wall interactions give rise to a variety of phenomena including the formation of cell-depleted layers, the Fåhraeus effect, and the Fåhraeus–Lindqvist effect. Cell–wall hydrodynamic interactions play a critical role in leukocyte adhesion, a key process in inflammatory response. In this chapter, a front-tracking method for simulating cell motion is presented, accounting for the flow-induced deformation. The salient features of the algorithm are described and the hydrodynamics of isolated capsules, vesicles, and erythrocytes in a dilute suspension are discussed. Simulations illustrate the hydrodynamic interception of a pair of cells and the lateral migration of isolated capsules in wall-bounded Poiseuille flow. In the most comprehensive simulations, the channel flow of 1096 capsules in a dense suspension is described. Combining the basic algorithm with a coarse-grain Monte-Carlo method for describing intermolecular forces allows us to study the molecular interaction between a cell and a vessel wall. The integrated algorithm is applied to illustrate leukocyte rolling under the influence of a shear flow.

5.1 Introduction

Recent advances in high-performance computing and the development of robust numerical methods have sparked interest in the microhydrodynamics of cells and particulate blood flow. Progress has been made in several directions toward the direct numerical simulation of blood flow in the microcirculation where the cellular nature of blood must be accurately resolved (e.g., Pivkin *et al.* 2006, Ding & Aidun 2006, Dupin *et al.* 2007, Munn & Dupin 2008, Doddi & Bagchi 2009). In spite of these advances, difficulties in developing accurate and efficient computational procedures are still outstanding.

Large deformation of erythrocytes in shear flow and squeezing motion through narrow capillary vessels have been noted for decades. In pathological conditions, such as sickle cell anemia, diabetes mellitus, and malignant malaria, erythrocytes are not able to deform, and this severely affects blood flow in the microcirculation. Due to the high volume fraction of red blood cells *in vivo* (40–45%), blood flow is dominated by strong hydrodynamic interactions among neighboring cells. Blood rheology in microvessels is primarily dictated by the deformability of the individual cells and the interaction between clusters of cells. Whole blood *in vivo* is a polydisperse suspension consisting of red blood cells, leukocytes (white blood cells), and thrombocytes (platelets). Although the last two components constitute only a small fraction of whole blood, their number density can increase significantly in disease. Interactions between different cells and between cells and vessel walls can significantly affect the local microhemodynamics (e.g., Helmke *et al.* 1997, Pappu & Bagchi 2007).

Erythrocytes contain a liquid, called the cytoplasma, and are enclosed by a lipid bilayer attached to a cytoskeleton, which is a network of filaments consisting of spectrin and actin. Cholesterol and other protein molecules are dispersed in the bilayer and cytoskeleton. The detailed molecular structure is neglected in the continuum approximation, and the erythrocyte is modeled as a viscous drop enclosed by a zero-thickness elastic membrane. The elastic nature of the membrane is characterized by a shear deformation modulus and an areal extensional modulus due to the collective stretch of entangled spectrin filaments (e.g., Boal 1994). The membrane also exhibits resistance to bending due to the lipid bilayer. A vesicle membrane enclosed by a lipid bilayer lacks in-plane elasticity and exhibits only flexural stiffness. Erythrocytes, capsules, and vesicles, exhibit two primary types of motion in shear flow: tank-treading where the interior liquid and membrane rotate while the cell maintains a fixed inclination with respect to the direction of the flow, and tumbling where the cell flips like a rigid body (e.g., Goldsmith & Marlow 1972, Fischer *et al.* 1978). Significant periodic deformation is observed in both types of motion.

Tank-treading and tumbling motion of a cell in simple shear flow with velocity $\mathbf{u} = (\dot{\gamma} y, 0, 0)$ was first described by Keller & Skalak (1982), where $\dot{\gamma}$ is the shear rate. In their analysis, the cell was modeled as a shape-preserving ellipsoid containing a viscous liquid and surrounded by a rotating membrane.

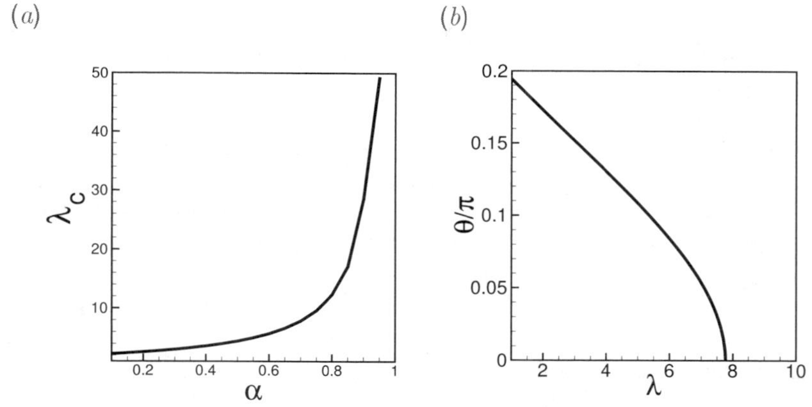

Figure 5.1.1 Illustration of the Keller & Skalak (1982) model for a nondeformable ellipsoid tank-treading in simple shear flow. Graphs of (*a*) the critical viscosity ratio λ_c against the aspect ratio α, and (*b*) inclination angle against the viscosity ratio λ for a tank-treading ellipsoid with axis ratio $\alpha = 0.7$.

Their analysis showed that the evolution of the inclination angle measured in counterclockwise direction from the x axis, θ, is governed by the differential equation

$$\frac{d\theta}{dt} = -\frac{\dot{\gamma}}{2}\Big(1 - \frac{4LB}{L^2+B^2}\,\frac{f_3}{f_2 - \lambda f_1}\cos 2\theta - \frac{L^2-B^2}{L^2+B^2}\cos 2\theta\Big), \tag{5.1.1}$$

where L and B are the ellipsoid semi-major and semi-minor axes, f_1, f_2 and f_3 are dimensionless functions of L and B, μ_0 is the viscosity of the suspending liquid, μ_1 is the viscosity of the interior liquid, and $\lambda = \mu_1/\mu_0$ is the viscosity ratio. For a given aspect ratio, $\alpha = B/L$, a steady solution is possible only when λ is less than a critical value, λ_c. For lower values of λ, the inclination angle decreases with increasing λ, as shown in figure 5.1.1. For $\lambda > \lambda_c$, a steady solution does not exist and tumbling motion arises.

One important assumption underlying the Keller & Skalak (1982) model is that the cell is nondeformable. In fact, large cell deformation coexists with tank treading and tumbling motion. Recent experiments by Abkarian *et al.* (2007), numerical simulations by Sui *et al.* (2008*a*, *b*), and theoretical models by Skotheim & Secomb (2007) have shown that a transition from tumbling to tank-treading can be triggered by changing the shear rate for a fixed viscosity ratio. The effect of the shear rate is not captured in the analysis of Keller & Skalak (1982) due to the absence of deformation. Skotheim &

Secomb (2007) demonstrated that swinging dynamics and the effect of the shear rate can be incorporated in terms of an elastic membrane energy (see Chapter 7).

Capsule and vesicle dynamics can be described analytically in the limit of small deformation. Asymptotic solutions were developed by Barthès-Biesel and coworkers for small capsule deformation (Barthès-Biesel 1980, Barthès-Biesel & Rallison 1981, Barthès & Sgaier 1985), and by others for small vesicle deformation (Seifert 1999, Misbah 2006). The analysis successfully predicts mode transition and confirms the occurrence of a vacillating–breathing mode. Analytical solutions are not possible in the case of large deformation. Numerical studies of the flow-induced deformation in simple shear flow were presented by Pozrikidis (1995) and Ramanujan & Pozrikidis (1998) based on the boundary-integral formulation for Stokes flow. The axisymmetric deformation of a capsule in a straining flow, through a hyperbolic constriction, and across a cylindrical tube was studied by others (Li *et al.* 1988, Leyrat-Maurin & Barthés-Biesel 1994, and Queguiner & Barthés-Biesel 1997). The effect of membrane bending stiffness was considered by Pozrikidis (2001) and Kwak & Pozrikidis (2001). Diaz *et al.* (2001) investigated the effect of membrane viscosity. Barthés-Biesel *et al.* (2002) and Lac *et al.* (2004) the effect of the membrane constitutive laws.

In further studies, Diaz *et al.*(2000) investigated the effect of the internal capsule viscosity, and Pozrikidis (2003, 2005) presented three-dimensional boundary-integral simulation of erythrocyte deformation in linear shear flow and tube flow. Breyiannis & Pozrikidis (2000) simulated the shear flow of an idealized two-dimensional suspensions of elastic capsules in doubly periodic flow. More recently, Lac *et al.* (2007) examined the hydrodynamic interaction of two capsules in shear flow and Freund (2007) simulated the margination of leukocytes surrounded by multiple red blood cells in two-dimensional microvessels (see also Chapter 3). Dodson & Dimitrakopoulos (2008) developed a high-accuracy spectral boundary-element method and studied the large deformation of capsules in straining flow. Kessler *et al.* (2008) used a similar approach to study the swinging and tumbling of elastic capsules in shear flow. The boundary-integral method was further applied to study vesicle deformation in shear flow (Kraus *et al.* 1996), near a wall (Sukumaran & Seifert 2001), and in bounded parabolic flow (Coupier *et al.* 2008).

Alternative numerical methods include the immersed-boundary method, front-tracking methods, phase-field methods, lattice Boltzmann methods, and dissipative particle dynamics. Eggleton & Popel (1998), Li & Sarkar (2008), Bagchi (2007), Doddi & Bagchi (2008, 2009), and Bagchi & Kalluri (2009) applied the front-tracking method to study capsule deformation in linear shear flow and Poiseuille flow. Ding & Aidun (2006) and Dupin *et al.* (2007)

developed a three-dimensional lattice-Boltzmann method and simulated the flow of a large number of erythrocytes. Sui *et al.* (2008*a*,*b*) combined the immersed-boundary method with the lattice-Boltzmann method and investigated mode-switching due to the shear rate for capsules and red blood cells. Pivkin *et al.* (2006) and Pivkin & Karniadakis (2008) used dissipative particle dynamics to study erythrocyte motion in shear flow and investigated the role of erythrocyte-platelet interaction on thrombus formation (see also Chapter 6). Biben & Misbah (2003), Beaucourt *et al.* (2004), and Biben *et al.* (2005) implemented an advected-field method and studied vesicle dynamics in shear flow.

In this chapter, the implementation of the front-tracking method is discussed and simulations are presented for several flow configurations. The general goal is to delineate the extent and physiological significance of the flow-induced deformation of erythrocytes, capsules, and vesicles. The unique properties of the erythrocyte membrane, including elastic resistance to shear, area-incompressibility, and elastic resistance to bending, and viscosity differences between the plasma and hemoglobin, are taken into consideration in the mathematical formulation.

Results will be presented first for isolated capsules, vesicles, and erythrocytes in shear flow to illustrate the transition from tank-treading to tumbling. The interaction of a pair of different cells and the lateral migration of an isolated capsule in wall-bounded Poiseuille flow will then be discussed. Large-scale simulations with up to 1096 capsules will be presented to illustrate the dynamics of suspension flow in microchannels. The main algorithm will be coupled with a Monte-Carlo scheme implementing a coarse-grain model of molecular interaction between a cell and a vessel wall. The integrated model will allow us to study the adhesive rolling motion of leukocytes in shear flow.

5.2 Numerical method

The front-tracking method was initially developed for simulating multi-fluid flow with applications in drop and bubble dynamics (e.g., Unverdi & Tryggvason 1992, Tryggvason *et al.* 2001). The basic procedure is a variation of the immersed-boundary method developed by Peskin (1977) for cardiac blood flow. The main idea is to describe the flow in the entire flow domain in terms of a generalized equation of motion. Interfaces between different fluids contribute concentrated force fields represented by distributed delta functions.

In the case of incompressible flow, the governing equations are the continuity equation, $\nabla \cdot \mathbf{u} = 0$, and a generalized Navier–Stokes equation involving the effect of the interface,

$$\rho\left(\frac{\partial \mathbf{u}}{\partial t} + \mathbf{u}\cdot\nabla\mathbf{u}\right) = -\nabla p + \nabla\cdot[\mu(\nabla\mathbf{u} + (\nabla\mathbf{u})^T] - \iint_{\mathcal{D}} \mathbf{f}(\mathbf{x}', t)\,\delta_3(\mathbf{x}-\mathbf{x}')\,\mathrm{d}S(\mathbf{x}'), \qquad (5.2.1)$$

where ρ is the common density of the fluids, $\mathbf{u}$ is the fluid velocity, p is the pressure, $\mathbf{f}$ is the distribution density of an interfacial force field, δ_3 is the three-dimensional delta function, and $\mathbf{x}'$ is the position of a point at the interface, $\mathcal{D}$. The last term on the right-hand-side of (5.2.1) represents an interfacial body force term coupling the membrane to the surrounding fluid. The computation of $\mathbf{f}$ will be described later in this section.

The viscosity field is described in terms of an indicator function, $\mathcal{I}$, as

$$\mu = \mu_0\,[1 + (\lambda - 1)\mathcal{I}]\,, \qquad (5.2.2)$$

where $\mathcal{I} = 1$ in the interior of a cell and $\mathcal{I} = 0$ outside a cell. Unverdi & Tryggvason (1992) demonstrated that $\mathcal{I}$ can be obtained by solving a Poisson equation,

$$\nabla^2\mathcal{I} = \nabla\cdot\mathbf{G}\,, \qquad (5.2.3)$$

where

$$\mathbf{G} = \iint_S \delta_3(\mathbf{x}-\mathbf{x}')\,\mathbf{n}\,\mathrm{d}S(\mathbf{x}'), \qquad (5.2.4)$$

and $\mathbf{n}$ is the outward unit vector normal to an interface. To ensure continuity of velocity across the interface, the interfacial velocity is interpolated from the fluid velocity as

$$\mathbf{u}_S(\mathbf{x}', t) = \iiint \mathbf{u}(\mathbf{x}, t)\,\delta_3(\mathbf{x}-\mathbf{x}')\,\mathrm{d}V(\mathbf{x}), \qquad (5.2.5)$$

where the integration is performed over the whole domain of flow. The interface is then advected by integrating the differential equation

$$\frac{\mathrm{d}\mathbf{x}'}{\mathrm{d}t} = \mathbf{u}_S(\mathbf{x}', t) \qquad (5.2.6)$$

using standard numerical methods.

5.2.1 Navier–Stokes solver

In the front-tracking method, the Navier–Stokes equation is solved on a fixed Eulerian grid and the interfaces are tracked in terms of Lagrangian marker

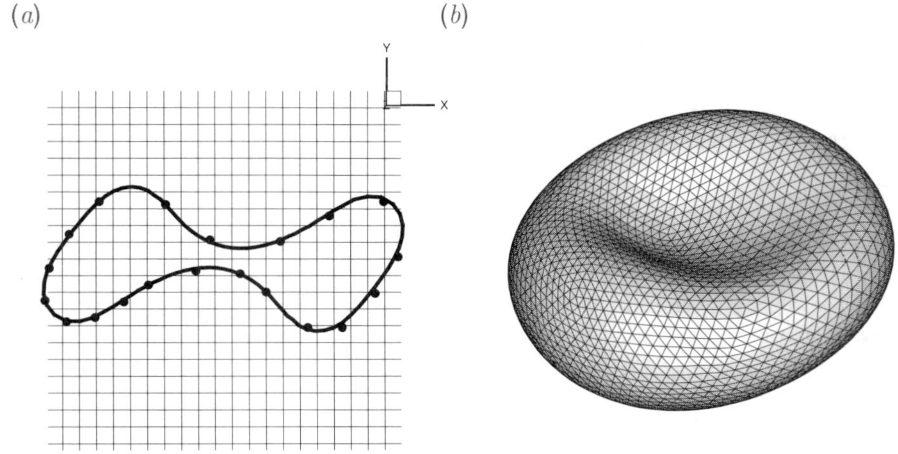

Figure 5.2.1 (*a*) Illustration of a fixed Eulerian mesh and accompanying moving Lagrangian mesh. (*b*) Illustration of the triangulated surface of a biconcave disk representing a deformed red blood cell.

points, as shown in figure 5.2.1. Any suitable fluid solver based on finite-volume, finite-difference, spectral, or spectral-element method can be employed. We have implemented a finite-difference–Fourier-transform method.

Following the general blueprint of projection methods for incompressible flow, we split the Navier–Stokes equation into an advection–diffusion part for the velocity and a Poisson equation part for the pressure. The body force term is included in the advection–diffusion part. A second-order finite difference scheme is used to spatially discretize the advection–diffusion equation. Inertial terms are treated explicitly using the second-order Adams-Bashforth method. Viscous terms are treated semi-implicitly using the second-order Crank-Nicolson method.

The discretization of the advection–diffusion equation yields the finite-difference equation

$$\rho\,\frac{\mathbf{u}^* - \mathbf{u}^n}{\Delta t} = \frac{1}{2}\Big(\mathbf{V}(\mathbf{u}^*) + \mathbf{V}(\mathbf{u}^n)\Big) - \frac{3}{2}\,(\mathbf{NL})^n + \frac{1}{2}\,(\mathbf{NL})^{n-1}, \qquad (5.2.7)$$

where $\mathbf{u}^*$ is the velocity at an intermediate time level between time levels n and $n+1$, Δt is the time step, $\mathbf{V}$ denotes the diffusion term, and $\mathbf{NL}$ denotes the nonlinear term. Terms involving the viscosity gradient are treated explicitly. The resulting discretized system can be solved iteratively or by direct inversion using, for example, the alternating direction implicit (ADI) method. The predicted velocity $\mathbf{u}^*$ obtained after integration is not necessarily solenoidal.

The pressure is computed by solving a Poisson equation that emerges by requiring a divergence-free velocity at the end of a complete time step,

$$\nabla^2 p^{n+1} = \frac{\rho}{\Delta t} \nabla \cdot \mathbf{u}^*. \tag{5.2.8}$$

Because in our simulations the flow domain is periodic in at least one direction, Fourier transforms are implemented to generate a decoupled system of two-dimensional partial differential equations, which can be easily inverted by fast algorithms. The predicted velocity is then projected onto the divergence-free space,

$$\frac{\mathbf{u}^{n+1} - \mathbf{u}^*}{\Delta t} = -\frac{1}{\rho} \nabla p^{n+1}. \tag{5.2.9}$$

To ensure numerical stability, the interfacial force and indicator function must vary smoothly across the interfaces. A smooth representation of the three-dimensional delta function must then be employed. Following Unverdi & Tryggvason (1992), we write

$$D(\mathbf{x} - \mathbf{x}') = \frac{1}{(4h)^3} \prod_{i=1}^{3} \left(1 + \cos \frac{\pi}{2h}(x_i - x_i')\right) \tag{5.2.10}$$

when $|x_i - x_i'| \leq 2h$ for $i = 1, 2, 3$, and $D(\mathbf{x} - \mathbf{x}') = 0$ otherwise, where h is the Eulerian grid size. As a result of this approximation, the interface is artificially smeared over four Eulerian points surrounding a central Lagrangian point.

5.2.2 Computation of the interfacial force

Skalak *et al.* (1973) proposed a strain-energy function to describe the shearing deformation and area dilatation of the erythrocyte membrane,

$$W = \frac{E_s}{8} \left[(\epsilon_1^2 + \epsilon_2^2 - 2)^2 + 2(\epsilon_1^2 + \epsilon_2^2 - \epsilon_1^2\epsilon_2^2 - 2)\right] + \frac{E_a}{8} \left(\epsilon_1^2\epsilon_2^2 - 1\right)^2, \tag{5.2.11}$$

where ϵ_1 and ϵ_2 are the principal stretch ratios, E_s is the shear elastic modulus and E_a is the area dilatation modulus. For a human erythrocyte, $E_s \approx 6 \times 10^{-3}$ dyn/cm, and $E_a \approx 500$ dyn/cm. Accordingly, the membrane is nearly incompressible but can easily be deformed. The interfacial force field arises from the principle of virtual work as

$$\mathbf{f}(\mathbf{x}', t) = -\frac{\partial W}{\partial \mathbf{x}'}. \tag{5.2.12}$$

The derivative is evaluated after discretization by standard numerical methods.

5.2.3 Membrane discretization

In numerical practice, the membrane of each cell is discretized into triangular elements based on the recursive subdivision of an icosahedron using the open-source GNU surface code (GTS). After discretization, the vertices of the generated triangles are projected onto a specified unstressed spherical, ellipsoidal, or biconcave shape, as illustrated in figure 5.2.1(*b*). In most cases, a node is surrounded by six triangular elements. Near few topological exceptions, a node is surrounded by only five elements.

Each vertex is now interpreted as a Lagrangian node and the interfacial force is computed by a finite-element method (Charrier *et al.* 1989, Shrivastava & Tang 1993). The displacement of the lth vertex of each triangle is first calculated,

$$\mathbf{v}_l = \mathbf{R} \cdot (\mathbf{X}'_l - \mathbf{X}'_1) - \mathbf{M} \cdot (\mathbf{x}'_l - \mathbf{x}'_1) \tag{5.2.13}$$

for $l = 1, 2, 3$, where $\mathbf{x}'$ is the position of a point on the deformed cell and $\mathbf{X}'$ is the position on the undeformed cell. The matrices $\mathbf{M}$ and $\mathbf{R}$ project deformed and undeformed triangles onto a common plane using rigid-body rotation. The nodal displacement arising due to the in-plane deformation may then be obtained.

To compute the deformation gradient tensor, $\mathbf{D}$, we assume that the displacement $\mathbf{v}$ varies linearly inside each triangular element. The in-plane stretch ratios ϵ_1 and ϵ_2 are related to $\mathbf{D}$ by

$$\epsilon_i^2 = \frac{1}{2}\Big(G_{11} + G_{22} \pm \Big[(G_{11} - G_{22})^2 + 4\, G_{12}^2\Big]^{1/2}\Big), \tag{5.2.14}$$

or $i = 1, 2$, where $\mathbf{G} = \mathbf{D}^T\mathbf{D}$ is the right Cauchy–Green tensor. The in-plane force at a vertex is computed from the equation

$$\mathbf{f}_l^P = -\frac{\partial W}{\partial \epsilon_1}\frac{\partial \epsilon_1}{\partial \mathbf{v}_l} - \frac{\partial W}{\partial \epsilon_2}\frac{\partial \epsilon_2}{\partial \mathbf{v}_l}. \tag{5.2.15}$$

Using the transformation $\mathbf{f} = \mathbf{R}^T\mathbf{f}^P$, we finally express the local force in global coordinates. The resulting force at a Lagrangian node, $\mathbf{f}(\mathbf{x}', t)$, is the vector sum of all forces contributed by surrounding elements.

5.3 Capsule deformation in simple shear flow

First, we consider the deformation of a capsule in linear shear flow with velocity $\mathbf{u} = [\dot{\gamma}y, 0, 0]$, where $\dot{\gamma}$ is a specified shear rate. The membrane response is governed by the constitutive equation (5.2.11). The capsule equivalent radius, a is chosen as the characteristic length scale, and the inverse shear rate, $1/\dot{\gamma}$,

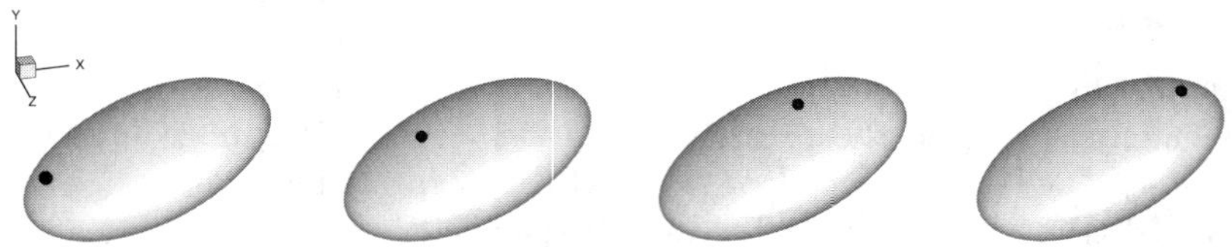

Figure 5.3.1 Steady shape of a deformed capsule with spherical unstressed shape for $\lambda = 1$ and $\mathrm{Ca} = 0.4$. The rotating position of a marker point demonstrates the membrane tank-treading motion.

is chosen as the characteristic time scale. Dimensionless parameters include the capillary number expressing the significance of viscous stresses relative to the elastic membrane tension, $\mathrm{Ca} = \dot{\gamma} a \mu_0 / E_s$, the ratio of the membrane area dilation modulus to shear deformation modulus, $C = E_a/E_s$, and the viscosity ratio, λ.

In the simulations reported in this section, the constant C is set to unity and the Reynolds number is set to $\mathrm{Re} = \rho \dot{\gamma} a^2 / \mu_0 \sim 10^{-2}$ so that inertial forces play a minor role. The computational domain is a cubic box with side length $2\pi a$ covered by a 80^3 Eulerian grid. The capsule membrane is discretized into 5120 triangular elements.

5.3.1 Spherical capsules

When $\mathrm{Ca} = 0.4$ and $\lambda = 1$, a spherical capsule deforms into a steady prolate shape aligned at a certain angle with respect to the direction of the incident simple shear flow, as shown in figure 5.3.1. The position of a marker point around the membrane is shown at different time intervals to demonstrate the occurrence of tank-treading motion. The bold lines in figure 5.3.2 originating from zero describe the evolution of the Taylor deformation parameter, $D = (L - B)/(L + B)$, where L and B are the maximum and minimum capsule dimensions. The thin lines show the evolution of the inclination angle, θ. Both are plotted against the scaled time, $t^* = t\dot{\gamma}$.

The results presented in figure 5.3.2(a) for $\lambda = 1$ show that, as the dimensionless shear rate Ca increases, the capsule deformation becomes more pronounced while the inclination angle becomes smaller. These predictions are consistent with boundary-integral simulation by Lac *et al.* (2004). The results presented in figure 5.3.2(b) illustrate the effect of the viscosity ratio, λ, for $\mathrm{Ca} = 0.05$. As λ increases, the deformation becomes less pronounced and the inclination angle becomes higher. Similar behavior is observed for capsules

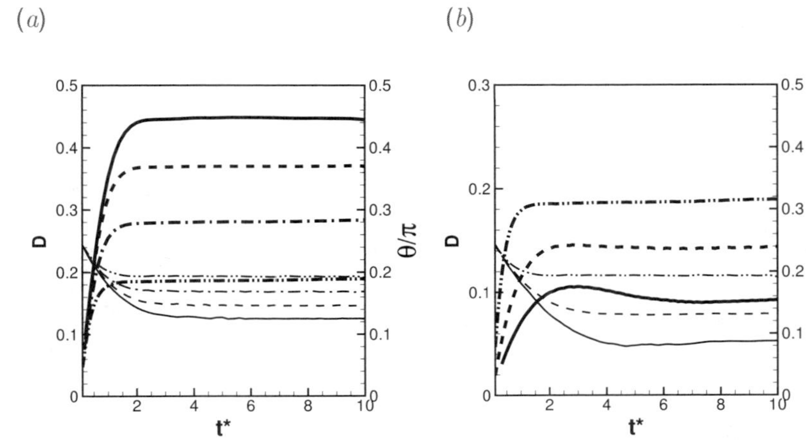

Figure 5.3.2 Evolution of the Taylor deformation parameter (bold lines) and inclination angle θ (thin lines) for a spherical unstressed capsule in shear flow. (*a*) Effect of Ca for $\lambda = 1$: $\mathrm{Ca} = 0.4$ (solid line), 0.2 (dashed line), 0.1 (dash-dotted line), and 0.05 (dash-double-dotted line). (*b*) Effect of the viscosity ratio for $\mathrm{Ca} = 0.05$: $\lambda = 1$ (dash-double-dotted line), 5 (dashed line), and 10 (solid line),

whose membrane is described by a neo-Hookean law (e.g., Sui *et al.* 2008*a*). The numerical results are consistent with the small-deformation analysis of Barthès-Biesel & Rallison (1981), the boundary-integral simulations of Ramanujan & Pozrikidis (1998) and Lac *et al.* (2004), and the experiments of Rehage *et al.* (2002) (Doddi & Bagchi 2008, 2009).

5.3.2 Ellipsoidal capsules

To illustrate the occurrence of tumbling, we consider the deformation of an oblate unstressed capsule with aspect ratio $\alpha = 0.7$ whose membrane is described by equation (5.2.11) with moduli ratio $C = 1$. Transition from tank-treading to tumbling can be triggered either by increasing the viscosity ratio, λ, or by decreasing the shear rate. The first possibility is illustrated in figure 5.3.3(*a, b*) for $\mathrm{Ca} = 0.075$ and $\lambda = 3, 7$, and 10. The graphs in this figure describe the evolution of the inclination angle, θ, and Taylor deformation parameter, D. For $\lambda = 3$, the inclination angle oscillates but remains positive, revealing a swinging mode. We recall that the Keller & Skalak theory is unable to capture this mode. For $\lambda = 10$, a tumbling motion is observed where θ varies between $\pm\pi/2$. Interestingly, for $\lambda = 7$, a full tumbling motion is not observed. Instead, θ periodically becomes positive and negative. This behav-

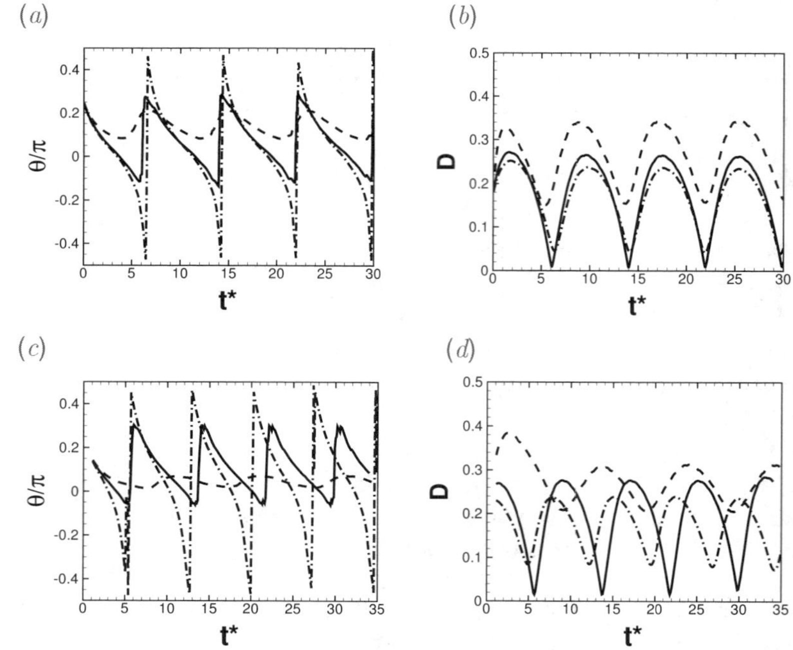

Figure 5.3.3 Swinging, tumbling, and vacillating–breathing modes of oblate capsules in simple shear flow. (a, b) Effect of λ for $\text{Ca} = 0.075$: $\lambda = 3$ (dashed line), 7 (solid line), and 10 (dash-dotted line). (c, d) Effect of Ca for $\lambda = 5$: $\text{Ca} = 0.4$ (dashed line), 0.05 (solid line), and 0.02 (dash-dotted line).

ior contradicts the results of the Keller & Skalak (1982) model predicting the onset of tumbling as soon as θ becomes zero. In all three cases described in figure 5.3.3(a, b), large oscillations in the deformation parameter arise.

A transition from tank-treading to tumbling also occurs by lowering the capillary number, as described in figure 5.3.3(c, d). Time sequences of evolving capsule profiles are shown in figure 5.3.4(a–c). A swinging or oscillatory mode, a vacillating–breathing mode, and a tumbling mode are observed, respectively, for $\text{Ca} = 0.4$, 0.05, and 0.2. Tank-treading is evident in the swinging and vacillating–breathing modes, as indicated by the position of a marker point around the membrane contour. In the tumbling mode, the marker point oscillates back and forth around a mean position.

Significant shape deformation is observed for all three modes, as shown in figure 5.3.4(a–c). It appears that the maximum shape deformation oc-

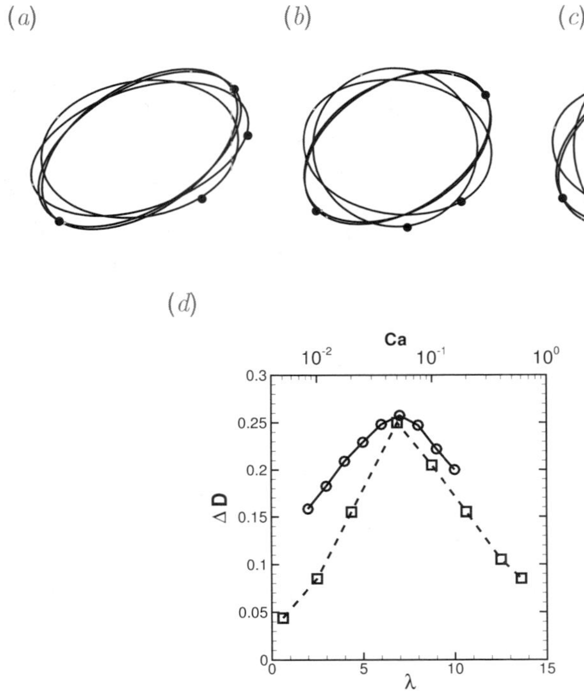

Figure 5.3.4 (*a*) Swinging, (*b*) vacillating-breathing, and (*c*) tumbling motion of an oblate capsule with aspect ratio $\alpha = 0.7$ and $\lambda = 5$. The capillary number is $\mathrm{Ca} = 0.4, 0.05$, and 0.02. (*d*) Amplitude of the shape deformation, ΔD: the solid line connecting circular symbols shows the effect of λ for $\mathrm{Ca} = 0.075$; the dashed line connecting square symbols shows the effect of Ca for $\lambda = 5$.

curs in the vacillating–breathing mode. To confirm this observation, in figure 5.3.4(*d*) we plot the oscillation amplitude, ΔD, against the capillary number or viscosity ratio. Increasing λ or decreasing Ca causes ΔD to first increase, then reach a maximum when the vacillating–breathing mode appears, and finally decrease with a further increase in λ or decrease in Ca when the tumbling mode appears. A detailed discussion of the transition process can be found in Bagchi & Murthy (2009). As λ increases, the mean inclination angle decreases at a rate that is much lower than that predicted by Keller & Skalak (1982) due to the appearance of the vacillating–breathing mode.

5.3.3 Vesicles

A vesicle is a liquid capsule enclosed by an incompressible membrane endowed with flexural stiffness but lacking resistance against shearing deformation. In this section, we describe the implementation of the front-tracking method for vesicles enclosed by nearly incompressible membranes. Area incompressibility is enforced by using a large value of E_a in the surface constitutive equation (5.2.11). A pertinent dimensionless parameter is the capillary number, $\mathrm{Ca} = \dot{\gamma}\mu_0 a/E_a$. For $\mathrm{Ca} = 0.001$, the increase in surface area due to the flow-induced deformation is less than 0.2%.

A bending energy functional for biological membranes was proposed by Helfrich (1973)

$$W_B = \frac{E_B}{2}\iint_S (2\kappa - c_0)^2\,\mathrm{d}S + E_G \iint_S \kappa_g \mathrm{d}S, \tag{5.3.1}$$

where E_B is the bending modulus associated with the mean curvature, κ, E_G is the bending modulus associated with the Gaussian curvature, κ_g, and c_0 is the spontaneous curvature. It is convenient to use an expression of the surface force density derived by Zhong-can & Helfrich (1989),

$$\mathbf{f}_b = \left[E_B\left(2\kappa + c_0\right)\left(2\kappa^2 - 2\kappa_g - c_0\kappa\right) + 2E_B\Delta_{LB}\kappa\right]\mathbf{n}, \tag{5.3.2}$$

where $\mathbf{n}$ is the unit normal vector to the membrane, $\Delta_{LB} = (\mathbf{I}_s \cdot \nabla) \cdot (\mathbf{I}_s \cdot \nabla)$ is the surface Laplace–Beltrami operator, and $\mathbf{I}_s = \mathbf{I} - \mathbf{nn}$ is the surface projection operator.

The curvature of the triangulated membrane is calculated by quadratic surface fitting,

$$z' = ax'^2 + bx'y' + cy'^2 + dx' + ey', \tag{5.3.3}$$

where (x', y', z') are local coordinates with origin at a Lagrangian point of interest, P_i, and the coordinate z' is aligned with the estimated normal vector. The technique is described in detail by Garimella & Swartz (2003) and Petitjean (2002). One-ring neighbor points are used to compute the coefficients using a least-square method, and iterations are performed to obtain an accurate fitting until satisfactory convergence to the estimated normal vectors has been secured. The curvatures κ and κ_g are expressed in terms of the fitted coefficients.

To discretize the Laplace–Beltrami operator, we adopt the framework of computational image reconstruction (e.g., Reuter *et al.* 2009). If $[P_i, P_j, P_{j+}]$ is a triangle sharing vertex P_i, then

$$\Delta_{LB}\kappa(P_i) = \frac{1}{2A}\sum_{j\in N(i)} n_j^{\mathrm{T}}\Big(\nabla_s\kappa(P_j) + \nabla_s\kappa(P_{j+})\,||P_j - P_{j+}||\Big), \tag{5.3.4}$$

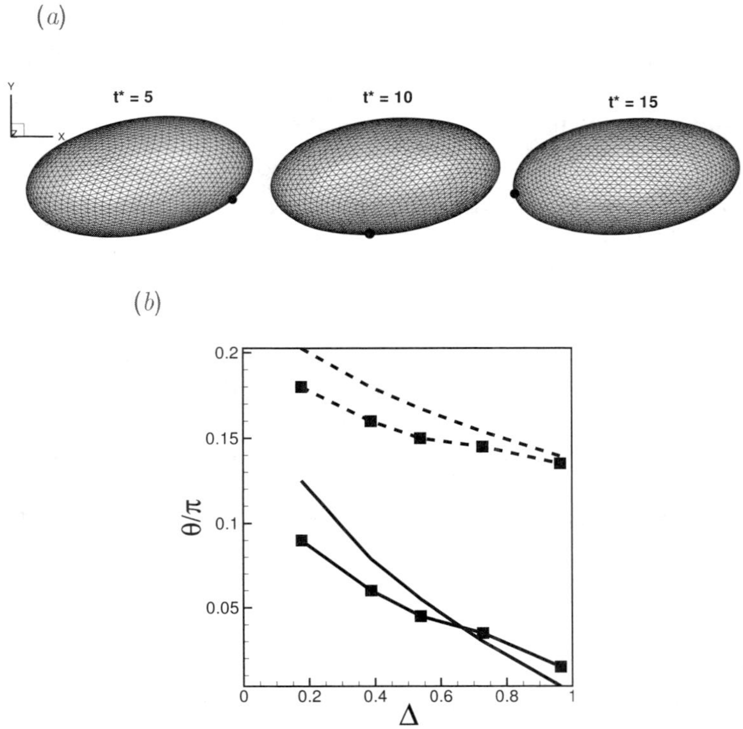

Figure 5.3.5 (*a*) Illustration of the transient deformation of a vesicle in simple shear flow with $\Delta = 0.5$ and $\lambda = 5.6$. (*b*) The computed inclination angle (lines with symbols), plotted against the vesicle excess area Δ, is compared with the experimental data (lines without symbols) of Kantsler & Steinberg (2005) for $\lambda = 1$ (dashed lines) and 5 (solid lines).

where $N(i)$ are one-ring neighbor vertices surrounding P_i and A is the area of all triangles sharing the vertex P_i (Xu 2004).

Relevant dimensionless parameters are the membrane excess area $\Delta = S/R_0^2 - 4\pi$, where $R_0 = (3V/4\pi)^{1/3}$ and V is the vesicle volume, the dimensionless bending rigidity $\chi = \dot{\gamma}\mu_0 R_0^3/E_B$ expressing the ratio of viscous stresses to bending resistance, and the viscosity ratio, λ. The spontaneous curvature is taken to be zero.

Figure 5.3.5(a) illustrates the evolution of a prolate vesicle with $\Delta = 0.5$, $\chi = 50$, and $\lambda = 5.6$. As time progresses, the vesicle aligns with the shear flow while the membrane rotates in a tank-treading mode. The computed inclination angle is successfully compared with experimental results by Kantsler & Steinberg (2005) in figure 5.3.5(*b*).

5.3.4 Red blood cells

A generalization of the numerical method allows us to simulate the deformation of erythrocytes (RBCs) taking into consideration the membrane resistance to shearing deformation, bending, and surface dilation. The resting shape of an RBC is a biconcave disk described by

$$z(r) = \left[1 - \left(\frac{r}{R_0}\right)\right]^{1/2} \left[C_0 + C_2\left(\frac{r}{R_0}\right)^2 + C_4\left(\frac{r}{R_0}\right)^4\right], \qquad (5.3.5)$$

where $0 \le r \le R_0$ and $0 \le \theta \le 2\pi$ (Evans & Fung 1972, Ramanujan & Pozrikidis 1998). The coefficients C_0, C_2, and C_4 depend on the cell osmolarity. For osmolarity of 300, $C_0/R_0 = 0.207$, $C_2/R_0 = 2.003$, and $C_4/R_0 = -1.123$ (Fung 1984). To discretize the biconcave disk, we follow a method described in Ramanujan & Pozrikidis (1998) and map the surface of a discretized sphere using the equations

$$x = R\eta, \qquad y = \frac{R}{2}\sqrt{1 - r^2}(C_0 + C_2 r^2 + C_4 r^4), \qquad z = R\zeta, \qquad (5.3.6)$$

where $\eta^2 + \zeta^2 = r^2$ and the length R is adjusted to preserve the cell volume. Relevant dimensionless parameters include the capillary number, $\text{Ca} = \dot{\gamma}\mu_0 R_0/E_s$, the ratio of the elastic moduli, $C = E_a/E_s$, the dimensionless bending rigidity, $E_B^* = E_s R_0^2/E_B$, and the viscosity ratio, λ. We take $E_B^* = 0.01$ and set the spontaneous curvature to $c_o R_0 = -2.09$. The ratio C varies between 50 and 400 so that the area dilatation is less than 0.5% in all cases.

Figure 5.3.6 shows snapshots of a deformed cell in shear flow. Tank-treading motion is observed in 5.3.6(*a*) for $\text{Ca} = 1.0$ and $\lambda = 1$, and tumbling motion is observed in 5.3.6(*b*) for $\text{Ca} = 0.1$ and $\lambda = 5$. In these simulations, the computational domain was covered by a 120^3 Eulerian grid, and the membrane was discretized into 20480 triangular elements. The dimensionless time step, $\dot{\gamma}\Delta$, varies between 10^{-3} and 10^{-4}. In most cases, we are able to continue the simulations up to scaled time $t\dot{\gamma} \approx 35$ without regridding. The average run time is 24 hours on four AMD Opteron Quad-Core processors.

5.4 Capsule interception

Having discussed the dynamics of individual capsules, vesicles, and erythrocytes in a dilute suspension, we proceed to consider intercellular interactions. The interception of two spherical rigid particles was discussed by Batchelor &

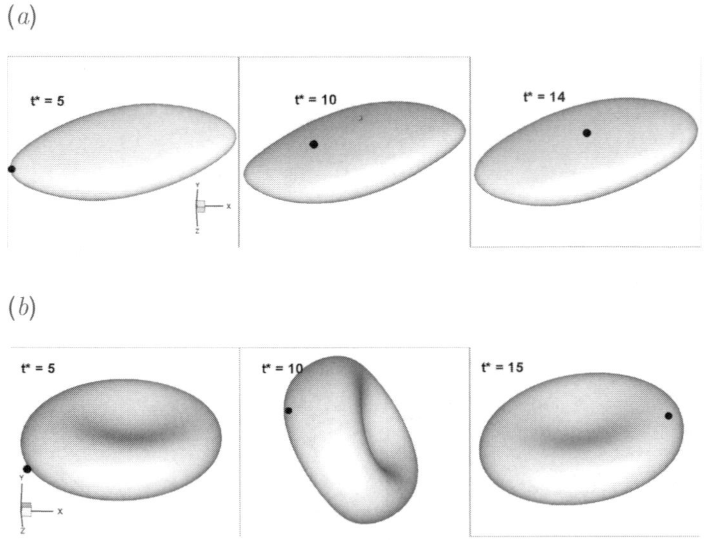

Figure 5.3.6 Erythrocyte deformation in shear flow illustrating (*a*) tank-treading motion for $\text{Ca} = 1.0$ and $\lambda = 1$, and (*b*) tumbling motion for $\text{Ca} = 0.1$ and $\lambda = 5$.

Green (1972). In Stokes flow, two rigid spheres with perfectly smooth surfaces return to their initial elevation after interception. In the case of deformable particles, such as liquid drops, interception causes a net migration and a shear-induced dispersion (e.g., Guido & Simeone 1998, Loewenberg & Hinch 1997, Charles & Pozrikidis 1998). Lac *et al.* (2007) simulated the interception of two identical deformable capsules in shear flow. We will presently extend their results to capsules with different physical properties.

First, we consider two spherical capsules, labeled 1 and 2, with the same volume but different membrane properties determined by the first capillary number, Ca_1, and the capillary number ratio, $\beta = \text{Ca}_2/\text{Ca}_1$. The membrane mechanics is governed by (5.2.11). Figure 5.4.1(*a*) illustrates the capsule interception for $\text{Ca}_1 = 0.02$, $\beta = 15$, and equal interior and exterior fluid viscosities. In the early stage of the motion, the capsules deform to obtain nearly steady shapes, as though they evolved in isolation. During interception, the capsules roll over each other but do not collide. The second capsule undergoes significant deformation due to the pronounced membrane deformability. After interception, the capsules separate and resume a steady shape in an effectively infinite flow.

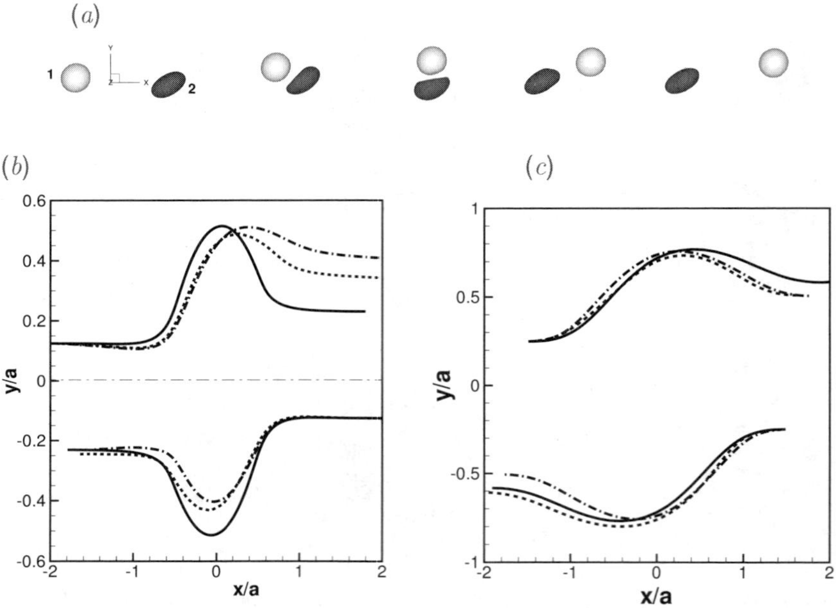

Figure 5.4.1 Interception of two capsules in simple shear flow. (a) Interaction sequence for $\mathrm{Ca}_1 = 0.02$ and $\beta = 15$. (b) Trajectories of the capsule centers showing the effect of the capillary number ratio β for $\mathrm{Ca}_1 = 0.02$, $\lambda_1 = \lambda_2$, and $\beta = 1$ (solid line), 5 (dashed line), and 30 (dash-dotted line). (c) Effect of λ for $\mathrm{Ca}_1 = \mathrm{Ca}_2 = 0.05$, and $\lambda_1 = \lambda_2 = 1$ (solid line), $\lambda_1 = \lambda_2 = 5$ (dash-dotted line), and $\lambda_1 = 1, \lambda_2 = 5$ (dashed line).

Figure 5.4.1(b) illustrates the effect of β on capsule center trajectories. The capsule separation after interception is larger than the initial separation due to shear-induced migration. Interestingly, for $\beta \neq 1$, the final separation is larger than that for $\beta = 1$. Thus, the shear-induced migration is more pronounced for different capsules, and this suggests the possibility of enhanced mixing in a polydisperse suspension. For $\beta = 1$, the lateral displacements of the two capsules are identical. For $\beta \neq 1$, the displacement of the more rigid capsule is significantly more pronounced.

To investigate the effect of the capsule viscosities, multiple indicator functions are introduced in the front-tracking method, one for each capsule or groups of capsules having the same viscosity. Figure 5.4.1(c) shows capsule-center trajectories for two cases with $\lambda_1 = \lambda_2 = 1$ and 5, and one case with $\lambda_1 = 1, \lambda_2 = 5$. The lateral shift after interception decreases as the capsule viscosity increases, reducing the interfacial deformation. Trajectory shifts are different for capsule pairs with different physical properties. However, the effect of the capsule viscosity is not as dramatic as the effect of the capillary number previously described.

5.5 Capsule motion near a wall

Deformable particles exhibit a flow-induced lateral drift described as migration away from a wall. The lateral migration of liquid drops has been studied extensively by experimental and theoretical methods (e.g., Chan & Leal 1981, Hiller & Kowalewski 1987, Smart & Leighton 1991). Migration causing the formation of particle-free zones near walls has been observed in the case of erythrocytes (e.g., Goldsmith 1971). An erythrocyte-free zone spontaneously appears near vessel walls in the microcirculation, reducing the local cell concentration and the apparent viscosity of the suspension. The Fåhraeus and Fåhraeus-Lindqvist effects describe the observed decrease in the hematocrit and apparent viscosity with decreasing vessel diameter.

In our simulations, we study the lateral migration of a capsule with spherical unstressed shape in pressure-driven flow through a channel confined between two parallel walls. The capsule diameter is 0.32 times the distance between the walls. The capsule deformation is determined by the capillary number, $\mathrm{Ca} = \mu_0 U_{cl}/E_s$, where U_{cl} is the centerline velocity of the unidirectional Hagen–Poiseuille flow. The capsule membrane is governed by the constitutive equation (5.2.11). At the initial instant, the capsule is released from a point near the lower wall.

Capsule migration over an extended period of time is evident in the results presented in figure 5.5.1(a) for unit viscosity ratio. A slipper shape initially develops due to variations in the shear rate between the channel walls. Over time, the capsule migrates toward the centerline obtaining a parachute shape due to the imposed pressure drop. The effect of the capillary number on the lateral trajectory and velocity is illustrated in figure 5.5.1(b). The migration velocity initially reaches a peak as the capsule rapidly adapts to the flow. As the capsule approaches the low-shear region near the centerline, the migration velocity declines. As Ca is increased, capsule deformation and migration become more pronounced.

Because the migration velocity is by orders of magnitude less than the streamwise velocity, the capsule deformation is quasi-steady. Approximate expressions for the migration velocity can be obtained when the capsule size is much smaller than the channel width and the capsule is sufficiently far from the walls. An asymptotic analysis for liquid drops shows that the migration velocity scales with $\mathrm{Ca}\,(1 - y/H)(a/H)^3$ (e.g., Helmy & Barthès-Biesel 1982,

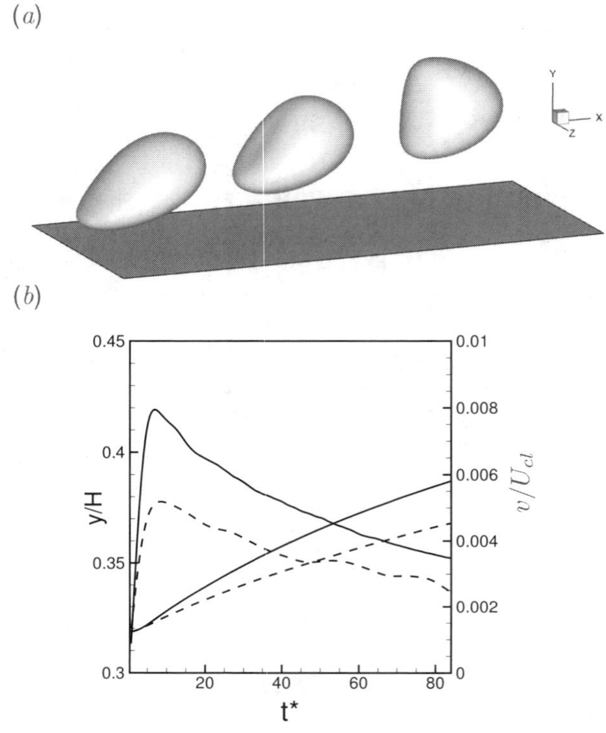

Figure 5.5.1 Lateral migration of a capsule in pressure-driven channel flow. (*a*) Transient capsule shapes are shown at three instances. (*b*) Effect of the capillary number on the lateral trajectory and migration velocity for $\mathrm{Ca} = 0.5$ (solid line) and 0.1 (dashed line).

Chan & Leal 1979). Doddi & Bagchi (2008) found that the asymptotic theory overpredicts the capsule deformation and the linear dependence of the migration velocity on Ca and $(1 - y/H)$ is not valid.

5.6 Suspension flow in a channel

Whereas capsule deformation in a dilute suspension has received a great deal of attention, only a few studies of suspension flow have been carried out. We have simulated the motion of 1096 capsules in pressure-driven flow between two parallel walls separated by distance H. Periodic boundary conditions are imposed in the streamwise (x) direction and spanwise (z) direction at the faces of a square computation domain with $H \approx 25a$, where a is the radius of an undeformed capsule.

(*a*) (*b*)

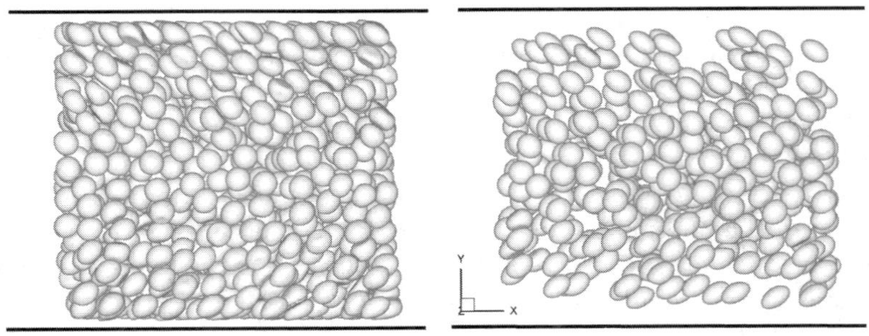

Figure 5.6.1 Instantaneous structure of a suspension in Poiseuille flow computed on a 300^3 Eulerian grid for $\mathrm{Ca} = 0.6$, $H/a \approx 25$, and (*a*) 29% or (*b*) 12% volume fraction.

The capillary number is defined as $\mathrm{Ca} = \mu_0 U_{cl}/E_s$, where U_{cl} is the centerline velocity of the Poiseuille flow in the absence of capsules. The capsule membrane obeys a neo-Hookean constitutive equation. A 300^3 Eulerian grid is employed, and each capsule membrane is divided into 1280 triangular elements. The simulations were performed on four AMD Opteron Quad-Core processors. The run time is approximately 3.5 h for each dimensionless time unit, tU_{cl}/a.

Figure 5.6.1(*a*) shows a snapshot of a suspension for 29% particle volume fraction and unit viscosity ratio after the motion has reached quasi-steady state. It is evident that the capsules near the wall deform more than those near the centerline due to the high local shear rate. Pairwise interceptions are responsible for transient flat contact surfaces developing during collision. Interceptions prohibit the continuous capsule migration toward the centerline. Figure 5.6.1(*b*) shows a snapshot of a moderately dense suspension with 12% volume fraction containing 454 capsules per periodic box. As the capsules migrate toward the centerline, a particle-free layer develops near the channel walls. Doddi & Bagchi (2009) present detailed information illustrating the dependence of the particle-free zone on the cell deformability, size ratio, and volume fraction.

Figure 5.6.2 shows time-averaged velocity profiles demonstrating the effect of capillary number and volume fraction. For the same pressure drop, the flow rate in the presence of the capsules is less than that of the unidirectional parabolic flow. A plug-flow velocity profile is established in the presence of capsules. The mean velocity decreases with increasing volume fraction and decreasing capillary number.

It is interesting to compare the numerical results with the the predictions of the core–annular two-phase model where the flow is divided into a concentrated central zone and a cell-depleted layer near the wall. Doddi & Bagchi (2009) showed that the model underpredicts the mean velocity and thus overpredicts the apparent viscosity of the suspension.The discrepancy is attributed to the jump in the local suspension viscosity across the interface of the two layers comprising the core–annular flow. In contrast, the numerical simulations show smooth variations of the local viscosity across the annular interface.

A three-layer model was proposed by Doddi & Bagchi (2009) where the channel is divided into a central region, a transition zone, and a cell-depleted region. The model parameters are extracted *a posteriori* from the simulations. The three-layer model accurately predicts the velocity profiles arising from the simulations. These results underscore the ability of particle-level simulations to provide new information on the hydrodynamics of dense suspensions and suggest improved low-dimensional models of blood flow in the microcirculation.

5.7 Rolling on an adhesive substrate

In the last case study, we consider the rolling of a deformable capsule on an adhesive substrate under the action of simple shear flow. This configuration serves primarily as a model of the rolling of a convected leukocyte on a blood vessel wall during inflammatory response. Adhesive rolling is important in the metastasis of cancer cells and in the scavenging of biomass, bacteria, and surface-bound particles. Other applications are found in targeted drug delivery with reference to the binding of a drug carrier to the vascular wall or diseased cell in an hemodynamic environment.

In the case of circulating leukocytes, adhesion is due to the binding of selectin molecules to respective ligands. Selectin molecules are mostly concentrated at the tips of microvilli protruding from the cell surface. When a leukocyte comes into close proximity with the vascular wall, the receptor–ligand interaction leads to the formation of adhesion bonds between the cell surface and the wall. The velocity of a captured cell is significantly reduced due to these bonds. Upon tethering, the cell spreads over the wall increasing the contact area and thus facilitating the formation of further bonds. Bonds stretch like springs under the action of hydrodynamic forces imparted from the cell to the bonds. The rate of randomly breaking bonds increases as the

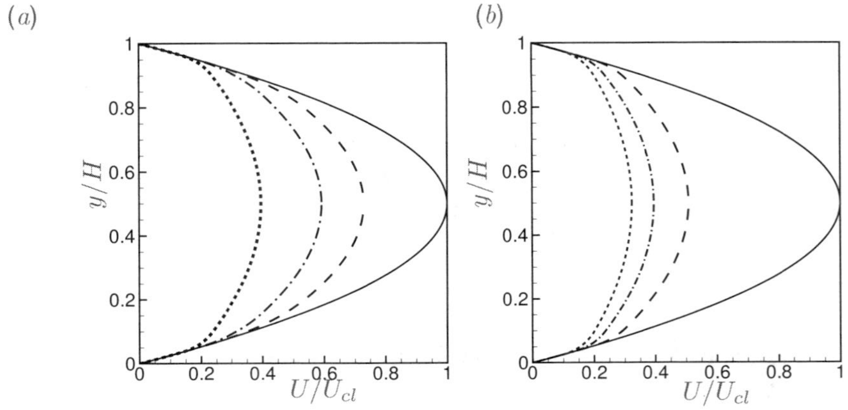

Figure 5.6.2 Mean velocity profile across a suspension of 122 capsules in a channel with width $H = 12.6a$ (Doddi & Bagchi 2009). (a) Effect of volume fraction for $\mathrm{Ca} = 0.05$: 0% (solid line), 12% (dashed line), 18% (dash-dotted line), and 26% (dashed line). (b) Effect of the capillary number at 26% volume fraction: $\mathrm{Ca} = 0.6$ (dashed line), 0.05 (dashed-dotted line), and 0.005 (dense dashed line). The Poiseuille flow profile is shown as a solid line.

bonds are increasingly stretched, resulting in tether breakup. As a result, a cell does not move in a continuous fashion but exhibits an intermittent stop-and-go translation (e.g., Springer 1995).

Hammer proposed an adhesive dynamics simulation method (ADS) to describe the motion of a leukocyte modeled as a rigid sphere (Hammer & Apte 1992, Chang & Hammer 2000, King & Hammer 2001; see also Chapter 8). The idea is to couple hydrodynamics to Monte-Carlo simulations for bond formation and breakup. Experiments have shown that cells may deform significantly during adhesion (e.g., Firrell & Lipowsky 1989). An improved computational model of leukocyte adhesion and rolling was developed by Khismatulin & Truskey (2004, 2005). Jadhav *et al.* (2005) combined the immersed-boundary method, the front-tracking method, and Monte Carlo simulations to predict the stop-and-go motion of deformable leukocytes. Their model is adopted in the present front-tracking simulations.

A leukocyte is modeled as a compound capsule with spherical undeformed shape. Bonds are allowed to develop at discrete but randomly distributed locations over the cell surface. Details of the implementation are given by Jadhav *et al.* (2005), Pappu & Bagchi (2008), and Pappu *et al.* (2008). Bonds are assumed to behave like Hookean springs. The force on a stretched bond is

$$f_b = k_b(l - l_0)\,, \tag{5.7.1}$$

where k_b is the spring constant and l, l_0 are the stretched and unstretched bond lengths (Dembo 1994). In the adhesive dynamics model, the probability of formation of a new bond, P_+, and the probability of breakage of an existing bond, P_-, during a time interval, Δt are

$$P_\pm = 1 - \exp(-k_\pm \Delta t)\,. \tag{5.7.2}$$

The rates of formation and breakage, k_+ and k_-, are related to the bond stretching by

$$k_+ = k_+^0 \exp(-\frac{k_{ts}\Delta l^2}{2K_B T}), \qquad k_- = k_-^0 \exp(\frac{(k_b - k_{ts})\Delta l^2}{2K_B T}), \tag{5.7.3}$$

where $\Delta l = l - l_0$, k_{ts} is the transition-state spring constant, K_B is the Boltzmann constant, T is the absolute temperature, and the superscript 0 indicates the reaction rates of the unstretched bonds.

Experimental values for the bond parameters are available. Typical values are $k_b = 1$ pN/nm, $l_0 = 0.1$ μm (Marshal *et al.* 2006), $k_{ts} = 0.99$ pN/nm, $k_-^0 = 1$ s^{-1} (Smith *et al.* 1999), and $k_+^0 = 1$ s^{-1} (Mehta *et al.* 1998). A bond is allowed to break up if $P_- > N_1$, and a new bond is allowed to form if $P_+ > N_2$, where N_1 and N_2 are random numbers in the range 0–1.

To transmit the adhesion force to the bulk hydrodynamics, a body force is added to the right-hand side of the Navier–Stokes equation,

$$\mathbf{F}(\mathbf{x}, t) = \iint \mathbf{f}_a(\mathbf{x}', t)\, \delta_3(\mathbf{x} - \mathbf{x}')\, \mathrm{d}S(\mathbf{x}')\,, \tag{5.7.4}$$

where $\mathbf{f}_a$ is the vector sum of all force bonds at a microvillus tip, δ_3 is the three-dimensional delta function, and $\mathbf{x}'$ is the location of a microvillus on the cell surface.

Figure 5.7.1(a) shows snapshots in the adhesive rolling of a cell on a flat substrate. The membrane mechanics is governed by a neo-Hookean law. Transient tethering of the cell is evident in the top and bottom panels where the cell is seen to spread over the substrate creating a flat contact zone. The midpanel shows a rolling cell with nearly zero contact area between two successive arrests. The cell trajectory and instantaneous velocity are shown in figure 5.7.1(b). The stop-and-go motion of the cell is clearly observed in these simulations. The role of cell deformation is discussed by Jadhav *et al.* (2005), Pappu & Bagchi (2008), and Pappu *et al.* (2008).

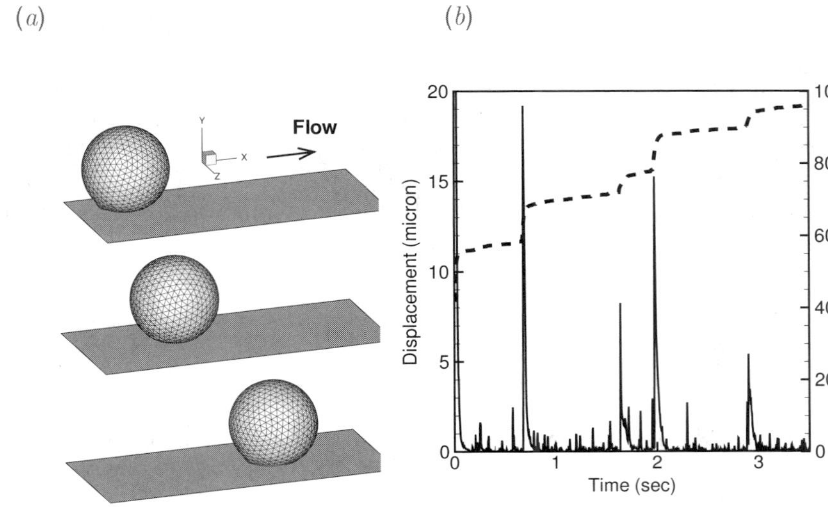

Figure 5.7.1 (*a*) Rolling of a capsule on an adhesive substrate under the influence of shear flow simulated by combining the front-tracking method with an adhesive dynamics scheme for stochastic bond formation and breakup. The top and bottom panels show a tethered cell with a flat contact area. The midpanel shows a rolling cell between two successive arrests. (*b*) History of cell displacement (dashed line) and velocity (solid line), showing discontinuous behavior.

5.8 Summary

We have discussed the implementation of a front-tracking immersed-boundary method for solving a variety of problems in microhemodynamics. Results were presented for capsules, vesicles, and erythrocytes in dilute and dense suspensions in infinite, wall-bounded, and pressure-driven flow.

The front-tracking immersed-boundary methodology is versatile in that the basic algorithm can be combined with finite-difference, finite-volume, spectral, pseudospectral, and other methods for computing the fluid flow. Monodisperse and polydisperse suspensions with a large number of cells can be accommodated, and coarse-grain models accounting for molecular interactions between neighboring cells and between cells and the vessel wall can be implemented by straightforward modifications.

The computational framework presented in this chapter can be extended to investigate a number of problems in microhydrodynamics involving deformable cells. Examples include studies of the effect of tank-treading to tumbling transition on the rheology of dilute suspension of red blood cells, the hydrodynamic interception between a pair of red blood cells, shear-induced self-diffusion in a dense suspension, the effect of red-blood cells on the adhesive rolling motion of leukocytes, and platelet margination. The general framework can be adapted to study the dispersion of nano- and micron-size particles regarded as models of drug carriers in the presence of blood cells with application in targeted drug delivery. In the most ambitious extension, the method can be applied to study the flow of blood particulates in microvascular networks.

References

ABKARIAN, M., FAIVRE, M. & VIALLAT, A. (2007) Swinging of red blood cells under shear flow. *Phys. Rev. Lett.* **98**, 188302.

BAGCHI, P. (2007) Mesoscale simulation of blood flow in small vessels. *Biophys. J.* **92**, 1858–1877.

BAGCHI, P. & KALLURI, R. M. (2009) Dynamics of nonspherical capsules in shear flow. *Phys. Rev. E* **80**, 016307.

BARTHÈS-BIESEL, D. (1980) Motion of a spherical microcapsule freely suspended in a linear shear flow. *J. Fluid Mech.* **100**, 831–853.

BARTHÈS-BIESEL, D. & RALLISON, J. M. (1981) The time-dependent deformation of a capsule freely suspended in a linear shear flow. *J. Fluid Mech.* **113**, 251–267.

BARTHÉS-BIESEL, D., & SGAIER, H. (1985) Role of membrane viscosity in the orientation and deformation of a spherical capsule suspended in shear flow. *J. Fluid Mech.* **160**, 119–135.

BARTHÉS-BIESEL, D., DIAZ, A. & DHENIN, E. (2002) Effect of constitutive laws for two-dimensional membranes on flow-induced capsule deformation. *J. Fluid Mech.* **460**, 211–222.

BATCHELOR, G. K. & GREEN, J. T. (1972) The hydrodynamic interaction of two small freely-moving spheres in a linear flow field. *J. Fluid Mech.* **56**, 375–400.

BEAUCOURT, J., RIOUAL, F., SEON, T. & MISBAH, C. (2004) Steady to unsteady dynamics of a vesicle in a flow. *Phys. Rev. E* **69**, 011906.

BIBEN, T. & MISBAH, C. (2003) Tumbling of vesicles under shear flow within an advected-field approach. *Phys. Rev. E* **67**, 031908.

BIBEN, T., KASSNER, K. & MISBAH, C. (2005) Phase-field approach to three-dimensional vesicle dynamics. *Phys. Rev. E* **72**, 041921.

BOAL, D. H. (1994) Computer simulation of a model network for the erythrocyte cytoskeleton. *Biophys. J.* **67**, 521–529.

BREYIANNIS, G. & POZRIKIDIS, C. (2000) Simple shear flow of suspensions of elastic capsules. *Theor. Comp. Fluid Dyn.* **13**, 327–347.

CHAN, P. C.-H. & LEAL, L. G. (1979) The motion of a deformable drop in a second-order fluid. *J. Fluid Mech.* **92**, 131–170.

CHAN, P. C.-H. & LEAL, L. G. (1981) An experimental study of drop migration in shear flow between concentric cylinders. *Int. J. Multiph. Flow* **7**, 83–94.

CHANG, K. C. & HAMMER, D. A. (2000) Adhesive Dynamics Simulations of Sialyl-Lewisx/E-selectin-Mediated Rolling in a Cell-Free System. *Biophys. J.* **79**, 1891-1902.

CHARLES, R. & POZRIKIDIS, C. (1998) Effect of the dispersed phase viscosity on the simple shear flow of suspensions of liquid drops. *J. Fluid Mech.* **365**, 205–233.

CHARRIER, J. M., SHRIVASTAVA, S. & WU, R. (1989) Free and constrained inflation of elastic membranes in relation to thermoforming-non-axisymmetric problems. *J. Strain Anal.* **24**, 55–74.

COULLIETTE, C. & POZRIKIDIS, C. (1998) Motion of an array of drops through a cylindrical tube. *J. Fluid Mech.* **358**, 1–28.

COUPIER, G., KAOUI, B., PODGORSKI, T. & MISBAH, C. (2008) Noninertial lateral migration of vesicles in bounded Poiseuille flow. *Phys. Fluids* **20**, 111702.

DACUNHA, F. R. & HINCH, E. J. (1996) Shear-induced dispersion in a dilute suspension of rough spheres. *J. Fluid Mech.* **309**, 211–223.

DEMBO, M. (1994) On peeling an adherent cell from a surface. *Lectures on Mathematics in the Life Sciences, Some Mathematical Problems in Biology* **26**, 51–77, American Mathematical Society, Providence, RI.

DIAZ, A., PELEKASIS, N. & BARTHÉS-BIESEL, D. (2000) Transient response of a capsule subjected to varying flow conditions: Effect of internal fluid viscosity and membrane elasticity. *Phys. Fluids* **12**, 948–958.

DIAZ, A., BARTHÉS-BIESEL, D. & PELEKASIS, N. (2001) Effect of membrane viscosity on the dynamic response of an axisymmetric capsule. *Phys. Fluids* **13**, 3835–3839.

DING, E.-J. & AIDUN, C. K. (2006) Cluster size distribution and scaling for spherical particles and red blood cells in pressure-driven flows at small Reynolds number. *Phys. Rev. Lett.* **96**, 204502.

DODDI, S. K. & BAGCHI, P. (2008) Lateral migration of a capsule in a plane Poiseuille flow in a channel. *Int. J. Multiph. Flow* **34**, 966–986.

DODDI, S. K. & BAGCHI, P. (2009) Three-dimensional computational modeling of multiple deformable cells flowing in microvessels. *Phys. Rev. E.* **79**, 046318.

DODSON, W. R. & DIMITRAKOPOULOS, P. (2008) Spindles, cusps, and bifurcation for capsules in Stokes flow. *Phys. Rev. Lett.* **101**, 208102.

DUPIN, M. M., HALLIDAY, I., CARE, C. M., ALBOUL, L. & MUNN, L. L. (2007) Modeling the flow of dense suspensions of deformable particles in three dimensions. *Phys. Rev. E.*, **75**, 066707.

EGGLETON, C. D. & POPEL, A. S. (1998) Large deformation of red blood cell ghosts in simple shear flow. *Phys. Fluids* **10**, 1834–1845.

EVANS, E. A. & FUNG, Y. C. (1972) Improved measurements of the erythrocyte geometry. *Macrovasc. Res.* **4**, 335–347.

FIRRELL, J. C. & LIPOWSKY, H. H. (1989) Leukocyte margination and deformation in mesenteric venules of rat. *Am. J. Physiol.* **256**, H1667–1674.

FISCHER, T. M., STOHR-LIESEN, M. & SCHMID-SCHÖNBEIN, H. (1978) The red cell as a fluid droplet: Tank tread-like motion of the human erythrocyte membrane in shear flow. *Science* **202**, 894–896.

FREUND, J. B. (2007) Leukocyte margination in a model microvessel. *Phys. Fluids* **19**, 023301.

FUNG, Y. C. (1984) *Biodynamics: Circulation.* Springer–Verlag, New York.

GARIMELLA, R. V. & SWARTZ, B. K. (2003) Curvature estimation for unstructured triangulation of surfaces. *Tech. Rep. LA-UR-03-8240*, Los Alamos National Laboratory.

GOLDSMITH, H. L., (1971) Red cell motions and wall interactions in tube flow. *Fed. Proc.* **30**, 1578–1590.

GOLDSMITH, H. L. & MARLOW, J. (1972) Flow behavior of erythrocytes. I. Rotation and deformation in dilute suspensions. *Proc. R. Soc. Lond. B* **182**, 351–384.

GUIDO, S. & SIMEONE, M. (1998) Binary collisions of drops in simple shear flow by computer-assisted video optical microscopy. *J. Fluid Mech.* **357**, 1–20.

HAMMER, D. A. & APTE, S. M. (1992) Simulation of cell rolling and adhesion on surfaces in shear flow: General results and analysis of selectin-mediated neutrophil adhesion. *Biophys. J.* **63**, 35-57.

HELFRICH, W. (1973) Elastic properties of lipid bilayers: Theory and possible experiments. *Naturforsch* **28**, 693–703.

HELMKE, B. P, BREMNER, S. N., ZWEIFACH, B. W., SKALAK, R. & SCHMID-SCHÖNBEIN, G.W. (1997) Mechanisms for increased blood flow resistance due to leukocytes. *Am. J. Physiol.* **273**, 2884–2890.

HELMY, A. & BARTHÈS-BIESEL, D. (1982) Migration of a spherical capsule freely suspended in an unbounded parabolic flow. *J. Mec. Theor. Appl.* **1**, 859–880.

HILLER, W. & KOWALEWSKI, T. A. (1987) An experimental study of the lateral migration of a droplet in a creeping flow. *Exper. Fluids.* **5**, 43–48.

HOCHMUTH, R. M. (1982) Solid and liquid behavior of red cell membrane. *Ann. Rev. Biophys. Bioeng.* **11**, 43–55.

JADHAV, S., EGGLETON, C. D. & KONSTANTOPOULOS, K. (2005) A 3-D computational model predicts that cell deformation affects selectin-mediated leukocyte rolling. *Biophys. J.* **88**, 96-104.

KANTSLER, V. & STEINBERG, V. (2005) Orientation and dynamics of a vesicle in tank-treading motion in shear flow. *Phys. Rev Lett.* **95**, 258101.

KANTSLER, V. & STEINBERG, V., (2006) Transition to tumbling and two regimes of tumbling motion of a vesicle in shear flow. *Phys. Rev. Lett.* **96**, 036001.

KELLER, S. R. & SKALAK, R. (1982) Motion of a tank-treading ellipsoidal particle in a shear flow. *J. Fluid Mech.* **120**, 27–47.

KESSLER, S., FINKEN, R. & SEIFERT, U. (2008) Swinging and tumbling of elastic capsules in shear flow. *J. Fluid Mech.* **605**, 207–226.

KHISMATULLIN, D. B. & TRUSKEY, G. A. (2004) A three-dimensional numerical study of the effect of channel height on leukocyte deformation and adhesion in parallel-plate flow chambers. *Microvasc. Res.* **68**, 188–202.

KHISMATULLIN, D. B. & TRUSKEY, G. A. (2005) Three-dimensional numerical simulation of receptor-mediated leukocyte adhesion to surfaces: Effects of cell deformability and viscoelasticity. *Phys. Fluids* **17**, 031505.

KING, M. R. & HAMMER, D. A. (2001) Multiparticle adhesive dynamics. Interactions between stably rolling cells. *Biophys. J.* **81**, 799-813.

KRAUS, M., WINTZ, W., SEIFERT, U. & LIPOWSKY, R. (1996) Fluid vesicles in shear flow. *Phys. Rev. Lett.* **77**, 3685.

KWAK, S. & POZRIKIDIS, C. (2001) Effect of membrane bending stiffness on the axisymmetric deformation of capsules in uniaxial extensional flow. *Phys. Fluids* **13**, 1234–1244.

LAC, E., BARTHÉS-BIESEL, D., PELEKASIS, N. A. & TSAMOPOULOS, J. (2004) Spherical capsules in three-dimensional unbounded Stokes flows: effect of the membrane constitutive law and onset of buckling. *J. Fluid Mech.* **516** 303–334.

LAC, E. & BARTHÉS-BIESEL, D. (2005) Deformation of a capsule in simple shear flow: Effect of membrane prestress. *Phys. Fluids* **17**, 072105.

LAC, E., MOREL, A. & BARTHÈS-BIESEL, D. (2007) Hydrodynamic interaction between two identical capsules in shear flow. *J. Fluid Mech.* **573**, 149–169.

LEIGHTON, D. T. & ACRIVOS, A. (1987) Measurement of shear-induced self-diffusion in a concentrated suspension of spheres. *J. Fluid Mech.* **177**, 109–131.

LEYRAT-MAURIN, A. & BARTHÈS-BIESEL, D. (1994) Motion of a deformable capsule through a hyperbolic constriction. *J. Fluid Mech.* **279**, 135–163.

LI, X. Z., BARTHÈS-BIESEL, D. & HELMY, A. (1988) Large deformations and burst of a capsule freely suspended in an elongational flow. *J. Fluid Mech.* **187**, 179–196.

LI, X. & POZRIKIDIS, C. (2000) Wall-bounded and channel flow of suspensions of liquid drops. *Int. J. Multiph. Flow* **26**, 1247–1279.

LI, X. & SARKAR, K. (2008) Front-tracking simulation of deformation and buckling instability of a liquid capsule enclosed by an elastic membrane. *J. Comp. Phys.* **227**, 4998–5018.

LOEWENBERG, M. & HINCH, E. J. (1997) Collision of two deformable drops in shear flow. *J. Fluid Mech.* **338**, 299–315.

MARSHALL, B. T., SARANGAPANI, K. K., WU, J., LAWRENCE, M. B., MCEVER, R. P. & ZHU, C. (2006) Measuring molecular elasticity by atomic force microscope cantilever fluctuations. *Biophys. J.* **90**, 681–692.

MEHTA, P., CUMMINGS, R. D. & MCEVER, R. P. (1998) Affinity and kinetic analysis of P-selectin binding to P-selectin glycoprotein ligand-1. *J. Biol. Chem.* **273**, 32506–32513.

MISBAH, C. (2006) Vacillating, breathing and tumbling of vesicles under shear flow. *Phys. Rev. Lett.* **96**, 028104.

MUNN, L. L. & DUPIN, M. M. (2008) Blood cell interactions and segregation in flow. *Ann. Biomed. Eng.* **36**, 534–544.

PAPPU, V. & BAGCHI, P. (2007) Hydrodynamic interaction between erythrocytes and leukocytes affects rheology of blood in microvessels. *Biorheology* **44**, 191–215.

PAPPU, V. & BAGCHI, P. (2008) Three-dimensional computational modeling and simulation of leukocyte rolling adhesion and deformation. *Comp. Bio. Med.* **38**, 738-753.

PAPPU, V., DODDI, S. K. & BAGCHI, P. (2008) A computational study of leukocyte adhesion and its effect on flow pattern in microvessels. *J. Theor. Biol.* **254**, 483-498.

PIVKIN, I. V., RICHARDSON, P. D. & KARNIADAKIS, G. E. (2006) Blood flow velocity effects and role of activation delay time on growth and form of platelet thrombi. *Proc. Natl. Acad. Sci.* **103**, 17164–17169.

PIVKIN, I. V. & KARNIADAKIS, G. E. (2008) Accurate coarse-grained modeling of red blood cells. *Phys. Rev. Lett.* **101**, 118105.

PESKIN, C. S. (1977) Numerical analysis of blood flow in the heart. *J. Comp. Phys.* **25**, 220–233.

PETITJEAN, S. (2002) A survey of methods for recovering quadrics in triangle meshes. *ACM Computing Surveys* **34**, 211–262.

POZRIKIDIS, C. (1995) Finite deformation of liquid capsules enclosed by elastic membranes in simple shear flow. *J. Fluid Mech.* **297**, 123–152.

POZRIKIDIS, C. (2001) Effect of bending stiffness on the deformation of liquid capsules in simple shear flow. *J. Fluid Mech.* **440**, 269–291.

POZRIKIDIS, C. (2003) Numerical simulation of the flow-induced deformation of red blood cells. *Ann. Biomed. Eng.* **31**, 1194–1205.

POZRIKIDIS, C. (2003) *Modeling and Simulation of Capsules and Biological Cells.* Chapman & Hall/CRC, Boca Raton.

POZRIKIDIS, C. (2005) Numerical simulation of cell motion in tube flow. *Ann. Biomed. Eng.* **33**, 165–178.

POZRIKIDIS, C. (2005) Axisymmetric motion of a file of red blood cells through capillaries. *Phys. Fluids* **17**, 031503.

QUEGUINER, C. & BARTHÈS-BIESEL, D. (1997) Axisymmetric motion of capsules through cylindrical channels. *J. Fluid Mech.* **348**, 349–376.

RAMANUJAN, S. & POZRIKIDIS, C. (1998) Deformation of liquid capsules enclosed by elastic membranes in simple shear flow: large deformations and the effect of fluid viscosities. *J. Fluid Mech.* **361**, 117–143.

REUTER, M., BIASOTTI, S., GIORGI, D., PATANE, G. & SPAGNUOLO, M., (2009) Discrete Laplace-Beltrami operators for shape analysis and segmentation. *Comput. Graphics* **33**, 381–390.

RIOUAL, F., BIBEN, T. & MISBAH, C. (2004) Analytical analysis of a vesicle tumbling under a shear flow. *Phys. Rev. E* **69**, 061914.

SECOMB, T. W. (2003) Mechanics of red blood cells and blood flow in narrow tubes. In: *Modeling and Simulation of liquid capsules and biological cells*, Pozrikidis, C. (Ed.), Chapman & Hall /CRC, Boca Raton.

SECOMB, T. W., STYP-REKOWSKA, B. & PRIES, A. R. (2007) Two-dimensional simulation of red blood cell deformation and lateral migration in microvessels. *Ann. Biomed. Eng.* **35**, 755–765.

SHRIVASTAVA, S., & TANG, J. (1993) Large deformation finite element analysis of non-linear viscoelastic membranes with reference to thermoforming. *J. Strain Anal.* **28**, 31–43.

SKALAK R., TOZEREN A., ZARDA R. P. & CHIEN S. (1973) Strain energy function of red blood cell membrane. *Biophys. J.* **13**, 245–264.

SKOTHEIM, J. M. & SECOMB, T. W. (2007) Red blood cells and other non-spherical capsules in shear flow: Oscillatory dynamics and the tank-treading-to-tumbling transition. *Phys. Rev. Lett.* **98**, 078301.

SMART, J. R. & LEIGHTON, D. T. (1991) Measurement of the drift of a droplet due to the presence of a plane. *Phys. Fluids* **3**, 21–31.

SMITH, M. J., BERG, E. L. & LAWRENCE, M. B. (1999) A direct comparison of selectin-mediated transient adhesive events using high temporal resolution. *Biophys. J.* **77**, 3371-3383.

SPRINGER, T. A. (1995) Traffic signals on endothelium for lymphocyte recirculation and leukocyte emigration. *Annu. Rev. Physiol.* **57**, 827–872.

SUI, Y., LOW, H. T., CHEW, Y. T. & ROY, P. (2008*a*) Tank-treading, swinging, and tumbling of liquid-filled elastic capsules in shear flow. *Phys. Rev. E* **77**, 016310.

SUI, Y., CHEW, Y. T., ROY, P., CHENG, Y. P. & LOW, H.T. (2008b) Dynamic motion of red blood cells in simple shear flow *Phys. Fluids* **20**, 112106.

SUKUMARAN, S. & SEIFERT, U. (2001) Influence of shear flow on vesicles near a wall: A numerical study. *Phys. Rev. E* **64**, 011916.

TRYGGVASON, G., BUNNER, B., ESMAEELI, A., JURIC, I., AL-RAWAHI, N., TAUBER, W., HAN, J., NAS, S. & JAN., Y.-J. (2001) A front-tracking method for the computations of multiphase flow. *J. Comp. Phys.* **169**, 708–759.

UNVERDI, S. O. & TRYGGVASON, G. (1992) A front-tracking method for viscous, incompressible multi-fluid flows. *J. Comp. Phys.* **100**, 25–37.

WALTER, A., REHAGE, H. & LEONHARD, H. (2001) Shear induced deformation of micro-capsules: Shape oscillations and membrane folding. *Colloids Surf. A* **123**, 183–185.

XU, G. (2004) Convergent discrete Laplace–Beltrami operators over triangular surfaces. *Geom. Model. Proc.*, Proceedings of the IEEE Computer Society, Washington, DC.

ZHANG, J., JOHNSON, P. C. & POPEL, A.S. (2008) Red blood cell aggregation and dissociation in shear flows simulated by lattice Boltzmann method. *J. Biomech.* **41**, 47–55.

ZHONG-CAN, O.-Y. & HELFRICH, W. (1989) Bending energy of vesicle membranes: General expressions for the first, second, and third variation of the shape energy and applications to spheres and cylinders. *Phys. Rev. A* **39**, 5280–5288.

6

Dissipative particle dynamics modeling of red blood cells

D. A. Fedosov, B. Caswell, G. E. Karniadakis

Division of Applied Mathematics
Division of Engineering
Brown University
Providence

Capsules and red blood cells suspended in flow exhibit a rich dynamics due to the deformability of the enclosing membranes. To accurately capture statics and dynamics, mechanical models of the membrane must be available incorporating shear elasticity, bending rigidity, and membrane viscosity. In the approach described in this chapter, the membrane of a red blood cell is modeled as a network of interconnected nonlinear springs emulating the cytoskeleton spectrin network. Dissipative forces in the network mimic the effect of the lipid bilayer. The macroscopic elastic properties of the network are analytically related to the spring parameters, circumventing *ad-hoc* adjustment. Chosen parameter values yield model membranes that reproduce optical tweezer stretching experiments. When probed with an attached oscillating microbead, predicted viscoelastic properties are in good agreement with experiments using magnetic optical twisting cytometry. In shear flow, red blood cells respond by tumbling at low shear rates and tank-treading at high shear rates. In transitioning between these regimes, the membrane exhibits substantial deformation controlled largely by flexural stiffness. Raising the membrane or internal fluid viscosity shifts the transition threshold to higher shear rates and reduces the tank-treading frequency. Simulations reveal that a purely elastic membrane devoid of viscous properties cannot adequately capture the cell dynamics. Results are presented to demonstrate the dependence of the transition thresholds from biconcave to parachute shapes in capillary flow on the cell properties and mean flow velocity.

6.1 Introduction

Red blood cells are soft biconcave capsules with average diameter 7.8 μm containing a viscous liquid and enclosed by a viscoelastic membrane. The cell membrane consists of a nearly incompressible lipid bilayer attached to a spectrin protein network, comprising the cytoskeleton, held together by short actin filaments. This membrane structure ensures the integrity of the cell in narrow capillaries whose cross-section is smaller than the size of the undeformed biconcave disk (e.g., Fung 1993). Consistent with the spectrin cytoskeleton structure, the membrane can be modeled as a network of viscoelastic springs exhibiting elastic and viscous response. Bending stiffness can be introduced in terms of a network bending energy and constraints on the cell surface area and volume can be imposed to ensure the area incompressibility of the lipid bilayer and the volume incompressibility of the interior fluid.

Theoretical and numerical analyses have sought to describe cell behavior and deformation in a variety of flows. Examples include models of ellipsoidal cells enclosed by viscoelastic membranes (e.g., Abkarian *et al.* 2007, Skotheim & Secomb 2007), numerical models based on shell theory (e.g., Fung 1993, Eggleton & Popel 1998, Pozrikidis 2005), and discrete descriptions at the spectrin protein level (e.g., Discher *et al.* 1998, Li *et al.* 2005) or at a mesoscopic level (e.g., Noguchi & Gompper 2005, Dupin *et al.* 2007, Pivkin & Karniadakis 2008). The membranes of healthy red blood cells exhibit nonlinear elastic response in steady stretching and viscous response in dynamic testing. Most existing membrane models incorporate only elastic response. Fluid-and solid-like models demand high computational cost due to the strong coupling of solid mechanics and fluid flow.

Semicontinuum models of deformable cells employ boundary-element, immersed-boundary, and front-tracking methods combined with a discrete membrane representation (e.g., Eggleton & Popel 1998, Pozrikidis 2005). The membrane is described by a set of point particles whose motion is coupled to the fluid flow on an Eulerian grid. Most models assume that all fluid viscosities are equal and ignore thermal fluctuations. Modeling a cell at the spectrin–protein level is constrained by high computational cost.

Mesoscopic models of viscoelastic capsules and red blood cells have been developed to describe three-dimensional motion (e.g., Noguchi & Gompper 2005, Dupin *et al.* 2007, Pivkin & Karniadakis 2008). Noguchi & Gompper (2005) simulated the deformation of vesicles enclosed by viscoelastic membranes using the method of multi-particle collision dynamics (e.g., Malevanets & Kapral 1999). Dupin *et al.* (2007) combined a lattice-Boltzmann method with a discrete membrane representation neglecting the membrane viscosity and the presence of thermal fluctuations (e.g., Succi 2001). Their implementation smears the sharp interface between the external and internal fluids into a diffuse transition zone.

The elasticity of the red blood cell membrane is attributed to a spectrin cytoskeleton network of approximately 27×10^3 nodes. The population number was reduced (coarse-grained) by Pivkin & Karniadakis (2008) using the dissipative particle dynamics (DPD) approach into 500 DPD particles connected by springs (e.g., Hoogerbrugge & Koelman 1992). The model of Pivkin & Karniadakis (2008) is the starting point for the work discussed in this chapter.

First, a theoretical analysis will be presented for a membrane network exhibiting specified macroscopic properties without parameter adjustment. The predicted cell mechanical properties will be compared with optical tweezers stretching experiments by Suresh *et al.* (2005). The predicted rheological properties will be compared with magnetic optical twisting cytometry experiments by Puig-de-Morales-Marinkovic *et al.* (2007). Red blood dynamics in shear flow exhibiting tumbling and tank-treading will be studied in detail with a view to delineating the effect of the membrane shear modulus, bending rigidity, external, internal, and membrane viscosities. Simulations of cell motion in Poiseuille flow will confirm that the biconcave-to-parachute transition depends on the flow strength and membrane properties. Comparison with available experiments will demonstrate that the computational model is able to accurately describe realistic red blood cell motion.

Comparison of the numerical simulations with theoretical predictions by Abkarian *et al.* (2007), Skotheim & Secomb (2007), and others will reveal discrepancies. It appears that the predictions of current theoretical models are only qualitatively accurate due to strong simplifications.

6.2 Mathematical framework

In the theoretical model, the membrane of a red blood cell is represented by a viscoelastic network. The motion of the internal and external fluids is described by the method of dissipative particle dynamics (DPD) (e.g., Hoogerbrugge & Koelman 1992). The membrane model is sufficiently general to be used with other simulation techniques, such as Brownian dynamics, lattice Boltzmann, multiparticle collision dynamics, and the immerse-boundary method.

6.2.1 Dissipative particle dynamics

Dissipative particle dynamics (DPD) is a mesoscopic simulation technique for computing the flow of complex fluids. A DPD system consists of N point particles, each representing a lump of atoms or molecules, described by their position, $\mathbf{r}_i$, velocity, $\mathbf{v}_i$, and mass m_i, where $i = 1, \ldots, N$. Particle interactions are mediated by conservative (C), dissipative (D), and random (R) interparticle forces given by

$$\mathbf{F}_{ij}^C = F_{ij}^C(r_{ij})\,\hat{\mathbf{r}}_{ij}, \qquad \mathbf{F}_{ij}^D = -\gamma\,\omega^D(r_{ij})\,(\mathbf{v}_{ij}\cdot\hat{\mathbf{r}}_{ij})\,\hat{\mathbf{r}}_{ij},$$

$$\mathbf{F}_{ij}^R = \sigma\,\omega^R(r_{ij})\,\frac{\xi_{ij}}{\sqrt{\mathrm{d}t}}\,\hat{\mathbf{r}}_{ij}, \tag{6.2.1}$$

where $\mathbf{r}_{ij}$ is the distance between the ith and jth particle, $\hat{\mathbf{r}}_{ij} = \mathbf{r}_{ij}/r_{ij}$ is a unit vector, $r_{ij} = |\mathbf{r}_{ij}|$, $\mathbf{v}_{ij} = \mathbf{v}_i - \mathbf{v}_j$, and $\mathrm{d}t$ is a small time interval. The coefficients γ and σ are the amplitudes of the dissipative and random forces, and the factors ω^D and ω^R are corresponding weights. The random force definition employs normally distributed random variables ξ_{ij} with zero mean, unit variance, and pairwise symmetry, $\xi_{ij} = \xi_{ji}$. The forces vanish beyond a cutoff radius defining the DPD length scale, r_c.

A typical conservative force is

$$F_{ij}^C(r_{ij}) = \begin{cases} a_{ij}(1 - r_{ij}/r_c) & \text{for } r_{ij} \le r_c, \\ 0 & \text{for } r_{ij} > r_c, \end{cases} \tag{6.2.2}$$

where a_{ij} is the conservative force coefficient between the ith and jth particles. The random force weight function $\omega^R(r_{ij})$ is chosen to be

$$\omega^R(r_{ij}) = \begin{cases} (1 - r_{ij}/r_c)^m & \text{for } r_{ij} \le r_c, \\ 0 & \text{for } r_{ij} > r_c. \end{cases} \tag{6.2.3}$$

In the original DPD method, m was set to unity. Different exponent values can be used to alter the fluid viscosity and increase the Schmidt number $Sc = \nu/D$, where D is the self-diffusivity and $\nu = \mu/\rho$ is the kinematic viscosity (e.g., Fan *et al.* 2006, Fedosov *et al.* 2008).

Temperature control is achieved by balancing random and dissipative forces according to the fluctuation–dissipation theorem,

$$\omega^D(r_{ij}) = \left[\omega^R(r_{ij})\right]^2, \qquad \sigma^2 = 2\gamma k_B T, \tag{6.2.4}$$

where T is the equilibrium absolute temperature and k_B is the Boltzmann constant (e.g., Espanol & Warren 1995).

The particles move in space according to the Newton's second law of motion,

$$\frac{\mathrm{d}\mathbf{r}_i}{\mathrm{d}t} = \mathbf{v}_i, \qquad \frac{\mathrm{d}\mathbf{v}_i}{\mathrm{d}t} = \frac{1}{m_i}\sum_{j\neq i}\mathbf{F}_{ij}, \tag{6.2.5}$$

where $\mathbf{r}_i$ is the particle position and $\mathbf{F}_{ij}$ is the force exerted on the ith by the jth particle. The particle equation of motion is integrated in time using the velocity Verlet algorithm (e.g., Allen & Tildesley 1987).

6.2.2 Mesoscopic viscoelastic membrane model

The cell membrane is represented by a curved triangulated network defined by N_v vertices, $\{\mathbf{x}_i\}$, connected by N_s springs (edges) forming N_t triangular faces, where $i = 1, \ldots, N_v$. The total energy of the network consists of an in-plane elastic energy, a viscous dissipation energy (IP), a bending energy (B), a surface area energy (A), and a volume energy (V),

$$V(\{\mathbf{x}_i\}) = V_{IP} + V_B + V_A + V_V. \tag{6.2.6}$$

The individual energy components are discussed in this section.

Elastic energy and viscous dissipation

The in-plane elastic energy is given by

$$V_{IP} = \sum_{j=1}^{N_s} [U_{IPS}(\ell_j) + U_{IPV}(\Delta v_j)] + \sum_{k=1}^{N_t} \frac{C_q}{A_k^q}, \tag{6.2.7}$$

where IPS stands for the in-plane spring energy and IPV stands for the in-plane viscous dissipation. The first sum in (6.2.7) expresses the contribution of viscoelastic springs; ℓ_j is the length of the jth spring and Δv_j is the relative velocity of the spring end points. The second sum expresses a stored elastic energy assigned to each triangular patch, where A_k is the area of the kth triangle. The constant C_q and exponent q will be defined.

We employ the worm-like chain (WLC) model in combination with a stored elastic energy (WLC-C), or a power function (POW) potential (WLC-POW). The WLC energy is given by

$$U_{WLC} = \frac{k_B T \ell_m}{4p} \frac{3x^2 - 2x^3}{1 - x}, \tag{6.2.8}$$

where $x = \ell/\ell_m \in (0, 1)$, ℓ_m is the maximum spring extension and p is the persistence length. The power-function energy is given by

$$U_{POW} = \frac{k_p}{(n-1)\ell^{n-1}}, \tag{6.2.9}$$

where k_p is a spring constant and n is a specified exponent.

Attractive forces exerted by WLC springs cause element compression. The second term in the WLC-C model (6.2.7) contributes an elastic energy that tends to expand the surface area. The equilibrium state of a single triangular plaquette with WLC-C energy defines an equilibrium spring length, ℓ_0. A relationship between the WLC spring parameters C_q and q can be obtained by setting the Cauchy stress derived from the virial theorem to zero, yielding

$$C_q^{WLC} = \frac{\sqrt{3}A_0^{q+1}}{4pq\ell_m} k_B T \frac{4x_0^2 - 9x_0 + 6}{1 - x_0}, \tag{6.2.10}$$

where $x_0 = \ell_0/\ell_m$ and $A_0 = \sqrt{3}l_0^2/4$ (e.g., Dao *et al.* 2006). Given the equilibrium length and spring parameters, this formula provides us with a value for C_q to be used in(6.2.7) for a chosen q.

Similar considerations apply to the WLC-POW model where an equilibrium spring length can be defined by balancing the WLC and POW forces. In this manner, p and k_p can be related to the WLC parameters and a chosen exponent, n. Since the POW term is able to mediate triangle area compression, the stored elastic energy is omitted and C_q is set to zero. The viscous component associated with each spring will be defined.

Bending energy

The bending energy is concentrated at the element edges according to the bending potential

$$V_B = \sum_{j=1}^{N_s} k_b \left[1 - \cos(\phi_j - \phi_0)\right], \tag{6.2.11}$$

where k_b is a bending modulus, ϕ_j is the instantaneous angle formed between two adjacent triangles sharing the jth edge, and ϕ_0 is the spontaneous angle. A schematic illustration of these angles is shown in figure 6.2.1.

Area and volume constraints

The last two terms in (6.2.6) enforce area conservation of the lipid bilayer and incompressibility of the interior fluid by way of area and volume constraints,

$$V_A = \frac{k_a}{2A_0^{tot}} (A - A_0^{tot})^2 + \frac{k_d}{2A_0} \sum_{j=1}^{N_t} (A_j - A_0)^2, \tag{6.2.12}$$

and

$$V_V = \frac{k_v^2}{2V_0^{tot}} (V - V_0^{tot})^2, \tag{6.2.13}$$

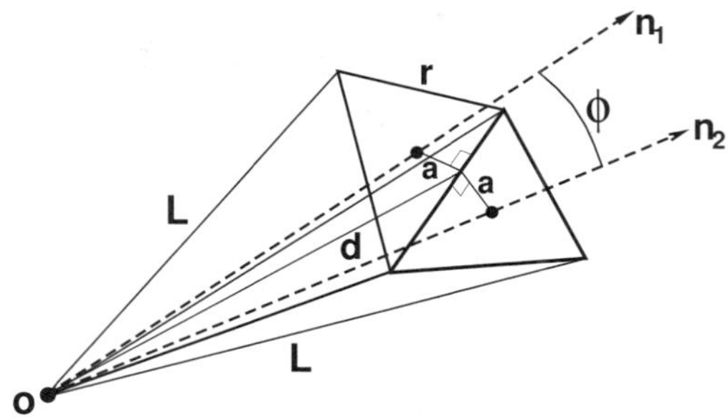

Figure 6.2.1 Schematic illustration of two equilateral triangles on a sphere of radius L.

where k_a, k_d, and k_v are constraint constants for global area, local area, and volume; A and V are the instantaneous membrane area and cell volume; and A_0^{tot} and V_0^{tot} are their respective specified values.

Nodal forces

Nodal forces $\mathbf{f}_i$ are derived from the elastic network energy by taking partial derivatives,

$$\mathbf{f}_i = -\frac{\partial V(\{\mathbf{x}_i\})}{\partial \mathbf{x}_i}, \tag{6.2.14}$$

for $i = 1, \ldots, N_v$. Exact expressions are outlined in the Appendix of this chapter.

6.2.3 Triangulation

The average shape of a normal red blood cell can be described by

$$z = \pm D_0 \left(1 - \frac{4(x^2+y^2)}{D_0^2}\right)^{1/2} \left(a_0 + a_1 \frac{x^2+y^2}{D_0^2} + a_2 \frac{(x^2+y^2)^2}{D_0^4}\right), \tag{6.2.15}$$

where $D_0 = 7.82$ μm is the cell diameter, $a_0 = 0.0518$, $a_1 = 2.0026$, and $a_2 = -4.491$ (Evans & Skalak 1980). The cell area and volume are, respectively, 135 μm^2 and 94 μm^3.

In our simulations, the membrane network structure is generated by triangulating the unstressed equilibrium shape described by (6.2.15). The cell shape is first imported into a commercial grid generation software to produce an initial triangulation using the advancing-front method. Subsequently, free-energy relaxation is performed by flipping the diagonal edges of quadrilateral

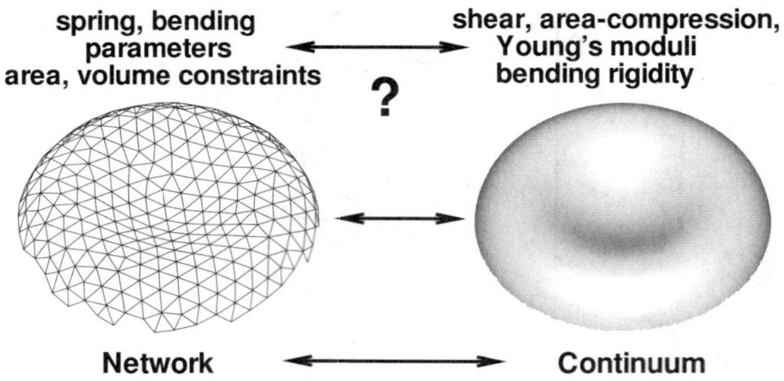

Figure 6.3.1 Illustration of a membrane network and corresponding continuum model.

elements formed by two adjacent triangles, while the vertices are constrained to move on the prescribed surface. The relaxation procedure includes only elastic in-plane and bending energy components.

6.3 Membrane mechanical properties

Several parameters must be chosen to ensure a desired mechanical response of the membrane network model. Figure 6.3.1 depicts a network and its continuum counterpart. To circumvent *ad-hoc* parameter adjustment, relationships are derived between local model parameters and network macroscopic properties for an hexagonal network. A similar analysis for a two-dimensional particulate sheet of equilateral triangles was presented by Dao *et al.* (2006).

Figure 6.3.2 illustrates an element in a hexagonal network with vertex $\mathbf{v}$ placed at the origin of a local Cartesian system. Using the virial theorem, we find that the Cauchy stress tensor at $\mathbf{v}$ is given by

$$\tau_{\alpha\beta} = -\frac{1}{S}\left[\frac{f(r_1)}{r_1} r_1^\alpha r_1^\beta + \frac{f(r_2)}{r_2} r_2^\alpha r_2^\beta + \frac{f(|\mathbf{r}_2 - \mathbf{r}_1|)}{|\mathbf{r}_2 - \mathbf{r}_1|}(r_2^\alpha - r_1^\alpha)(r_2^\beta - r_1^\beta)\right]$$
$$-\left(q\frac{C_q}{A^{q+1}} + \frac{k_a(A_0^{tot} - N_t A)}{A_0^{tot}} + \frac{k_d(A_0 - A)}{A_0}\right)\delta_{\alpha\beta}, \qquad (6.3.1)$$

where α and β stand for x or y, $f(r)$ is the spring force, N_t is the total number of triangles, $A_0^{tot} = N_t A_0$, $S = 2A_0$, $\delta_{\alpha\beta}$ is the Kronecker delta, and S is the area of the hexagonal element centered at $\mathbf{v}$ (e.g., Allen & Tildesley 1987).

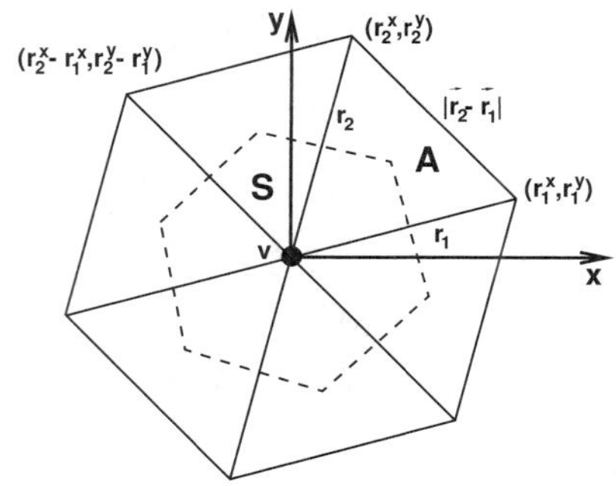

Figure 6.3.2 Illustration of an element in a hexagonal triangulation.

6.3.1 Shear modulus

The shear modulus is derived from the network deformation by applying an engineering shear strain, γ, to a material vector embedded in the surface, $\mathbf{r}_1$, so that the deformed material vector is

$$\mathbf{r}_1' = \mathbf{r}_1 \cdot \mathbf{J} = \begin{bmatrix} r_1^x + \frac{1}{2} r_1^y \\ \frac{1}{2} r_1^x \gamma + r_1^y \end{bmatrix}, \tag{6.3.2}$$

where

$$\mathbf{J} = \begin{bmatrix} 1 & \gamma/2 \\ \gamma/2 & 1 \end{bmatrix} + O(\gamma^2) \tag{6.3.3}$$

is the linear strain tensor and $\mathbf{r}_1 = (r_1^x; r_1^y)$, as shown in figure 6.3.2. Because the shear deformation is area preserving, only spring forces contribute to the membrane shear modulus.

Expanding τ_{xy}' in a Taylor series, we find that

$$\tau_{xy}' = \tau_{xy} + \left.\frac{\partial \tau_{xy}'}{\partial \gamma}\right|_{\gamma=0} \gamma + O(\gamma^2). \tag{6.3.4}$$

The linear shear modulus of the network is

$$\mu_0 = \left.\frac{\partial \tau_{xy}'}{\partial \gamma}\right|_{\gamma=0}. \tag{6.3.5}$$

Differentiating the first term of τ_{xy}' we obtain

$$\frac{\partial}{\partial \gamma}\left(\frac{f(r_1')}{r_1'} r_1^{x'} r_1^{y'}\right)_{\gamma=0} = \left(\frac{\partial \frac{f(r_1)}{r_1}}{\partial r_1} \frac{(r_1^x r_1^y)^2}{r_1} + \frac{f(r_1) r_1}{2}\right)_{r_1=\ell_0}. \tag{6.3.6}$$

Using the vector-product definition of the area of a triangle, we obtain

$$(r_1^x r_1^y)^2 + (r_2^x r_2^y)^2 + (r_2^x - r_1^x)^2 (r_2^y - r_1^y)^2 = 2A_0^2. \tag{6.3.7}$$

The linear shear modulus of the WLC-C model is

$$\mu_0^{WLC-C} = \frac{\sqrt{3}k_B T}{4p\ell_m x_0}\left(\frac{3}{4(1-x_0)^2} - \frac{3}{4} + 4x_0 + \frac{x_0}{2(1-x_0)^3}\right), \tag{6.3.8}$$

and the linear shear modulus of the WLC-POW model is

$$\mu_0^{WLC-POW} = \frac{\sqrt{3}k_B T}{4p\ell_m x_0}\left(\frac{x_0}{2(1-x_0)^3} - \frac{1}{4(1-x_0)^2} + \frac{1}{4}\right) + \frac{\sqrt{3}k_p(m+1)}{4\ell_0^{m+1}}. \tag{6.3.9}$$

6.3.2 Compression modulus

The linear elastic area compression modulus K is found from the in-plane pressure following a small area expansion,

$$p = -\frac{1}{2}(\tau_{xx} + \tau_{yy}) = \frac{3\,\ell}{4A} f(\ell) + q\frac{C_q}{A^{q+1}} + \frac{(k_a + k_d)(A_0 - A)}{A_0}. \tag{6.3.10}$$

Defining the compression modulus,

$$K = -\frac{\partial p}{\partial \log A}\bigg|_{A=A_0} = -\frac{1}{2}\frac{\partial p}{\partial \log \ell}\bigg|_{\ell=\ell_0} = -\frac{1}{2}\frac{\partial p}{\partial \log x}\bigg|_{x=x_0}, \tag{6.3.11}$$

and using equations (6.3.10) and (6.3.11), we obtain

$$K^{WLC-C} = \frac{\sqrt{3}k_B T}{4p\ell_m(1-x_0)^2}\left[\left(q+\frac{1}{2}\right)(4x_0^2 - 9x_0 + 6) + \frac{1+2(1-x_0)^3}{1-x_0}\right] + k_a + k_d \tag{6.3.12}$$

and

$$K^{WLC-POW} = 2\mu_0^{WLC-POW} + k_a + k_d. \tag{6.3.13}$$

For $q = 1$, we find that

$$K^{WLC-C} = 2\mu_0^{WLC-C} + k_a + k_d. \tag{6.3.14}$$

The compression modulus of the nearly constant-area membrane enclosing a red blood cell is much larger than the shear elastic modulus.

The Young's modulus and Poisson ratio of the two-dimensional sheet are given by

$$Y = 4\frac{K\mu_0}{K + \mu_0}, \qquad \nu = \frac{K - \mu_0}{K + \mu_0}. \tag{6.3.15}$$

As $K \to \infty$, we obtain $Y \to 4\mu_0$ and $\nu \to 1$, as required. To ensure a nearly constant area, we set $k_a + k_d \gg \mu_0$. In practice, the values $\mu_0 = 100$ and $k_a + k_d = 5000$ yield a nearly incompressible membrane with Young's modulus approximately 2% smaller than the asymptotic value $4\mu_0$.

The analytical expressions given in (6.3.8, 6.3.9, 6.3.12, 6.3.13, and 6.3.15) were verified by numerical tests on a regular two-dimensional sheet of springs. The two-dimensional sheet was confirmed to be isotropic for small shear strains and stretches, and anisotropic for large deformations (e.g., Fedosov 2009).

6.3.3 Bending rigidity

Helfrich (1973) proposed an expression for the bending energy of a lipid bilayer,

$$E_c = \frac{k_c}{2} \iint (C_1 + C_2 - 2C_0)^2 \, \mathrm{d}A + k_g \iint C_1 C_2 \, \mathrm{d}A, \tag{6.3.16}$$

where C_1 and C_2 are the principal curvatures, C_0 is the spontaneous curvature, and k_c, k_g are bending rigidities. The second term on the right-hand side of (6.3.16) is constant for any closed surface and thus inconsequential.

A relationship between the bending modulus, k_b, and the macroscopic membrane bending rigidity, k_c, can be derived for a spherical shell. Figure 6.2.1 shows two equilateral triangles with edge length r whose vertices lie on a sphere of radius L. The angle subtended between the triangle normals $\mathbf{n}_1$ and $\mathbf{n}_2$ is denoted by ϕ. In the case of a spherical shell, the total energy in (6.3.16) is found to be

$$E_c = 8\pi k_c \left(1 - \frac{C_0}{C_1}\right)^2 + 4\pi k_g = 8\pi k_c \left(1 - \frac{L}{L_0}\right)^2 + 4\pi k_g, \tag{6.3.17}$$

where $C_1 = C_2 = 1/L$ and $C_0 = 1/L_0$. In the network model, the energy of the triangulated sphere is

$$V_B = N_s \, k_b \, [1 - \cos(\phi - \phi_0)]. \tag{6.3.18}$$

Expanding $\cos(\phi - \phi_0)$ in a Taylor series provides us with the leading term

$$V_B = \frac{1}{2} N_s k_b (\phi - \phi_0)^2 + O\big((\phi - \phi_0)^4\big). \tag{6.3.19}$$

With reference to figure 6.3.2, we find that $2a \approx \phi L$ or $\phi = r/(\sqrt{3}L)$, and $\phi_0 = r/(\sqrt{3}L_0)$.

For a sphere, $A = 4\pi L^2 \approx N_t A_0 = \sqrt{3}N_t r^2/4 = \sqrt{3}N_s r^2/6$, and $r^2/L^2 = 8\pi\sqrt{3}/N_s$. Finally, we obtain

$$V_B = \frac{1}{2} N_s k_b \Big(\frac{r}{\sqrt{3}L} - \frac{r}{\sqrt{3}L_0}\Big)^2 = \frac{N_s k_b r^2}{6L^2}\,\Big(1 - \frac{L}{L_0}\Big)^2 = \frac{4\pi k_b}{\sqrt{3}}\,\Big(1 - \frac{L}{L_0}\Big)^2. \qquad (6.3.20)$$

Setting the macroscopic bending energy, E_c, equal to the membrane bending energy, V_B, for $k_g = -4k_c/3$ and $C_0 = 0$, we obtain $k_b = 2k_c/\sqrt{3}$, in agreement with the limit of a continuum approximation (e.g., Lidmar *et al.* 2003).

The spontaneous angle ϕ_0 is set according to the total number of vertices on the sphere, N_v. It can be shown that $\cos\phi = 1 - 1/[6(L^2/r^2 - 1/4)]$ and the number of edges is $N_s = 2N_v - 4$. The bending stiffness, k_b, and spontaneous angle, ϕ_0, are given by

$$k_b = \frac{2}{\sqrt{3}}\,k_c, \qquad \phi_0 = \arccos\Big(\frac{\sqrt{3}(N_v - 2) - 5\pi}{\sqrt{3}(N_v - 2) - 3\pi}\Big). \qquad (6.3.21)$$

6.3.4 Membrane viscosity

Since interparticle dissipative interaction is an intrinsic part of the formulation, incorporating dissipative and random forces into springs fits naturally into the DPD scheme. Straightforward implementation of standard DPD dissipative and random interactions expressed by (6.2.1) is insufficient. The reason is that, when projected onto the connecting vector, the contribution of the inter-particle relative velocity, $\mathbf{v}_{ij}$, is negligible for small dissipative coefficients, γ. Large values promote numerical instability.

Best performance is achieved by assigning to each spring a viscous dissipation force $-\gamma\mathbf{v}_{ij}$, where γ is a scalar coefficient. However, any alteration of the dissipative forces requires a corresponding change in fluctuating forces consistent with the fluctuation–dissipation balance to ensure a constant membrane temperature, $k_B T$. The general framework of the fluid-particle model (e.g., Espanol 1998) is employed with

$$\mathbf{F}^D_{ij} = -\mathbf{T}_{ij} \cdot \mathbf{v}_{ij}, \qquad \mathbf{T}_{ij} = A(r_{ij})\,\mathbf{I} + B(r_{ij})\,\mathbf{e}_{ij}\mathbf{e}_{ij}, \qquad (6.3.22)$$

and

$$\mathbf{F}^R_{ij}\,\mathrm{d}t = \sqrt{2k_B T}\,\Big(\tilde{A}(r_{ij})\,\mathrm{d}\overline{\mathbf{W}^S_{ij}} + \tilde{B}(r_{ij})\frac{1}{3}\,\mathrm{tr}[\mathrm{d}\mathbf{W}_{ij}]\,\mathbf{I} + \tilde{C}(r_{ij})\,\mathrm{d}\mathbf{W}^A_{ij}\Big) \cdot \mathbf{e}_{ij}, \qquad (6.3.23)$$

where the superscripts R and D stand for "random" and "dissipative," $\mathbf{I}$ is the identity matrix, $\mathrm{tr}[\mathrm{d}\mathbf{W}_{ij}]$ is the trace of a random matrix of independent Wiener increments $\mathrm{d}\mathbf{W}_{ij}$ whose symmetric and antisymmetric parts are denoted with superscripts S and A, and

$$d\overline{\mathbf{W}_{ij}^{S}} \equiv d\mathbf{W}_{ij}^{S} - \frac{1}{3}\mathrm{tr}[d\mathbf{W}_{ij}^{S}]\,\mathbf{I} \tag{6.3.24}$$

is the traceless symmetric part. The scalar weight functions $A(r)$, $B(r)$, $\tilde{A}(r)$, $\tilde{B}(r)$, and $\tilde{C}(r)$ are related by

$$\begin{aligned} A(r) &= \frac{1}{2}\left[\tilde{A}^2(r) + \tilde{C}^2(r)\right], \\ B(r) &= \frac{1}{2}\left[\tilde{A}^2(r) - \tilde{C}^2(r)\right] + \frac{1}{3}\left[\tilde{B}^2(r) - \tilde{A}^2(r)\right]. \end{aligned} \tag{6.3.25}$$

The standard forms of the dissipative and random forces are recovered by setting $\tilde{A}(r) = \tilde{C}(r) = 0$ and $B(r) = \gamma\omega^D(r)$. We employ spatially constant weight functions $A(r) = \gamma^T$, $B(r) = \gamma^C$, and $\tilde{C}(r) = 0$, where γ^T and γ^C are dissipative coefficients. Accordingly,

$$\mathbf{T}_{ij} = \gamma^T \mathbf{I} + \gamma^C \mathbf{e}_{ij}\mathbf{e}_{ij} \tag{6.3.26}$$

and the dissipative interaction force becomes

$$\mathbf{F}_{ij}^{D} = -\left(\gamma^T \mathbf{1} + \gamma^C \mathbf{e}_{ij}\mathbf{e}_{ij}\right)\cdot \mathbf{v}_{ij} = -\gamma^T \mathbf{v}_{ij} - \gamma^C (\mathbf{v}_{ij}\cdot \mathbf{e}_{ij})\,\mathbf{e}_{ij}. \tag{6.3.27}$$

The first term on the right-hand side provides us with the main viscous contribution. The second term is identical in form to the central dissipative force of the standard DPD introduced in (6.2.1). To satisfy the fluctuation–dissipation balance, the following random interaction force ensuring $3\gamma^C > \gamma^T$ is used,

$$\mathbf{F}_{ij}^{R} dt = (2k_B T)^{1/2}\left((2\gamma^T)^{1/2} d\overline{\mathbf{W}_{ij}^{S}} + (3\gamma^C - \gamma^T)^{1/2}\frac{1}{3}\,\mathrm{tr}[d\mathbf{W}_{ij}]\,\mathbf{I}\right)\cdot \mathbf{e}_{ij}. \tag{6.3.28}$$

These stipulations for the dissipative and random forces in conjunction with an elastic response constitute a mesoscopic viscoelastic spring.

To relate the membrane shear viscosity η_m to the model dissipative parameters γ^T and γ^C, an element of the hexagonal network shown in figure 6.3.2 is subjected to a constant shear rate, $\dot{\gamma}$. The shear stress τ_{xy} at short times can be approximated from the contribution of the dissipative force in (6.3.27),

$$\begin{aligned} \tau_{xy} &= -\frac{1}{2A_0}\left[\gamma^T\dot{\gamma}\left((r_y^1)^2 + (r_y^2)^2 + (r_y^2 - r_y^1)^2\right)\right. \\ &\quad \left. + \frac{\gamma^C\dot{\gamma}}{l_0^2}\left((r_x^1 r_y^1)^2 + (r_x^2 r_y^2)^2 + (r_x^2 - r_x^1)^2(r_y^2 - r_y^1)^2\right)\right] \\ &= \dot{\gamma}\sqrt{3}\,(\gamma^T + \frac{1}{4}\gamma^C). \end{aligned} \tag{6.3.29}$$

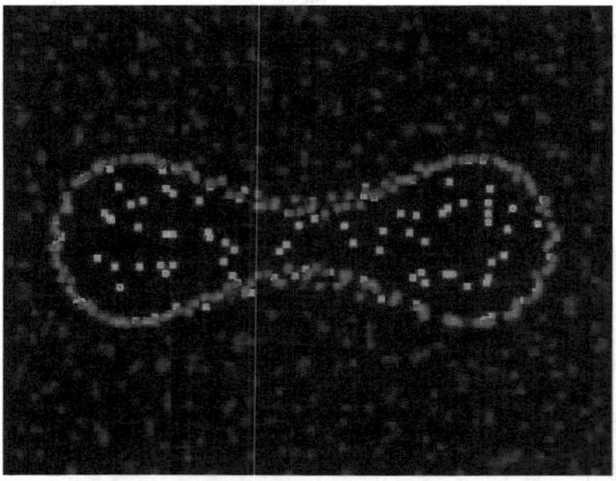

Figure 6.4.1 A slice through a sample equilibrium simulation. Red particles are membrane vertices, blue particles represent the external fluid, and green particles represent the internal fluid. *(Color in the electronic file.)*

The membrane viscosity is

$$\eta_m = \frac{\tau_{xy}}{\dot{\gamma}} = \sqrt{3}\,(\gamma^T + \frac{1}{4}\,\gamma^C). \tag{6.3.30}$$

This equation indicates that γ^T accounts for the largest portion of the membrane dissipation. Accordingly, the numerical results are nearly insensitive to the value of γ^C. Since large values lead to numerical instability, γ^C is set to its minimum value, $\frac{1}{3}\,\gamma^T$, in the simulations.

6.4 Membrane-solvent interfacial conditions

The cell membrane encloses a viscous fluid and is surrounded by a liquid solvent. Figure 6.4.1 shows a snapshot of a simulation at equilibrium. Red particles are membrane vertices, blue particles represent the external fluid, and green particles represent the internal fluid. To prevent mixing between the internal and external fluid, membrane impenetrability is required. Fluid adherence to the membrane or no-slip is implemented in terms of pairwise interactions between fluid particles and membrane nodes.

Bounce-back reflection of fluid particles at the triangular plaquettes satisfies membrane impenetrability and better enforces no-slip compared to specular reflection. However, bounce-back reflection alone does not guarantee no-slip on the membrane. Fluid particles whose centers are located at a distance less than the cutoff radius, r_c, require special treatment to account for dissipative interactions with the membrane. In practice, this necessitates that the DPD dissipative force coefficient between fluid particles and membrane vertices be properly set (e.g., Fedosov 2009).

An analogy with linear shear flow over a flat plate is used to determine the dissipative force coefficient γ for the fluid in the vicinity of the membrane. In the continuum approximation, the total shear force on an area A of the plate is $A\eta\dot{\gamma}$, where η is the fluid viscosity and $\dot{\gamma}$ is the local shear rate. To mimic the membrane surface, wall particles are distributed over the plate matching the configuration of the cell network model. The force on a single wall particle in this system exerted by the surrounding fluid under shear can be expressed as

$$F_v = \iiint_{V_h} n\, g(r)\, F^D \,\mathrm{d}V, \tag{6.4.1}$$

where F^D is the DPD dissipative force between fluid and wall particles, n is the fluid number density, $g(r)$ is the radial distribution function of fluid particles relative to the wall particles, and V_h is the half-sphere volume of fluid above the plate. Thus, the total shear force on the area A is equal to $N_A F_v$, where N_A is the number of plate particles residing inside the area A. When conservative interactions between fluid and wall particles are neglected, the radial distribution function simplifies to $g(r) = 1$.

Setting $N_A F_v = A\eta\dot{\gamma}$ yields an expression for the dissipative force coefficient γ in terms of the fluid density and viscosity and the wall density, N_A/A. Near a wall where the half-sphere lies inside the range of the linear shear flow, the shear rate cancels out. This formulation has been confirmed to enforce satisfactory no-slip boundary conditions for linear shear flow over a flat plate, and is an excellent means for enforcing no-slip at the membrane surface.

6.5 Numerical and physical scaling

The dimensionless constants and variables in the DPD model must be scaled with physical units. The characteristic length scale r^M is based on the cell diameter at equilibrium, D_0^M, where $[D_0^M] = r^M$ and the superscript M denotes model units. The length scale adopted in the present work is

$$r^M = \frac{D_0^P}{D_0^M}\,[m], \tag{6.5.1}$$

where the superscript P denotes physical units, and $[m]$ stands for meters. The Young's modulus is used as an additional scaling parameter. An energy unit scale can be derived by setting the model and physical Young's moduli equal,

$$Y^M \frac{(k_BT)^M}{(r^M)^2} = Y^P \frac{(k_BT)^P}{m^2}, \tag{6.5.2}$$

yielding the model energy scale,

$$(k_BT)^M = \frac{Y^P}{Y^M} \frac{(r^M)^2}{m^2} (k_BT)^P = \frac{Y^P}{Y^M} \left(\frac{D_0^P}{D_0^M}\right)^2 (k_BT)^P. \tag{6.5.3}$$

Once the model energy unit has been defined, the membrane bending rigidity can be expressed in energy units. With the above length and energy scales, the force scale for membrane stretching is given by

$$N^M = \frac{(k_BT)^M}{r^M} = \frac{Y^P}{Y^M} \frac{D_0^P}{D_0^M} \frac{(k_BT)^P}{m} = \frac{Y^P}{Y^M} \frac{D_0^P}{D_0^M} N^P. \tag{6.5.4}$$

Membrane rheology and dynamics require a time scale in addition to the scales previously defined. A general model time scale is defined as

$$\tau = \frac{t_i^P}{t_i^M} s = \left(\frac{D_0^P}{D_0^M} \frac{\eta^P}{\eta^M} \frac{Y_0^M}{Y_0^P}\right)^\alpha, \tag{6.5.5}$$

where η is a characteristic viscosity and α is a chosen scaling exponent similar to the power-law exponent in rheology.

6.6 Membrane mechanics

The mechanical properties of cell membranes are typically measured by deformation experiments using either micropipette aspiration techniques or optical tweezers (e.g., Evans 1983, Discher *et al.* 1994, Henon *et al.* 1999, Suresh *et al.* 2005). The shear modulus μ_0 of a healthy RBC lies in the range 2 – 12 μN/m, and the bending rigidity k_c lies in the range $1 \times 10^{-19} - 7 \times 10^{-19}$ J corresponding to 25–171 k_BT at room temperature 23°C.

To set the mechanical properties of the network model, triangulation of the cell shape described by equation (6.2.15) is first performed, yielding the equilibrium spring length

$$\ell_0 = \frac{1}{N_s} \sum_{i=1}^{N_s} \ell_0^i. \tag{6.6.1}$$

A shear modulus of a healthy cell provides us with a scaling base, $\mu_0 = \mu_0^M$. The WLC spring model requires a choice for the maximum extension length,

ℓ_m^M. It is more convenient to set the ratio $x_0 = \ell_0^M/\ell_m^M$ governing the cell nonlinear response at large deformation. The ratio x_0 is fixed at 2.2 in all simulations (e.g., Fedosov 2009).

Necessary model parameters can be calculated from (6.3.8 and 6.3.9) for given values of ℓ_0^M, μ_0^M, and x_0, thereby circumventing manual adjustment. The calculation of the area compression modulus K^M and Young's modulus Y^M follows from equations (6.3.12, 6.3.13, and 6.3.15) for specified area constraint parameters k_a and k_d. In the simulations, we use $\mu_0^M = 100$, $k_a = 4900$, $k_d = 100$, and $k_v = 5000$. We note that the global area compression and volume constraints are strong, while the local area constraint is weak. The bending rigidity k_c is set to $58(k_BT)^M$ corresponding to physical units 2.4×10^{-19} J at room temperature. The exponent n in relation (6.2.9) is set to 2.

6.6.1 Equilibrium shape and the stress-free model

After initial setup, an equilibrium simulation is run to confirm that the cell retains the biconcave shape. Figure 6.6.1(a) shows an equilibrated shape computed with the WLC-C or WLC-POW model using typical red blood cell parameters. A network on a nondevelopable surface cannot be constructed with triangles having the same edge lengths. Consequently, if all springs have the same equilibrium length, the cell surface will necessarily develop local bumps manifested as local stress anomalies in the membrane.

In fact, the potential energy relaxation performed during the triangulation process produces triangles with a narrow distribution of spring lengths around a specified equilibrium value. Accordingly, a network constructed without annealing implemented by energy relaxation of the equilibrium shape would still display pronounced bumps and fail to relax to an equilibrium stress-free axisymmetric shape.

The relaxed cell shape is affected by the ratio of the membrane modulus of elasticity to the bending rigidity expressed by the Föppl-von Kármán number

$$\kappa = \frac{Y_0 R_0^2}{k_c}, \qquad (6.6.2)$$

where $R_0 = \sqrt{A_0/(4\pi)}$.

Figure 6.6.1(b) displays an equilibrated shape computed with the WLC-C or WLC-POW model. The bending rigidity is ten times lower than that of the red blood cell membrane, $k_c = 2.4 \times 10^{-19}$ J. Membrane stress fluctuations are significantly pronounced under these conditions.

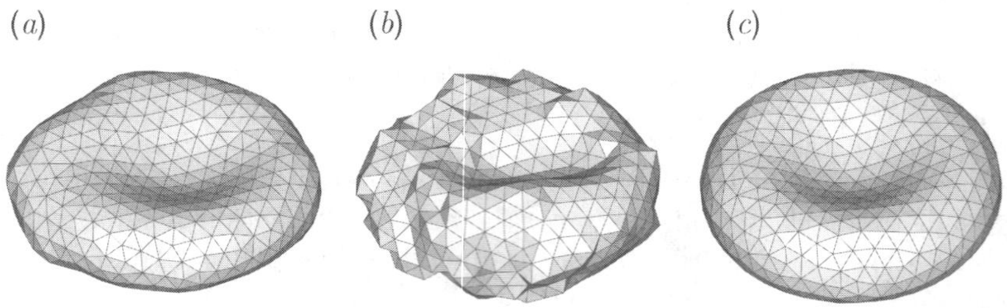

Figure 6.6.1 Equilibrium shape of a cell computed with the WLC-C or WLC-POW model for (a) $k_c = 2.4 \times 10^{-19}$ J and (b) $k_c = 2.4 \times 10^{-20}$ J. (c) Equilibrium shape with the WLC-POW stress-free model for $k_c = 2.4 \times 10^{-20}$ J.

Shape regularization

A stress-free shape eliminating membrane stress anomalies was produced by computational annealing. For each spring, the equilibrium spring length ℓ_0^i is adjusted to be the edge length after triangulation, while the ratio x_0 is kept constant at 2.2, for $i = 1, \ldots, N_s$. The maximum spring extension is then set individually to $\ell_m^i = l_0^i \times x_0$. The initial cell network defines local areas for each triangular plaquette, A_0^j, for $j = 1, \ldots, N_t$. The total cell surface area,

$$A_0^{tot} = \sum_{j=1}^{N_t} A_0^j, \tag{6.6.3}$$

and the total cell volume, V_0^{tot}, are calculated from the triangulation. After this modification, a new network that is virtually free of irregularities appears.

The annealing process disqualifies the WLC-C model. The reason is that the assumed isotropic in-plane area-expansion potential expressed by the last term in (6.2.7) is not able to accommodate individual equilibrium spring lengths for each triangle side. Because the POW potential is defined in terms of spring length, it is endowed with the necessary degrees of freedom for equilibrium length adjustment. The individual spring parameters p^i and k_s^i of the WLC-POW model are recalculated based on ℓ_0^i, ℓ_m^i, μ_0^M using (6.3.9) in conjunction with the relation $f_{WLC} = f_{POW}$ for the given spring equilibrium length.

Figure 6.6.1(c) shows an equilibrium shape computed with the WLC-POW stress-free model for bending rigidity $k_c = 2.4 \times 10^{-20}$ J. Because mem-

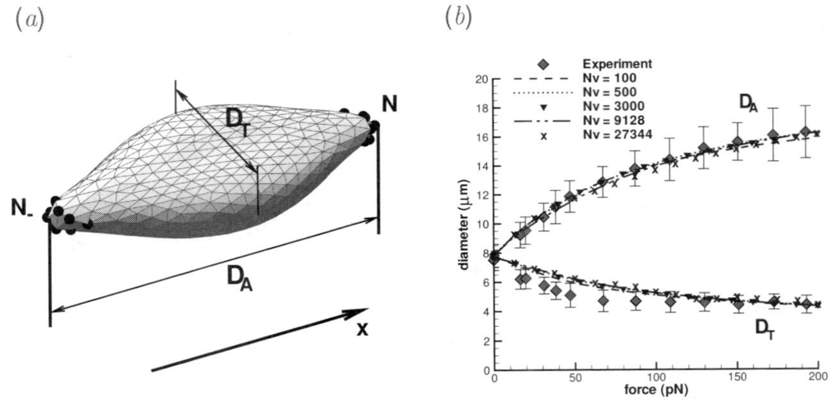

Figure 6.6.2 (a) Schematic illustration of cell deformation. (b) Stretching response with the WLC-POW stress-free model for different coarse-graining levels or number of vertices N_v in the network representation. The diamonds represent experimental results by Suresh *et al.* (2005).

brane stress artifacts are eliminated, arbitrary surface networks can be employed even for small flexural stiffness. However, if the generated network departs too much from a regular hexagonal triangulation, the analytic formulas used to estimate the network macroscopic properties are no longer reliable.

Stretching test

The reconstructed cell was subjected to stretching analogous to that imposed on cells in optical tweezers experiments (Suresh *et al.* 2005). A stretching force F_s^P up to 200 pN is applied at the outermost $N_+ = \epsilon N_v$ vertices with the largest x coordinates in the positive x direction, and at the outermost $N_- = N_+$ vertices with the smallest x coordinates in the negative x direction, as shown in figure 6.6.2(a). The vertex fraction ϵ is set to 0.02, corresponding to the contact diameter of an attached silica bead $d_c = 2$ μm used in the experiments.

For each external force, the cell is allowed to relax to an equilibrium stretched state. The axial diameter, D_A (defined as the maximum distance between the sets of points N_+ and N_-), and the transverse diameter, D_T (defined as the maximum distance between two points from the set of all vertices projected onto a plane normal to the axial diameter) are averaged during a specified simulation time. Results presented in figure 6.6.2(b) for $\mu_0^P = 6.3 \times 10^{-6}$ N/m are in good agreement with experimental data for all levels of coarse graining. Noticeable discrepancies for the transverse diameter are likely due to experimental error. The optical measurements were performed from a single observation angle. Numerical simulations show that stretched cells may rotate in the yz plane. Consequently, measurements from a single observation angle are likely to underpredict the maximum transverse diameter.

6.7 Membrane rheology from twisting torque cytometry

Early measurements of cell relaxation time employed a micropipette technique to study cell extension and recovery (e.g., Hochmuth *et al.* 1979). The relaxation time extracted from an exponential fit of cell recovery after deformation is on the order of 0.1 s. However, since the deformation is inherently nonuniform in these experiments, it is doubtful that the global technique produces an accurate characteristic membrane time scale (e.g., Yoon *et al.* 2008, Fedosov 2009).

In recent experiments, Puig-de-Morales-Marinkovic *et al.* (2007) applied optical magnetic twisting cytometry (OMTC) to infer a dynamic complex modulus of the cell membrane. In this procedure, the cell membrane response is measured locally by observing the motion of an attached ferromagnetic microbead driven by an oscillating magnetic field. The experiments have confirmed that the membrane is a viscoelastic material. Our viscoelastic membrane model will be tested against the results of the optical magnetic twisting cytometry. The numerical simulations emulate the aforementioned experiments where the motion of a microbead attached to the top of the biconcave cell due an oscillating torque is studied, as shown in figure 6.7.1(a). The data allow us to infer rheological membrane properties, including the complex modulus.

In the numerical model, the microbead is represented by a set of vertices deployed on a rigid sphere. A group of cell vertices near the bottom of the microbead simulates the area of attachment. The torque on the microbead is applied only to the bead vertices. Figure 6.7.1(b) shows a typical response to an oscillating torque. The bead motion, monitored by the displacement of the center of mass, oscillates with the applied torque frequency. The oscillation is shifted by a phase angle, ϕ, that depends on the applied frequency. In the case of a purely elastic material and in the absence of inertia, the phase angle ϕ is zero for any torque frequency.

The linear complex modulus of a viscoelastic material can be extracted from the phase angle and torque frequency using the relations

$$g'(\omega) = \frac{\Delta T}{\Delta d} \cos\phi, \qquad g''(\omega) = \frac{\Delta T}{\Delta d} \sin\phi, \tag{6.7.1}$$

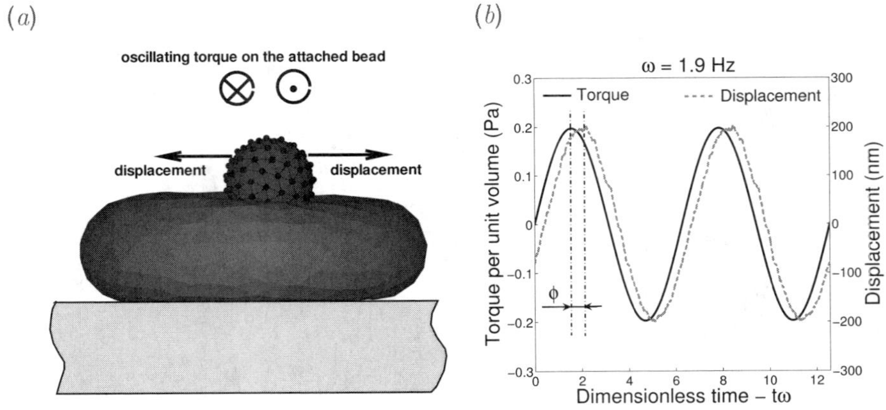

Figure 6.7.1 (*a*) Illustration of the numerical setup in twisting torque cytometry. (*b*) Response of an attached microbead subject to an oscillating torque exerted on the bead.

where $g'(\omega)$ and $g''(\omega)$ are two-dimensional storage and loss moduli and ΔT and Δd are the torque and bead displacement amplitudes. In the absence of inertia, the phase angle ϕ ranges between 0 and $\pi/2$.

The computed complex modulus is compared with experimental data by Puig-de-Morales-Marinkovic *et al.* (2007) in figure 6.7.2. Good agreement is found for bending rigidity $k_c = 4.8 \times 10^{-19}$ J and membrane viscosity $\eta_m = 0.022$ Pa s. Numerical twisting cytometry suggests that the storage modulus behaves as

$$g'(\omega) \sim (k_c Y_0)^{0.65}. \tag{6.7.2}$$

Since the Young's modulus of healthy cells is fixed by the cell stretching test, figure 6.7.2 essentially illustrates the dependence of g' on the membrane bending rigidity. To ensure good agreement with experiments, the bending rigidity of a healthy cell must be in the range $4 - 5 \times 10^{-19}$ J, which is twice the widely adopted value, $k_c = 2.4 \times 10^{-19}$ J.

For small displacements, the loss modulus, g'', depends mainly on the surface viscosity and is insensitive to the membrane elastic properties. The simulated loss modulus follows a power law in frequency with exponent $\alpha = 0.85$ to be used in (6.5.5) as a time scale. In the experiments, the exponent is approximately 0.75. The agreement is fair in light of fitting errors over two frequency decades in simulations and experiments. The data shown in inset of figure 6.7.2 suggest that inertia affects g' at high frequencies. Decreasing the bead mass would allow us to obtain rheological data for higher

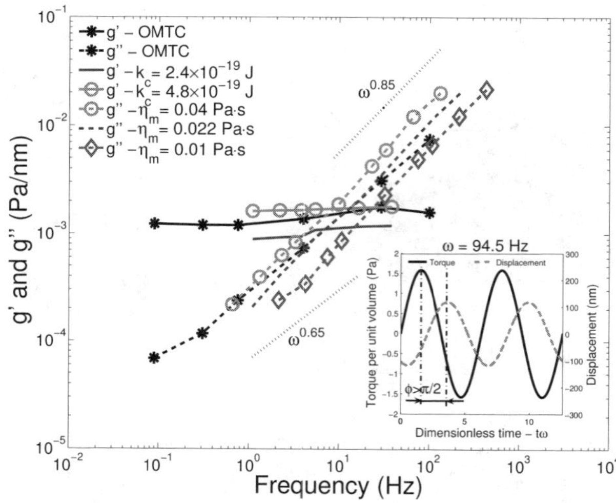

Figure 6.7.2 Graphs of the functions g' and g'' obtained from simulations with different membrane viscosities and bending rigidities. The numerical results are compared with experimental data by Puig-de-Morales-Marinkovic *et al.* (2007). The inset illustrates the effect of inertia for high frequencies of the driving torque. (*Color in the electronic file.*)

torque frequencies, but the computational cost is high due to the small time step. When the loss modulus dominates the storage modulus, the bead displacement amplitude at fixed torque amplitude is extremely small and hard to measure in the laboratory. In contrast, bead displacement in simulations can be successfully detected on a length scale as small as several nanometers.

6.8 Cell deformation in shear flow

Experiments have shown that red blood cells tumble at low shear rates and exhibit a tank-treading motion at high shear rates (e.g., Tran-Son-Tay *et al.* 1984, Fischer 2004, 2007, Abkarian *et al.* 2007). Fischer (2004) attributed this behavior to a minimum elastic energy state of the cell membrane. Cells have been made to tank-tread in the laboratory for several hours. When the flow is stopped, the cells relax to their original biconcave shape and attached microbeads recover their original relative position. It appears that tank-treading is possible only when a certain elastic energy barrier has been surpassed. In theoretical analyses, ellipsoidal cell models tank-treading along a fixed ellipsoidal path have been considered (e.g., Abkarian *et al.* 2007,

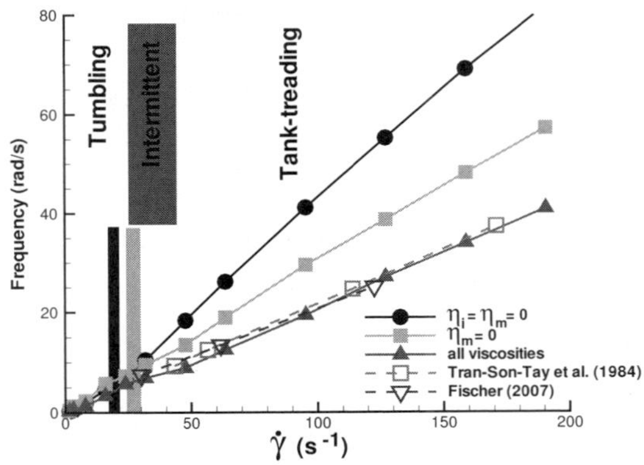

Figure 6.8.1 Tumbling and tank-treading frequency of a RBC in shear flow for $\eta_o = 0.005$ Pa s, $\eta_i = \eta_m = 0$ (circles); $\eta_o = \eta_i = 0.005$ Pa s, $\eta_m = 0$ (squares); $\eta_o = \eta_i = 0.005$ Pa s, $\eta_m = 0.022$ Pa s (triangles). (*Color in the electronic file.*)

Skotheim & Secomb 2007, see also Chapter 7). The theory predicts that the dynamics depends on the membrane shear modulus, shear rate, and viscosity ratio $\lambda = (\eta_i + \eta_m)/\eta_o$, where η_i, η_m, and η_o are the interior, membrane, and exterior fluid viscosities.

For viscosity ratio $\lambda < 3$, the theory predicts tumbling at low shear rates and tank-treading at high shear rates (e.g., Skotheim & Secomb 2007). The cells exhibit an unstable behavior in a narrow intermittent region around the tumbling-to-tank-treading transition where tumbling can be followed by tank-treading and *vice versa*. For $\lambda > 3$, a broad region of intermittent behavior instead of stable tank-treading is predicted. Red blood cells with viscosity ratio $\lambda > 3$ are observed to tank tread while exhibiting a swinging motion with a certain frequency and amplitude about an average tank-treading axis. The reliability of the theoretical predictions will be judged by comparison with the results of our simulations.

In the first simulation, a cell is suspended in a linear shear flow between two parallel walls. The viscosities of the external solvent and internal cytosol fluid are set to $\eta_o = \eta_i = 0.005$ Pa s. Consistent with results of twisting torque cytometry, the membrane viscosity is set to $\eta_m = 0.022$ Pa s. Figure 6.8.1 presents information on the cell tumbling and tank-treading frequency under different conditions. Experimental observations by Tran-Son-Tay *et al.* (1984) and Fischer (2007) are included for comparison.

In the case of a purely elastic membrane with or without inner solvent (circles and squares), the numerical results significantly overpredict the tank-treading frequency compared with experimental measurements. The internal solvent viscosity could be further increased to improve agreement with experimental data. However, since the cytosol is a hemoglobin solution with a well-defined viscosity of about 0.005 Pa s, excess viscous dissipation must occur inside the membrane. (e.g., Cokelet & Meiselman 1968). The data plotted with triangles in figure 6.8.1 show good agreement with experimental data for increased membrane viscosity.

The tumbling frequency is nearly independent of the medium viscosities. Increasing the viscosity of the internal fluid or raising the membrane viscosity only slightly shifts the tumbling-to-tank-treading threshold into higher shear rates. We estimate that the tank-treading energy barrier of a cell is approximately $E_c = 3 - 3.5 \times 10^{-17}$ J. In a theoretical model proposed by Skotheim & Secomb (2007), the energy barrier was set to $E_c = 10^{-17}$ J to ensure agreement with experimental data (Chapter 7). Membrane deformation during tank treading is indicated by an increase in the elastic energy difference with increasing shear rate to within approximately 20% of E_c.

An intermittent regime is observed with respect to the shear rate in all cases. Consistent with experiments, the width of the transition zone broadens as the membrane viscosity increases. Similar results regarding intermittency were reported by Kessler *et al.* (2008) for viscoelastic vesicles. We conclude that theoretical predictions of cell dynamics in shear flow are qualitatively correct, at best, due to the assumption of ellipsoidal shape and fixed ellipsoidal tank-treading path. Experiments by Abkarian *et al.* (2007) have shown and the present simulations have confirmed that the cell deforms along the tank-treading axis with strains of order $0.1 - 0.15$.

Cell deformation in shear flow depends on the ratio of the membrane elastic to bending modulus expressed by the Föppl-von Kármán number κ defined in (6.6.2). Figure 6.8.2 shows several snapshots of tumbling and tank-treading cells with bending rigidity ten times that commonly adopted for red blood cells, $k_c = 2.4 \times 10^{-18}$ J, corresponding to Föppl-von Kármán number $\kappa = 85$. Tumbling to tank-treading transition occurs at shear rates 20–25 s^{-1}. The results show negligible deformation during tumbling and small deformation during tank-treading following the transition.

Figure 6.8.3 presents analogous results for tumbling and tank-treading cells with bending rigidity $k_c = 2.4 \times 10^{-19}$ J corresponding to $\kappa = 850$. Significant shape deformation is observed during tumbling and tank-treading. However, the frequency of the motion is hardly changed from that corresponding to $\kappa = 85$. A further decrease of the bending rigidity results in membrane

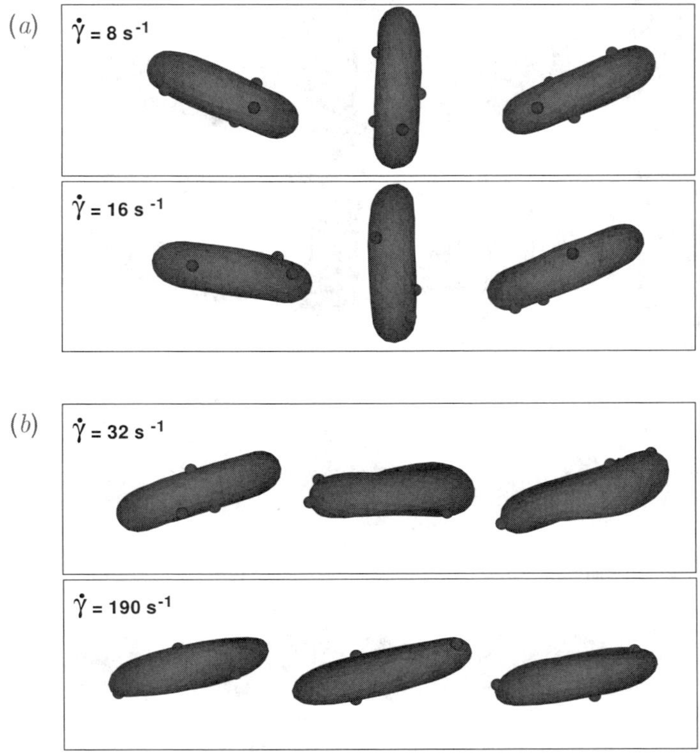

Figure 6.8.2 Snapshots of a (a) tumbling and (b) tank-treading cell at different shear rates for viscosities $\eta_o = \eta_i = 0.005$ Pa s and $\eta_m = 0.022$ Pa s, bending rigidity $k_c = 2.4 \times 10^{-18}$ J, and Föppl-von Kármán number $\kappa = 85$. Blue particles are added as tracers during post-processing for visual clarity. (*Color in the electronic file.*)

buckling. The discrete network cannot adequately capture the membrane bending on length scales comparable to the element size. To screen out the effect of membrane discretization, simulations were performed with 1000 and 3000 membrane network vertices. Similar results were obtained for corresponding Föppl-von Kármán numbers.

The simulations suggest that the membrane bending rigidity is several times higher than the widely accepted value, $k_c = 2.4 \times 10^{-19}$ J. Simulations of twisting torque cytometry presented previously in this chapter corroborate this assertion. An increase in the membrane shear modulus raises the Föppl-von Kármán number and the tank-treading energy barrier, E_c, and thus also shifts the tumbling-to-tank-treading transition to higher shear rates.

(a)

(b)

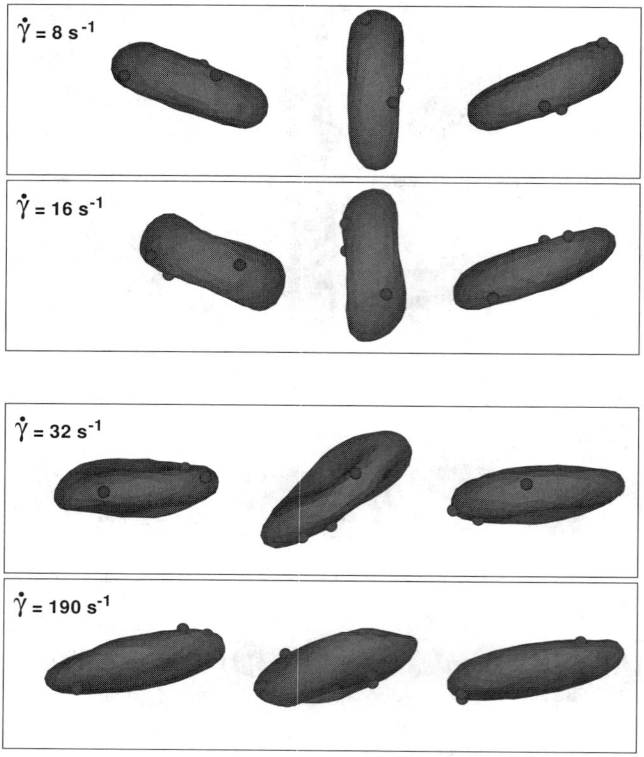

Figure 6.8.3 Snapshots of a (*a*) tumbling and (*b*) tank-treading cell at different shear rates for viscosity $\eta_o = \eta_i = 0.005$ Pa s, $\eta_m = 0.022$ Pa s, bending rigidity $k_c = 2.4 \times 10^{-19}$ J, and Föppl-von Kármán number $\kappa = 850$. Blue particles are added as tracers during post-processing for visual clarity. (*Color in the electronic file.*)

We have seen that a cell oscillates or swings around tank-treading axes with a certain frequency and amplitude, as shown in figures 6.8.2 and 6.8.3. Figure 6.8.4 presents graphs of the average tank-treading angle and swinging amplitude. The numerical results are consistent with experimental data by Abkarian *et al.* (2007). The average swinging angle is larger for a purely elastic membrane without inner cytosol. The inclination angle is independent of the internal fluid and membrane viscosities, and the swinging amplitude is insensitive to the fluid and membrane properties. The swinging frequency is exactly twice the tank-treading frequency.

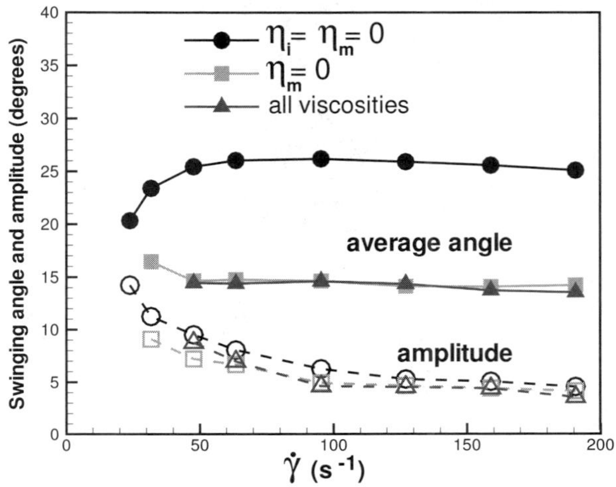

Figure 6.8.4 Graphs of the swinging average angle in degrees (filled symbols) and amplitude (open symbols) for (a) $\eta_o = 0.005$ Pa s and $\eta_i = \eta_m = 0$ (circles); (b) $\eta_o = \eta_i = 0.005$ Pa s and $\eta_m = 0$ (squares); (c) $\eta_o = \eta_i = 0.005$ Pa s and $\eta_m = 0.022$ Pa s (triangles).

6.9 Tube flow

The mean velocity of pressure-driven flow in a circular tube is defined as

$$\bar{v} = \frac{1}{A} \iint v(r)\, \mathrm{d}S, \tag{6.9.1}$$

where A is the cross-sectional area and $v(r)$ is the axial velocity. For a Newtonian fluid, $\bar{v} = v_c/2$, where v_c is the centerline velocity.

At low mean velocities, a cell suspended in tube flow retains its biconcave shape. As the driving pressure gradient becomes higher, the cell obtains the parachute-like shape, as shown in figure 6.9.1 for tube diameter $9\mu m$, in agreement with experimental observations (e.g., Tsukada *et al.* 2001). To identify the biconcave-to-parachute transition, we compute the gyration tensor

$$G_{mn} = \frac{1}{N_v} \sum_i (r_m^i - r_m^C)(r_n^i - r_n^C), \tag{6.9.2}$$

where $\mathbf{r}^i$ are the membrane vertex coordinates, $\mathbf{r}^C$ is the membrane center of mass, and m, n stand for x, y, or z (e.g., Mattice & Suter 1994). The

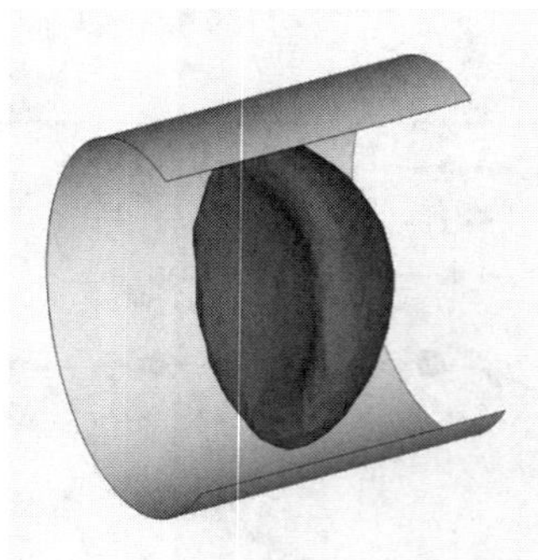

Figure 6.9.1 Parachute shape of a cell suspended in Poiseuille flow through a $9\mu\text{m}$ diameter tube.

(*a*) (*b*)

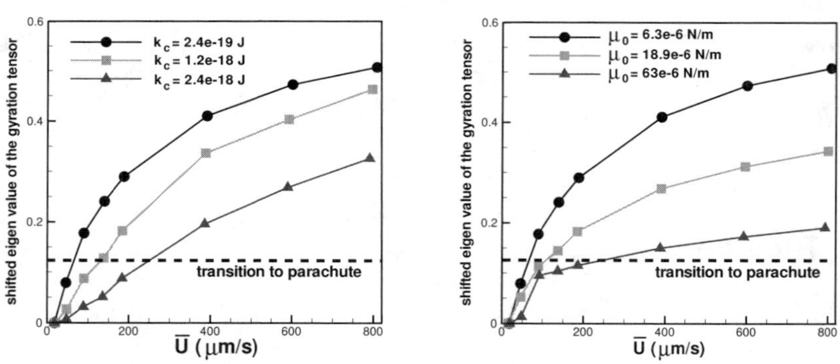

Figure 6.9.2 Excess axial eigenvalue of the gyration tensor above that for a biconcave disk for (*a*) different bending rigidities and (*b*) different membrane shear moduli. The cell volume fraction is $C = 0.05$.

eigenvalues of the gyration tensor allow us to accurately characterize the cell shape. For the equilibrium biconcave shape, the gyration tensor has two large eigenvalues corresponding to the midplane of the biconcave disk, and one small eigenvalue corresponding to the disk thickness. At the biconcave-to-parachute transition, the small eigenvalue increases, indicating that the cell elongates along the tube axes.

Figure 6.9.2 illustrates the dependence of the axial eigenvalue on the mean flow velocity for different membrane bending rigidities and shear moduli.

(a) (b)

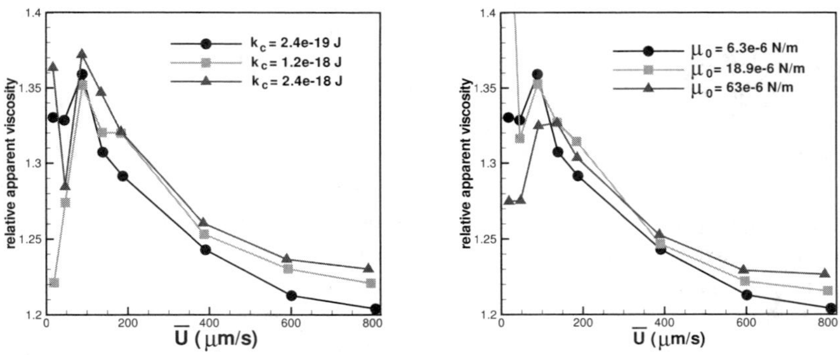

Figure 6.9.3 Relative apparent viscosity for (*a*) different bending rigidities and (*b*) different membrane shear moduli. The cell volume fraction is $C = 0.05$.

The dashed line describes the biconcave-to-parachute transition. For healthy cells, the transition occurs at a mean velocity of about 65 μm/s. Transition for larger bending rigidity or membrane shear modulus occurs at stronger flows. The critical mean velocity changes almost linearly with the bending rigidity, k_c, and shear modulus, μ_0. These results are consistent with numerical simulations by Noguchi & Gompper (2005). Stiffer capsules suffer a smaller elongation along the tube axis at the same mean velocity. The results in figure 6.9.2 corroborate the notion that stiffer cells exhibit stronger resistance to flow.

The relative apparent viscosity of the suspension is defined as

$$\lambda_{app} = \frac{\eta_{app}}{\eta_o}, \qquad \eta_{app} = \frac{n f R_0^2}{8\bar{u}}, \tag{6.9.3}$$

where n is the suspension number density, f is the force exerted on each particle, R_0 is the tube radius, η_o is the solvent viscosity, and $\bar{u}$ is the bulk velocity calculated using equation (6.9.1). The product nf is the streamwise pressure gradient, $\Delta P/L$, where L is the tube length. Figure 6.9.3 reveals a slight increase in the apparent viscosity with cell stiffening due to the increased flow resistance. The effect is small even for a tenfold increase in the membrane elastic modulus due to the low cell concentration, $C = 0.05$. A stronger effect is expected at higher cell volume fractions.

6.10 Summary

We have presented a mesoscopic model of red blood cells implemented by the dissipative particle dynamics method. The spectrin cytoskeleton is represented by a network of interconnected viscoelastic springs. The surface network accounts for bending resistance attributed to the lipid bilayer and incorporates local and global area constraints to ensure constant volume and surface area. The model was validated by several tests on membrane mechanics, rheology, and cell dynamics in shear and Poiseuille flow. The macroscopic properties of the membrane were related to the network parameters by theoretical analysis. The predicted mechanical properties of the cells agree with optical tweezers experiments even for a highly coarse-grained membrane representation with respect to the number of vertices in the spring network. Cell rheology was probed by numerical simulations mimicking optical magnetic twisting cytometry in experiments. The predicted membrane viscosity is equal to 0.022 Pa s, which is about twenty times that of water. The numerical results indicate that the bending rigidity of the the membrane can be two to three times higher than the widely accepted value $k_c = 2.4 \times 10^{-19}$ J.

Red blood cell deformation was simulated in shear and Poiseuille flow. In shear flow, a cell exhibits tumbling at low shear rates and tank-treading at high shear rates. A narrow intermittent region appears where these modes interchange. The theoretical model is able to quantitatively capture cell dynamics in shear flow. Comparison of the numerical results with existing theoretical predictions suggest that the latter suffers from oversimplification. Near the tumbling-to-tank-treading transition, simulated cells exhibit a strong deformation. Further experimental data on cell deformation around the tumbling-to-tank-treading transition could confirm the complex dynamics observed in the simulations. Simulations of cell motion in Poiseuille flow through a 9 μm diameter tube demonstrated a transition to a parachute shape at a mean velocity of about 65 μm/s. The threshold occurs at higher mean velocities for stiffer cells with a higher bending rigidity or shear modulus.

Most current cell models assume that the cell membrane is a purely elastic incompressible medium. The simulations described in this chapter show that membrane viscosity is essential for capturing single cell rheology and dynamics. The presented model is general enough to be used with other simulation methods, including Lattice-Boltzmann, Brownian dynamics, the immersed-boundary method, and multiparticle collision dynamics.

Acknowledgment

This work was supported by the NSF grant CBET-0852948 and by the NIH grant R01HL094270. Computations were performed at the NSF NICS facility.

Appendix

The modeled membrane is described by the potential energy $V(\{\mathbf{x}_i\})$ given in (6.2.6) with contributions defined in (6.2.7) – (6.2.13). Nodal forces corresponding to these energies are derived according to equation (6.2.14) and then divided into three parts: two-point interactions mediated by springs defined in (6.2.8) and (6.2.9), three-point interactions representing stored elastic energy and mediating area and volume conservation constraints according to (6.2.7), (6.2.12), and (6.2.13), and four-point interactions implementing flexural stiffness between adjacent faces.

Figure 6A.1(a) shows a sample triangular element of a membrane network. It is useful to introduce a distance matrix, $\mathbf{a}_{ij} = \mathbf{p}_i - \mathbf{p}_j$, where $i, j = 1, 2, 3$, and the normal vector $\boldsymbol{\xi} = \mathbf{a}_{21} \times \mathbf{a}_{31}$. The area of the triangle is

$$A_k = \frac{1}{2}\,|\boldsymbol{\xi}| = \frac{1}{2}\,(\xi_x^2 + \xi_y^2 + \xi_z^2)^{1/2}. \tag{6A.1}$$

The stored elastic energy for a single triangle generates the following nodal forces according to (6.2.7),

$$f_{s_i} = -\frac{\partial\,(C_q/A_k^q)}{\partial s_i} = \alpha\big(\xi_x \frac{\partial \xi_x}{\partial s_i} + \xi_y \frac{\partial \xi_y}{\partial s_i} + \xi_z \frac{\partial \xi_z}{\partial s_i}\big), \tag{6A.2}$$

where

$$\alpha = 2^q \,\frac{q\,C_q}{(\xi_x^2 + \xi_y^2 + \xi_z^2)^{q/2+1}} = \frac{q\,C_q}{4A_k^{q+2}}, \tag{6A.3}$$

s_i stands for x, y, and z, and $i = 1, 2, 3$. Explicitly,

$$(f_{x_1}, f_{y_1}, f_{z_1}) = \alpha\,(\boldsymbol{\xi} \times \mathbf{a}_{32}), \qquad (f_{x_2}, f_{y_2}, f_{z_2}) = \alpha\,(\boldsymbol{\xi} \times \mathbf{a}_{13}),$$
$$(f_{x_3}, f_{y_3}, f_{z_3}) = \alpha\,(\boldsymbol{\xi} \times \mathbf{a}_{21}). \tag{6A.4}$$

The global area conservation constraint represented by the first term on the right-hand side of (6.2.12) produces the nodal forces

$$f_{s_i} = -\frac{\partial}{\partial s_i}\Big[\frac{k_a(A - A_0^{tot})^2}{2A_0^{tot}}\Big] = -\frac{k_a(A - A_0^{tot})}{A_0^{tot}}\frac{\partial A}{\partial s_i} = \beta_a \sum_{k=1}^{N_t} \frac{\partial A_k}{\partial s_i}$$
$$= \beta_a \sum_{k=1}^{N_t} \frac{1}{4A_k}\big(\xi_x^k \frac{\partial \xi_x^k}{\partial s_i} + \xi_y^k \frac{\partial \xi_y^k}{\partial s_i} + \xi_z^k \frac{\partial \xi_z^k}{\partial s_i}\big), \tag{6A.5}$$

where $\beta_a = -k_a(A - A_0^{tot})/A_0^{tot}$, the superscript k denotes the kth triangle, and $i = 1, \ldots, N_v$. For a single triangle, the nodal forces have the functional form shown in (6A.4) with $\alpha = \beta_a/(4A_k)$.

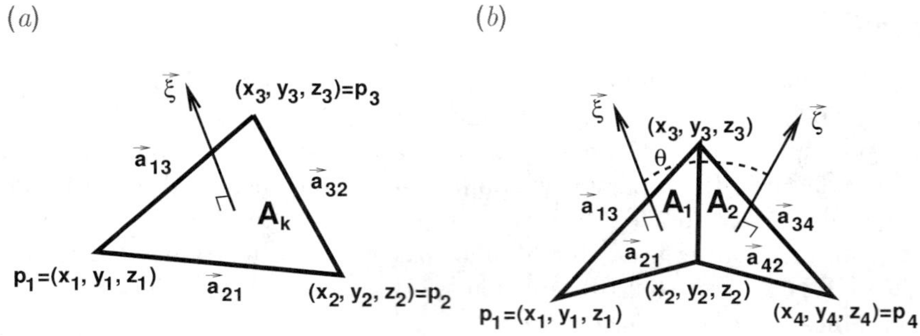

Figure 6A.1 (a) Illustration of a triangular element of the surface network, and (b) Sketch of two adjacent triangular elements of the network.

For a single triangle, the local area conservation constraint expressed by the second term on the right-hand side of (6.2.12) produces nodal forces given by (6A.4) with $\alpha = -k_d(A_k - A_0)/(4A_0A_k)$.

Global volume conservation expressed by (6.2.13) produces the nodal forces

$$f_{s_i} = -\frac{\partial}{\partial s_i}[\frac{k_v(V - V_0^{tot})^2}{2V_0^{tot}}] = -\frac{k_v(V - V_0^{tot})}{V_0^{tot}}\frac{\partial V}{\partial s_i} = \beta_v \sum_{k=1}^{N_t} \frac{\partial V_k}{\partial s_i}, \qquad (6A.6)$$

where $V_k = \frac{1}{6}(\boldsymbol{\xi}^k \cdot \mathbf{t}_c^k)$, and $\mathbf{t}_c^k = (p_1^k + p_2^k + p_3^k)/3$ is the center of mass of the kth triangle shown in figure 6.A.1. The nodal forces for a single triangle arise from the volume constraint as

$$\begin{aligned}
(f_{x_1}, f_{y_1}, f_{z_1}) &= \frac{\beta_v}{6}(\frac{1}{3}\boldsymbol{\xi} + \mathbf{t}_c \times \mathbf{a}_{32}), \\
(f_{x_2}, f_{y_2}, f_{z_2}) &= \frac{\beta_v}{6}(\frac{1}{3}\boldsymbol{\xi} + \mathbf{t}_c \times \mathbf{a}_{13}), \\
(f_{x_3}, f_{y_3}, f_{z_3}) &= \frac{\beta_v}{6}(\frac{1}{3}\boldsymbol{\xi} + \mathbf{t}_c \times \mathbf{a}_{21}).
\end{aligned} \qquad (6A.7)$$

Four-point interactions are encountered in the bending energy between two adjacent faces expressed by (6.2.11). Figure 6A.2(a) shows an arrangement of two adjacent triangular elements in the network. The triangle normal vectors are $\boldsymbol{\xi} = \mathbf{a}_{21} \times \mathbf{a}_{31}$ and $\boldsymbol{\zeta} = \mathbf{a}_{34} \times \mathbf{a}_{24}$, and the corresponding areas are $A_1 = |\boldsymbol{\xi}|/2$, and $A_2 = |\boldsymbol{\zeta}|/2$. Bending energy produces the nodal forces

$$f_{s_i} = -\frac{\partial}{\partial s_i}[k_b[1 - \cos(\theta - \theta_0)]] = -k_b \sin(\theta - \theta_0)\frac{\partial \theta}{\partial s_i}, \qquad (6A.8)$$

where θ is the angle subtended between the normals $\boldsymbol{\xi}$ and $\boldsymbol{\zeta}$, given by

$$\cos\theta = (\frac{\boldsymbol{\xi}}{|\boldsymbol{\xi}|} \cdot \frac{\boldsymbol{\zeta}}{|\boldsymbol{\zeta}|}). \tag{6A.9}$$

We write $\sin(\theta-\theta_0) = \sin\theta\cos\theta_0 - \cos\theta\sin\theta_0$, where $\sin\theta = \pm(1-\cos^2\theta)^{1/2}$ taken with the plus sign if $([\boldsymbol{\xi} - \boldsymbol{\zeta}] \cdot [\mathbf{t}_c^1 - \mathbf{t}_c^2]) \geq 0$ and with the minus sign otherwise, where $\mathbf{t}_c^1$ and $\mathbf{t}_c^2$ are the centers of mass vectors of the first and second triangle. The derivative of θ with respect to s_i is given by

$$\frac{\partial\theta}{\partial s_i} = \frac{\partial}{\partial s_i}\arccos(\frac{\boldsymbol{\xi}}{|\boldsymbol{\xi}|} \cdot \frac{\boldsymbol{\zeta}}{|\boldsymbol{\zeta}|}) = -\frac{1}{\sqrt{1-\cos^2\theta}}\frac{\partial}{\partial s_i}(\frac{\boldsymbol{\xi}}{|\boldsymbol{\xi}|} \cdot \frac{\boldsymbol{\zeta}}{|\boldsymbol{\zeta}|}). \tag{6A.10}$$

Analytical calculation of the derivatives produces the following nodal forces due to four-point interactions,

$$\begin{aligned}
(f_{x_1}, f_{y_1}, f_{z_1}) &= b_{11}\,(\boldsymbol{\xi} \times \mathbf{a}_{32}) + b_{12}\,(\boldsymbol{\zeta} \times \mathbf{a}_{32}),\\
(f_{x_2}, f_{y_2}, f_{z_2}) &= b_{11}\,(\boldsymbol{\xi} \times \mathbf{a}_{13}) + b_{12}\,(\boldsymbol{\xi} \times \mathbf{a}_{34} + \boldsymbol{\zeta} \times \mathbf{a}_{13}) + b_{22}(\boldsymbol{\zeta} \times \mathbf{a}_{34}),\\
(f_{x_3}, f_{y_3}, f_{z_3}) &= b_{11}\,(\boldsymbol{\xi} \times \mathbf{a}_{21}) + b_{12}\,(\boldsymbol{\xi} \times \mathbf{a}_{42} + \boldsymbol{\zeta} \times \mathbf{a}_{21}) + b_{22}(\boldsymbol{\zeta} \times \mathbf{a}_{42}),\\
(f_{x_4}, f_{y_4}, f_{z_4}) &= b_{12}\,(\boldsymbol{\xi} \times \mathbf{a}_{23}) + b_{22}\,(\boldsymbol{\zeta} \times \mathbf{a}_{23}),
\end{aligned} \tag{6A.11}$$

where

$$b_{11} = -\beta_b\cos\theta/|\boldsymbol{\xi}|^2, \quad b_{12} = \beta_b/(|\boldsymbol{\xi}||\boldsymbol{\zeta}|), \quad b_{22} = -\beta_b\cos\theta/|\boldsymbol{\zeta}|^2, \tag{6A.12}$$

and

$$\beta_b = k_b(\sin\theta\cos\theta_0 - \cos\theta\sin\theta_0)/\sqrt{1-\cos^2\theta}. \tag{6A.13}$$

References

ABKARIAN, M., FAIVRE, M. & VIALLAT, A. (2007) Swinging of red blood cells under shear flow. *Phys. Rev. Lett.* **98**, 188302.

ALLEN, M. P. & TILDESLEY, D. J. (1987) *Computer Simulation of Liquids.* Clarendon Press, New York.

COKELET, G. R. & MEISELMAN, H. J. (1968) Rheological comparison of hemoglobin solutions and erythrocyte suspensions. *Science* **162**, 275–277.

DAO, M., LI, J. & SURESH, S. (2006) Molecularly based analysis of deformation of spectrin network and human erythrocyte. *Materials Sci. Engin. C* **26**, 1232–1244.

DISCHER, D. E., MOHANDAS, N. & EVANS, E. A. (1994) Molecular maps of red cell deformation: Hidden elasticity and in situ connectivity. *Science* **266**, 1032–1035.

DISCHER, D. E., BOAL, D. H. & BOEY, S. K. (1998) Simulations of the erythrocyte cytoskeleton at large deformation. II. Micropipette aspiration. *Biophys. J.* **75**, 1584–1597.

DUPIN, M. M., HALLIDAY, I., CARE, C. M., ALBOUL, L. & MUNN, L. L. (2007) Modeling the flow of dense suspensions of deformable particles in three dimensions. *Phys. Rev. E* **75**, 066707.

EGGLETON, C. D. & POPEL, A. S. (1998) Large deformation of red blood cell ghosts in simple shear flow. *Phys. Fluids* **10**, 1834–1845.

ESPANOL, P. (1998) Fluid particle model. *Phys. Rev. E* **57**, 2930–2948.

ESPANOL, P. & WARREN, P. (1995) Statistical mechanics of dissipative particle dynamics. *Europhys. Lett.* **30**, 191–196.

EVANS, E. A. & SKALAK, R. (1980) *Mechanics and Thermodynamics of Biomembranes.* CRC Press, Boca Raton.

EVANS, E. A. (1983) Bending elastic modulus of red blood cell membrane derived from buckling instability in micropipette aspiration tests. *Biophys. J.* **43**, 27–30.

FAN, X., PHAN-THIEN, N., CHEN, S., WU, X. & NG, T. Y. (2006) Simulating flow of DNA suspension using dissipative particle dynamics. *Phys. Fluids* **18**, 063102.

FEDOSOV, D. A., KARNIADAKIS, G. E. & CASWELL, B. (2008) Dissipative particle dynamics simulation of depletion layer and polymer migration in micro- and nanochannels for dilute polymer solutions. *J. Chem. Phys.* **128**, 144903.

FEDOSOV, D. A. (2009) *Multiscale Modeling of Blood Flow and Soft Matter.* PhD thesis, Brown University, Providence.

FISCHER, T. M. (2004) Shape memory of human red blood cells. *Biophys. J.* **86**, 3304–3313.

FISCHER, T. M. (2007) Tank-tread frequency of the red cell membrane: Dependence on the viscosity of the suspending medium. *Biophys. J.* **93**, 2553–2561.

FUNG, Y. C. (1993) *Biomechanics: Mechanical Properties of Living Tissues.* Springer-Verlag, New York.

HELFRICH, W. (1973) Elastic properties of lipid bilayers: Theory and possible experiments. *Z. Naturforschung C* **28**, 693–703.

HENON, S., LENORMAND, G., RICHERT, A. & GALLET, F. (1999) A new determination of the shear modulus of the human erythrocyte membrane using optical tweezers. *Biophys. J.* **76**, 1145–1151.

HOCHMUTH, R. M., WORTHY, P. R. & EVANS, E. A. (1979) Red cell extensional recovery and the determination of membrane viscosity. *Biophys. J.* **26**, 101–114.

HOOGERBRUGGE, P. J. & KOELMAN, J. M. V .A. (1992) Simulating microscopic hydrodynamic phenomena with dissipative particle dynamics. *Europhys. Lett.* **19**, 155–160.

KESSLER, S., FINKEN, R. & SEIFERT, U. (2008) Swinging and tumbling of elastic capsules in shear flow. *J. Fluid Mech.* **605**, 207–226.

LI, J., DAO, M., LIM, C. T. & SURESH, S. (2005) Spectrin-level modeling of the cytoskeleton and optical tweezers stretching of the erythrocyte. *Biophys. J.* **88**, 3707–3719.

LIDMAR, J., MIRNY, L. & NELSON, D. R. (2003) Virus shapes and buckling transitions in spherical shells. *Phys. Rev. E* **68**, 051910.

MALEVANETS, A. & KAPRAL, R. (1999) Mesoscopic model for solvent dynamics. *J. Chem. Phys.* **110**, 8605–8613.

MATTICE, W. L. & SUTER, U. W. (1994) *Conformational Theory of Large Molecules: The Rotational Isomeric State Model in Macromolecular Systems.* Wiley Interscience, New York.

NOGUCHI, H. & GOMPPER, G. (2005) Shape transitions of fluid vesicles and red blood cells in capillary flows. *Proc. Natl. Acad. Sci. USA* **102**, 14159–14164.

PIVKIN, I. V. & KARNIADAKIS, G. E. (2008) Accurate coarse-grained modeling of red blood cells. *Phys. Rev. Lett.* **101**, 118105.

POZRIKIDIS, C. (2005) Numerical simulation of cell motion in tube flow. *Ann. Biomed. Eng.* **33**, 165–178.

PUIG-DE-MORALES-MARINKOVIC, M., TURNER, K. T., BUTLER, J. P., FREDBERG, J. J. & SURESH, S. (2007) Viscoelasticity of the human red blood cell. *Am. J. Physiol.-Cell Physiol.* **293**, 597–605.

SKOTHEIM, J. M. & SECOMB, T. W. (2007) Red blood cells and other non-spherical capsules in shear flow: Oscillatory dynamics and the tank-treading-to-tumbling transition. *Phys. Rev. Lett.* **98**, 078301.

SUCCI, S. (2001) *The Lattice Boltzmann Equation for Fluid Dynamics and Beyond.* Oxford University Press, Oxford.

SURESH, S., SPATZ, J., MILLS, J. P., MICOULET, A., DAO, M., LIM, C. T., BEIL, M. & SEUFFERLEIN, T. (2005) Connections between single-cell biomechanics and human disease states: Gastrointestinal cancer and malaria. *Acta Biomaterialia* **1**, 15–30.

TRAN-SON-TAY, R., SUTERA, S. P. & RAO, P. R. (1984) Determination of RBC membrane viscosity from rheoscopic observations of tank-treading motion. *Biophys. J.* **46**, 65–72.

TSUKADA, K., SEKIZUKA, E., OSHIO, C. & MINAMITANI, H. (2001) Direct measurement of erythrocyte deformability in diabetes mellitus with a transparent microchannel capillary model and high-speed video camera system. *Microvasc. Res.* **61**, 231–239.

YOON, Y. Z., KOTAR, J., YOON, G. & CICUTA, P. (2008) The nonlinear mechanical response of the red blood cell. *Phys. Biol.* **5**, 036007.

7 Simulation of red blood cell motion in microvessels and bifurcations

T. W. Secomb

Department of Physiology
University of Arizona
Tucson

The constitution and salient mechanical properties of mammalian red blood cells are well established. In principle, this information provides us with a basis for understanding and predicting blood flow, both in bulk and in narrow passages occurring in the microcirculation. However, the particular mechanical properties of the red blood cell membrane, the asymmetry of three-dimensional cell shapes in flow, and the strong cell-to-cell interactions occurring under typical physiological concentrations present us with substantial difficulties in analyzing and simulating red blood cell motion and deformation in flow. Progress can be made by making simplifying geometric assumptions to reduce the dimensionality of the mathematical problem. When blood flows through capillaries, the red blood cells often move in single files with approximately axisymmetric shapes. Computational approaches for axisymmetric configurations have led to predictions of flow resistance in narrow tubes in quantitative agreement with experimental data. A two-dimensional model was recently developed where a red blood cell is represented by a two-dimensional assembly of viscoelastic elements. The model has been used to describe several aspects of red blood cell motion in microvessels discussed in this chapter, including transverse cell migration, cell-to-cell interactions, and partition of cells in diverging bifurcations.

7.1 Introduction

Mammalian red blood cells (RBCs), also called erythrocytes, consist of a fluid interior enclosed by a flexible membrane. The flexibility of the red blood cells is crucial in allowing blood to flow readily through the circulation, even at a relatively high concentration (hematocrit) of 40–45% found in healthy individuals, and to traverse the small pathways of the microcirculation. The salient mechanical properties of red blood cells discussed in Chapter 1 have been known for nearly thirty years. The cell interior is an incompressible, nearly Newtonian fluid. The membrane exhibits viscoelastic response to in-plane shearing deformation described by a Voigt solid model (Evans & Hochmuth 1976), strong resistance to areal dilation, and a relatively small resistance to bending. This behavior arises from the mechanical properties of the lipid bilayer and protein cytoskeleton composing the membrane. In principle, these properties provide us with sufficient data for understanding and predicting the motion and interaction of RBCs in specified microvascular geometries. However, progress has been relatively slow and several fundamental aspects remain poorly resolved.

Difficulties in computing flow-induced RBC motion and deformation stem from the unique mechanical properties of the membrane. This can be understood by introducing a set of scalings for an individual cell suspended in plasma, and then considering the range of the resulting dimensionless parameters describing the mechanical properties. Distance is expressed in μm, time in ms, and velocity in mm/s, which is typical of blood flow in microvessels. Force is expressed in units of 10^{-7} dyn. The unit of viscosity is 1 cP (10^{-2} dyn s/cm^2), which is close to the viscosity of normal blood plasma, and the unit of pressure and stress is 10 dyn/cm^2.

The shear elastic modulus of the membrane, 0.006 dyn/cm, is equal to 6 in this scaling, while the area modulus, 500 dyn/cm, is equal to 5×10^5 (Hochmuth & Waugh 1987). The bending modulus of the membrane, approximately equal to 1.8×10^{-12} dyn cm, corresponds to a scaled value of 0.18 (Evans 1983). These disparate magnitudes lead to numerically difficult problems. For example, the large disparity between the shear and bulk modulus leads to stiff problems when membrane dynamics is considered. An alternative is to regard the membrane motion as strictly area-preserving (Secomb & Skalak 1982*b*). However, this approach leads to difficulties analogous to those encountered in the numerical computation of incompressible viscous flow. The small value of the scaled bending modulus allows red blood cells to develop local regions of high curvature where bending resistance plays an important role (Secomb *et al.* 1986).

The shear viscosity of the cell membrane is estimated to be approximately 10^{-3} dyn s/cm (Hochmuth & Waugh 1987), corresponding to a scaled value of 1000. Several studies have indicated that the actual value may be lower, closer to 10^{-4} dyn s/cm, corresponding to a scaled value of 100 (Tran-Son-Tay *et al.* 1984, Hsu & Secomb 1989, Secomb *et al.* 2007). A scaled value in the range 100–1000 implies that the membrane exhibits a large viscous resistance to flow-induced deformation when a cell is suspended in a low-viscosity medium such as plasma. Accordingly, rapid cell deformation is possible only when a cell strongly interacts closely with other cells or solid boundaries creating strong lubrication-type forces. These effects are difficult to simulate using conventional numerical schemes based on meshes of computational nodes with fixed spacing. Despite its importance, membrane viscosity has been neglected in most computational studies of flow-induced deformation.

The mechanics of blood flow in narrow tubes has been discussed in previous reviews (Secomb 1995, Secomb 2003). The apparent viscosity of blood is defined as the viscosity of a homogeneous fluid that would produce the same flow rate according to Poiseuille's law under given conditions. Measured in glass tubes, the apparent viscosity exhibits a strong decline as the tube diameter decreases below 1 mm according to the Fåhraeus-Lindqvist effect (Fåhraeus & Lindqvist 1931, Pries *et al.* 1992). The volume fraction of RBCs in narrow tubes (tube hematocrit) is typically lower than the hematocrit of blood measured in bulk. The reduction of hematocrit, known as the Fåhraeus effect, arises because the mean velocity of RBCs is higher than the mean plasma velocity, so that RBCs have a shorter transit time than plasma in a given tube (Fåhraeus 1928). Although several factors contribute to these two effects, the most important factor is the formation of a cell-free or cell-depleted layer near a tube wall.

For tube diameters above 30 μm, the variation of the blood apparent viscosity with tube diameter at normal hematocrit can be accurately described by a two-layer continuum model where a central core region with fixed viscosity of about 3 cP is surrounded by a layer of plasma whose width is independent of the tube diameter (Secomb 1995). Some limitations of this model are apparent. Firstly, a sharp transition from a cell-free zone to a zone of fixed viscosity containing cells is assumed. In reality, the interface between the cell-filled core and the cell-free layer is irregular in space and time. Secondly, the plasma layer width (~ 1.8 μm) is an empirical parameter. Thirdly, the model fails for tube diameters below 30 μm where multiple cells are present and the two-layer continuum approach no longer applies. A central objective of microvascular blood rheology research is to understand the factors governing the radial distribution of RBCs in microvessels.

Progress has been made in the case of blood flow through narrow capillaries where RBCs are typically arranged in a single file and the motion of the individual cells can be analyzed by neglecting interactions with other cells, as discussed in Section 7.2. For vessels with diameter 7 μm or higher, we encounter several RBCs inside a single cross-section in a multi-file arrangement, except at very low hematocrits (Gaehtgens *et al.* 1980). Understanding blood flow in microvessels with diameter larger than 7 μm requires computational approaches to simulate the motion and interaction of multiple cells in a concentrated suspension. The geometrical and mechanical complexity of multiple and strongly interacting deformable particles is a main obstacle to the theoretical understanding of blood rheology in the microcirculation.

Sections 7.3–7 of this chapter are devoted to describing an approach recently developed to address this problem. The main idea is to represent RBCs as two-dimensional assemblages of viscoelastic elements suspended in flow, as discussed in Section 7.3. Following sections describe applications in cell tank-treading, motion of cells in channels and diverging bifurcations, and motion of multiple interacting cells in channel flow.

7.2 Axisymmetric models for single-file RBC motion

Red blood cells in narrow capillaries frequently assume bullet-like shapes with approximate rotational symmetry about the tube axis. Convex shapes are observed at the front and concave shapes are observed at the rear surface of the cells. Several theoretical analyses of RBC motion in capillaries have been conducted for axisymmetric configurations (Barnard *et al.* 1968, Lighthill 1968, Zarda *et al.* 1977, Secomb *et al.* 1986).

The cell shape can be described by a function, $r(s)$, specifying the radial position of a point in the membrane in terms of the arc length, s, measured from the nose of the cell, as shown in figure 7.2.1. Under the assumption that fluid flow in the space between the cell and vessel wall is described by lubrication theory, the governing equations for steady motion of a deformable cell in a uniform cylindrical tube reduce into a set of nonlinear ordinary differential equations. The computational approach and predicted results are summarized in this section.

In the axisymmetric configuration, the principal axes of the membrane stress and strain coincide with the coordinate directions s and φ, where φ is the meridional angle. The membrane strain is characterized by the stretch ratios

$$\lambda_s = \frac{\mathrm{d}s}{\mathrm{d}s_0} \qquad \lambda_\phi = \frac{r}{r_0}, \tag{7.2.1}$$

where r_0 and s_0 correspond to material point particles in the unstressed configuration. The membrane curvatures are $\kappa_s = \mathrm{d}\theta/\mathrm{d}s$ and $\kappa_\varphi = \sin\theta/r$, where θ is the angle subtended between the normal vector to the membrane and the tube axis. The components of the membrane tension are denoted by t_s and t_φ, the shear force per unit length is denoted by q_s, and the bending moments are denoted by m_s and m_φ, as shown in figure 7.2.1.

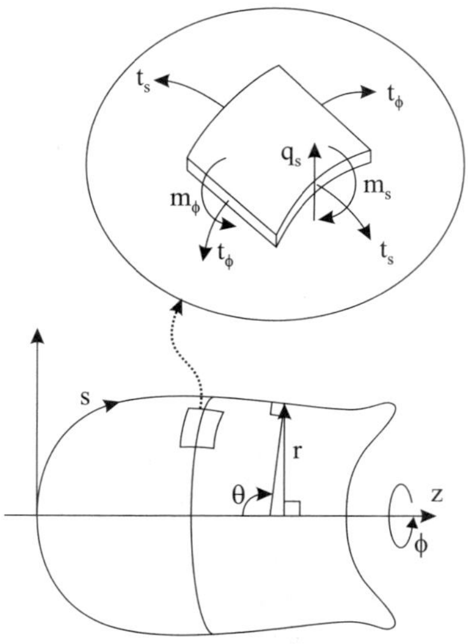

Figure 7.2.1 Schematic illustration of variables describing the geometry and stress resultants in an axisymmetric shell.

The principal tensions can be resolved into an isotropic (mean) component, t_m, and a deviatoric component, t_d, so that

$$t_s = t_m + t_d, \qquad t_\phi = t_m - t_d. \tag{7.2.2}$$

Because of its high areal modulus, the membrane can be treated as an area conserving medium, $\lambda_s \lambda_\phi \simeq 1$. Like the pressure in an incompressible fluid, the isotropic component of the tension is determined exclusively by forces acting on the membrane.

The bending moments are assumed to be isotropic and proportional to the change in the total curvature, defined as twice the mean curvature,

$$m_s = m_\phi = B\,[(\kappa_s + \kappa_\phi) - (\kappa_s + \kappa_\phi)_0], \tag{7.2.3}$$

where B is the bending resistance (Evans & Skalak 1980).

The deviatoric component of the membrane tension representing the in-plane membrane shear stress is given by

$$t_d = \frac{k}{2}\,(\lambda_s^2 - \frac{1}{\lambda_s^2}) - \frac{m_s}{2}(\kappa_s - \kappa_\varphi), \tag{7.2.4}$$

where k is the shear elastic modulus (Evans & Skalak 1980). The second term on the right-hand side results from the interaction between bending and tension forces in a membrane consisting of area-preserving leaflets (Secomb 1988).

Equilibrium of membrane normal stresses, tangential stresses, and bending moments requires

$$\frac{1}{r}\frac{\mathrm{d}(rq_s)}{\mathrm{d}s} = p + \kappa_s t_s + \kappa_\varphi t_\varphi, \qquad \frac{1}{r}\frac{\mathrm{d}(rt_s)}{\mathrm{d}s} = t_\varphi \frac{\cos\theta}{r} - \kappa_s q_s - t,$$
$$\frac{1}{r}\frac{\mathrm{d}(rm_s)}{\mathrm{d}s} = m_\varphi \frac{\cos\theta}{r} + q_s, \tag{7.2.5}$$

where p is the hydrostatic pressure difference across the membrane (external minus internal), and τ is the fluid shear stress acting on the membrane due to the motion of the external fluid (Timoshenko 1940).

Lubrication theory is used to relate p and τ to the cell shape and motion. The typical width of the gap between the cell and the vessel wall is assumed to be small compared to the typical length of the gap. The radial fluid velocity is then small compared to the axial velocity, and the pressure in the gap is independent of radial position. Details of the lubrication analysis can be found elsewhere (Secomb *et al.* 1986, Secomb 2003). In the end, the pressure gradient is expressed as a function of the radial position r,

$$\frac{\mathrm{d}p}{\mathrm{d}z} \equiv g(r), \tag{7.2.6}$$

where

$$g(r) = \frac{16\mu_p}{a^2 - r^2}\left[u_0\left(\frac{a^2}{2} + \frac{a^2 - r^2}{4\log(r/a)}\right) - aq_0\right]\left(a^2 + r^2 + \frac{a^2 - r^2}{\log(r/a)}\right)^{-1}, \tag{7.2.7}$$

μ_p is the plasma viscosity, u_0 is the negative of the cell velocity, a is the vessel radius, and q_0 is the leak-back, defined as the volumetric flow rate of fluid in the gap relative to the cell per unit circumferential length. The shear stress acting on the cell surface is given by

$$\tau(r) = \frac{g(r)}{4}\left(\frac{a^2 - r^2}{r\log(r/a) + 2r}\right) - \frac{\mu_p u_0}{r\log(r/a)}. \tag{7.2.8}$$

The preceding equations can be combined into a system of six ordinary differential equations for r, θ, κ_s, q_s, p, and τ_s with respect to arc length, s. The solution is subject to the constraints of fixed cell volume and surface area. The internal pressure must be found as part of the solution. The unstressed shape of the membrane is assumed to be a sphere with the same surface area as a red blood cell.

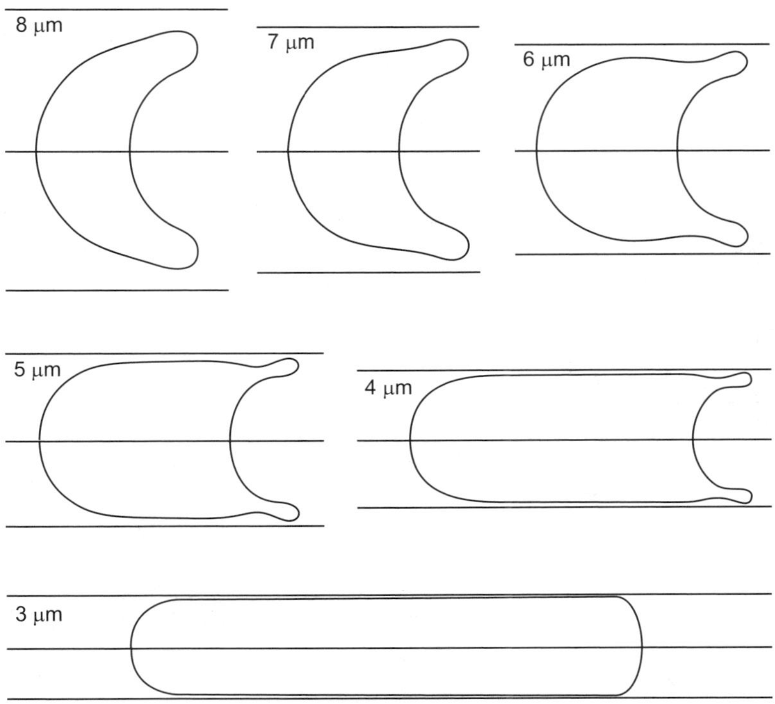

Figure 7.2.2 Computed shapes of axisymmetric red blood cells in cylindrical tubes. The flow direction is from right to left, and the cell velocity is $0.01\ \mathrm{cm/s}$. Tube diameters are indicated in each panel.

The differential equations were solved numerically using a multiple shooting method (Secomb *et al.* 1986). Predicted red blood cell shapes are shown in figure 7.2.2. Several characteristic features of red cells flowing in capillaries are evident, including a convex shape at the front, a concave shape at the rear, and an outward bulging shape near the trailing edge. The apparent viscosity estimated from the computed pressure drop across the cell agrees well with experimental values obtained from studies of blood flow through glass tubes with corresponding diameters (Secomb 1987).

7.3 Two-dimensional models for RBC motion

In reality, red blood cell shapes in capillaries are generally nonaxisymmetric. Even in narrow capillaries where the cells are constrained tightly to bullet-like shapes, the trailing edge of the cell is often asymmetric. As the capillary diameter increases, slipper-like shapes develop. A rigorous computational analysis of this motion requires solving a fully three-dimensional fluid-solid interaction problem, which is difficult for reasons already discussed.

In an early effort to assess the effect of asymmetry on the cell motion in capillaries, a two-dimensional model was introduced by Secomb & Skalak (1982*a*). In this model, the membrane is allowed to move with constant velocity around the perimeter of the cell in a tank-treading mode similar to that observed in capillary tubes (Gaehtgens & Schmid-Schönbein 1982). The analysis provides us with predictions for the dependence of the tank-treading velocity on the degree of asymmetry of the cell shape. Tank treading motion reduces the flow resistance relative to that for a symmetric cell configuration.

The three-dimensional mechanics of the cell tank-treading motion in capillary flow were further analyzed by Hsu & Secomb (1989). The cell shape and membrane motion were described in terms of a stream function that ensures area conservation during tank-treading (Secomb & Skalak 1982*b*). This model takes into account the viscous resistance of the membrane motion resulting from shearing deformation, neglected in the earlier analysis of Secomb & Skalak (1982*a*). The predicted tank-treading velocity is much lower than that seen in experiments, underscoring the important effect of membrane shear viscosity.

7.3.1 Element cell model

A novel two-dimensional model of cell motion and deformation was recently proposed (Secomb *et al.* 2007, Barber *et al.* 2008). The two-dimensional outline of a cell represents the cross-section of a three-dimensional cell in the mid-plane parallel to the flow direction. The model employs a number of interconnected viscoelastic elements, as shown in figure 7.3.1. A chain of elements around the perimeter of the cell represents the membrane. A set of internal viscous elements is introduced to represent viscous resistance to cell deformation.

The interior resistance has a component arising from the internal fluid viscosity, and another component expressing the effect of membrane shear viscosity due to membrane tank-treading or other three-dimensional deformation. Membrane flexural stiffness is implemented in terms of bending moments proportional to the angle subtended between neighboring elements at each node. Conservation of the cell interior area is enforced by introducing an interior pressure determined by the cell area, A,

$$p_{int} = k_p \, (1 - \frac{A}{A_{ref}}) \tag{7.3.1}$$

where A_{ref} and $k_p >> 1$ are two constants.

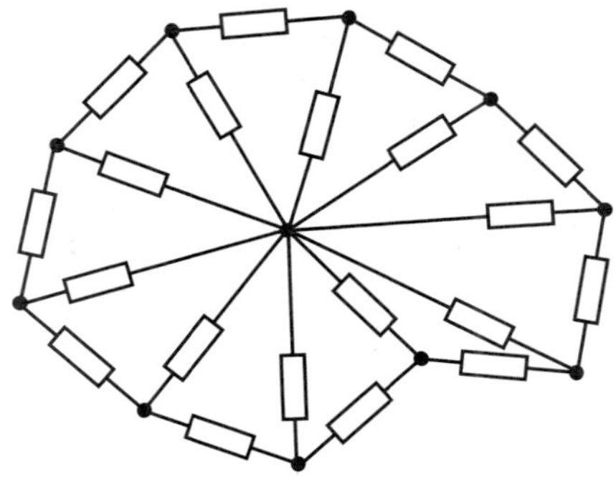

Figure 7.3.1 Schematic illustration of a two-dimensional red blood cell model. Rectangles denote viscoelastic elements connected at nodes.

Internal elements

The internal elements are purely viscous. The tension of the ith element is described by

$$T_i = \mu'_m \frac{1}{L_i} \frac{\mathrm{d}L_i}{\mathrm{d}t}, \tag{7.3.2}$$

where L_i is the element length and μ'_m is the element viscosity.

Membrane elements

Mechanical equilibrium of the ith membrane element requires

$$\frac{\mathrm{d}t_i}{\mathrm{d}s} = -g_i, \qquad \frac{\mathrm{d}q_i}{\mathrm{d}s} = -f_i, \qquad \frac{\mathrm{d}m_i}{\mathrm{d}s} = q_i. \tag{7.3.3}$$

The longitudinal (tension) force, t_i, the transverse (shear) force, q_i, the bending moment, m_i, the normal fluid loading, f_i, and the tangential fluid loading, g_i, are functions of position, s, along an element from node i to node $i+1$. The viscoelastic response of the ith membrane element is described by

$$\bar{t}_i = k_t \left(\frac{l_i}{l_0} - 1\right) + \mu_m \frac{1}{l_i} \frac{\mathrm{d}l_i}{\mathrm{d}t}, \tag{7.3.4}$$

where l_i is the element length, l_0 is a reference length, k_t is the elastic modulus, μ_m is the membrane viscosity, and $\bar{t}_i$ is the mean tension defined by

$$\bar{t}_i = \frac{1}{l_i} \int_0^{l_i} t_i(s)\,\mathrm{d}s. \tag{7.3.5}$$

The bending moments exerted on the ith element at nodes i and $i+1$ are given by

$$m_i(0) = -k_b \frac{\alpha_i}{l_0}, \qquad m_i(l_i) = -k_b \frac{\alpha_{i+1}}{l_0}, \tag{7.3.6}$$

where k_b is the bending modulus and $\alpha_i = \theta_i - \theta_{i-1}$ is the angle subtended between elements $i-1$ and i. Integrating (7.3.6), we find that

$$k_b \frac{\alpha_i - \alpha_{i+1}}{l_0 l_i} = \bar{q}_i, \tag{7.3.7}$$

where $\bar{q}_i$ is the mean shear in the element defined as in (7.3.5). With some manipulation, the tensions and shear forces at each element end-node can be expressed in terms of the mean values, $\bar{t}_i$ and $\bar{q}_i$, and loads, g_i and f_i, as

$$t_i(0) = \bar{t}_i + \frac{1}{l_i}\int_0^{l_i} g_i(s)(l_i - s)\,\mathrm{d}s, \qquad t_i(l_i) = \bar{t}_i - \frac{1}{l_i}\int_0^{l_i} g_i(s) s\,\mathrm{d}s,$$
$$q_i(0) = \bar{q}_i + \frac{1}{l_i}\int_0^{l_i} f_i(s)(l_i - s)\,\mathrm{d}s, \qquad q_i(l_i) = \bar{q}_i - \frac{1}{l_i}\int_0^{l_i} f_i(s) s\,\mathrm{d}s. \tag{7.3.8}$$

7.3.2 Governing equations and numerical method

The equations of force equilibrium at each membrane or internal node are expressed in terms of the element length, the time derivatives of the element length, the fluid loading generated by the external flow, and the internal pressure. The motion of the external fluid is assumed to satisfy the equations of Stokes flow, subject to the no-slip boundary condition at each membrane element. Other boundary conditions, such as pressure or velocity at flow boundaries, depend on the specific problem under consideration.

A finite-element method is used to solve the equations governing the fluid flow and cell motion. As already mentioned, the relatively large viscous resistance of RBCs to deformation allows cells to closely approach solid boundaries and other cells. The finite-element method is advantageous in that the flow inside narrow lubrication layers can be resolved with a fine mesh without requiring an equally fine mesh throughout the domain of flow.

The results presented in this chapter were obtained using the software package FlexPDE version 3.11 (PDE Solutions Inc., Antioch, CA). The code includes an automatic mesh generator that limits the element aspect ratio while ensuring adequate resolution in narrow regions between cells and other boundaries, as shown in figure 7.3.2.

In the simulations, the RBC shape is represented by 20 membranes nodes. The number of exterior fluid elements varies in the range 100–1000.

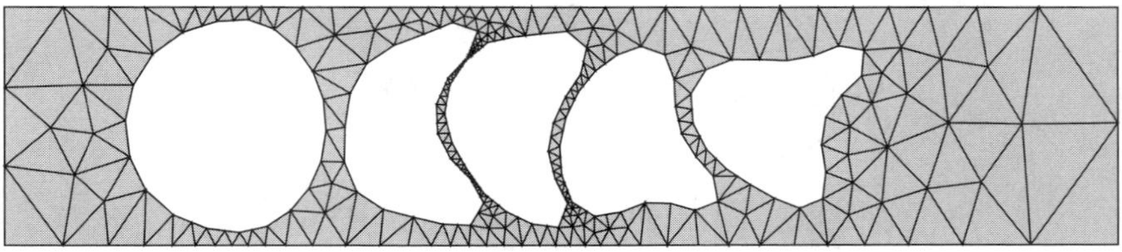

Figure 7.3.2 A typical finite-element mesh used to compute the fluid flow showing high mesh resolution in narrow gaps between cells.

The FlexPDE package allows scalar equations to be coupled with the partial differential equations solved. In our case, the equations of force equilibrium at each node are coupled with the equations of Stokes flow for the external fluid. The solution of the resulting system produces the velocity at each node, along with the velocity and pressure fields in the fluid. Further details are provided by Secomb *et al.* (2007).

The time integration is performed by an explicit method based on the computed nodal velocities. Either Euler integration or a trapezoidal predictor–corrector scheme is used (Barber *et al.* 2008). The computed sequence of cell shapes is visualized using still or moving images. In most cases, a time step of 1 ms is used. Smaller time steps are used in calculations at high shear rates and viscosities or when cells approach boundaries or other cells. The computational time depends on the mesh complexity. For the cases analyzed here, each time step requires 10–300 s of CPU time on a personal computer equipped with a 2 GHz processor.

The initial cell shape is a disk with area A_{ref}. At the initial instant, the length of each external element is less than the reference length, l_0. As the motion develops, the segments elongate and the cell perimeter increases leading to noncircular shapes with the same area. The use of a circular initial shape ensures that the emerging noncircular shapes represent true model predictions and are not the result of an assumed initial profile.

7.4 Tank-treading in simple shear flow

Red blood cells in a dilute suspension undergoing simple shear flow in a high-viscosity suspending fluid assume nearly steady elongated shapes aligned approximately with the direction of the flow. A small marker particle bound to the membrane exhibits periodic tank-treading motion around the cell perimeter (Fischer & Schmid-Schönbein 1977, Fischer *et al.* 1978).

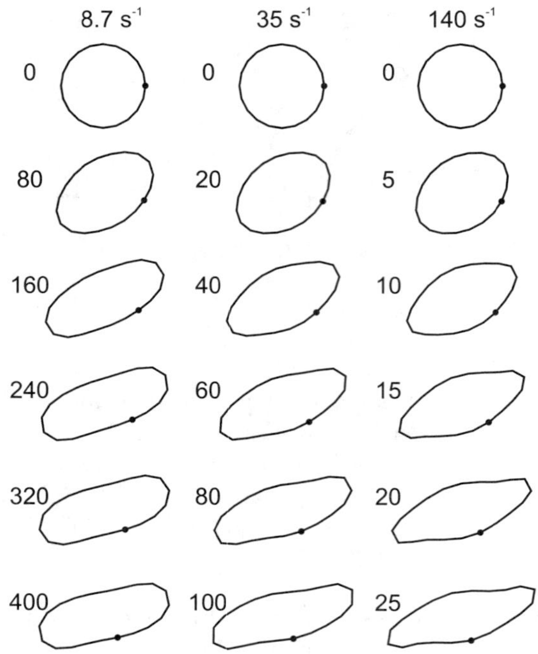

Figure 7.4.1 Predicted cell shapes for tank-treading motion in an ambient medium with viscosity 13 cP at shear rates 8.7, 35, and 140 s^{-1}. The fluid velocity is from left to right above each cell and from right to left below each cell. Time measured in ms since initiation of the motion is printed in each frame. A dot around each shape represents a point particle moving with the membrane velocity.

The two-dimensional model described in this chapter was used to simulate the cell motion. A comparison between experimental and predicted characteristics of the motion was made to refine the model parameters. Good agreement between the predicted and observed dependence of the cell length and tank-treading frequency on the imposed shear rate was obtained with dimensionless parameter values $k_t = 0.12$, $\mu_m = 200$, $\mu'_m = 100$, and $k_b = 0.9$. Predicted cell shapes computed with these parameter values are shown in figure 7.4.1.

The cells deform into elongated shapes aligned at a small positive angle with respect to the direction of the flow. This behavior is consistent with that predicted by earlier models assuming ellipsoidal cell shapes (Keller & Skalak 1982). As the shear rate is increased, the cell deformation becomes more pronounced. As the shear rate increases from 8.7 to 35 to 140 s^{-1}, the tank-treading frequency increases from 0.30 to 1.1 to 3.7 s^{-1}. At higher shear rates, the cell contour develops a counterclockwise twist at the extremities, as shown in figure 7.4.1. This feature is not evident in experimental studies of tank-treading motion, possibly because observations are typically made with an optical axis perpendicular to the moving surfaces used to drive the flow so that such deformations cannot be detected.

When a capsule undergoes stable tank-treading motion in simple shear flow, variations of the cell length and tank-treading frequency with shear rate are sensitive to the elastic and viscous properties of the capsule. The ability of a given model to produce close agreement with experimental findings is a stringent test of the model assumptions and parameters. In the case of RBCs, a detailed set of experimental data is available (Fischer & Schmid-Schönbein 1977, Fischer *et al.* 1978). It would be of interest to examine how well predictions of other published models agree with this data set.

7.5 Channel flow

The two-dimensional model can be used to simulate the motion of RBCs in microvessels. In the simplest case, a single RBC is convected in pressure-driven flow between two parallel walls, as show in figure 7.5.1. A cell initially placed at the centerline remains symmetric in shape and position, quickly developing a shape that is convex at the front and convex at the rear with a slight outward bulge of the membrane near the trailing edge (upper sequence). These features are consistent with observations of RBC shapes in capillaries (Skalak & Branemark 1969, Gaehtgens *et al.* 1980, Secomb 2003, Pries & Secomb 2008). The predicted cell shape resembles an observed shape described as parachute.

A cell initially placed near a channel wall undergoes a rapid transition to a tear-drop shape, as shown in the lower sequence of figure 7.5.1. The cell then migrates away from the wall toward the centerline and equilibrates to an asymmetric shape resembling a slipper shape often seen in the laboratory. The progression of shape change and migration are similar to those described for vesicles (Cantat & Misbah 1999). Transverse migration is of particular interest because it leads to the formation of a cell-free or cell-depleted layer near a wall. A particle with upstream to downstream symmetry cannot exhibit transverse migration due to reversibility of Stokes flow. The deformability of the cell and the developing shape are critical in determining the rate of transverse migration.

In the example shown in figure 7.5.1, migration toward the centerline occurs within the first few tens of μm, and is essentially completed within 100 μm of travel time. In a recent study, cell trajectories were predicted for a

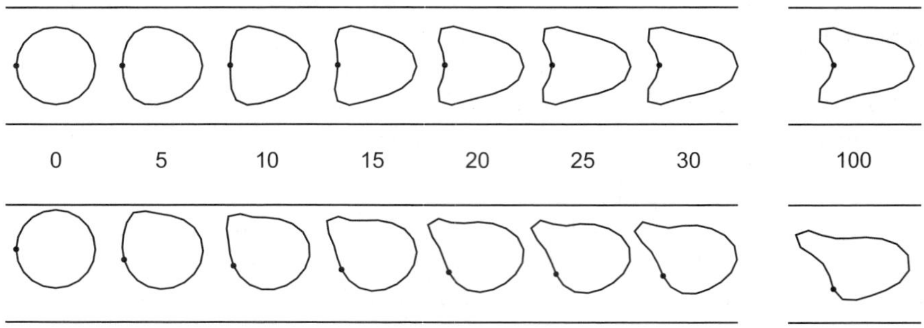

Figure 7.5.1 Simulated sequence of RBC shapes in an 8 μm channel. The suspending medium viscosity is 1 cP, and the flow velocity is approximately 1 mm/s. Time measured in ms since initiation of motion is shown for each frame. A dot on each shape indicates a point particle attached to the membrane. The distance between successive cell profiles is exaggerated for clarity. In the upper sequence, the cell is traveling along the centerline. In the lower sequence, the cell center is initially placed at a distance of 1 μm above the centerline.

specific microvessel geometry in the rat mesentery and compared with experimentally observed cell trajectories with good agreement (Secomb *et al.* 2007). Cells entering a microvascular segment at a bifurcation with asymmetric initial positions approach the centerline within about 30 μm of travel distance.

7.6 Motion through diverging bifurcations

Observations in the microcirculation reveal that the tube hematocrit, defined as the volume fraction of RBCs, varies greatly among vessel segments. A major reason for this variation is the nonproportional partition of cells and plasma in diverging microvessel bifurcations. The branch with the higher flow generally receives a larger hematocrit.

This behavior can be understood by considering the effect of a cell-free or cell-depleted layer near a vessel wall in the segment upstream from the bifurcation. RBCs clustered around the centerline of the upstream segment tend to follow the streamline of the centerline fluid, which normally enters the branch with the higher flow. The more strongly the cells are clustered, the more pronounced the unequal partitioning of plasma and RBCs. The transverse migration of RBCs in microvessels is a major determinant of hematocrit partition in bifurcations, as discussed in Section 7.5.

Once a red blood cell has entered a diverging bifurcation, it encounters a region of strongly varying flow. A cell may undergo large deformation, especially if the cell trajectory carries it near the dividing surface between the two daughter branches. A cell driven against this surface typically becomes flattened and may remain close to the surface for a long period of time. The process illustrated in figure 7.6.1 is commonly observed in blood flow in the microcirculation.

In the example illustrated in figure 7.6.1, a parent vessel of width 8 μm is divided into two daughter branches of equal width 6 μm. The flow rates are held constant, with a dimensionless flow rate of 8 entering the parent vessel, corresponding to mean velocity 1 mm/s, and outlet flow rates of 2 and 6 in the upper and lower daughter vessel. Simulations are presented for three initial cell positions, $y_0 = 1.0$, 1.1, and 1.2 μm from the centerline. These positions are close to the dividing streamline separating cells entering the upper and lower branches. When $y_0 = 1.0$ μm, the cell impinges on the flow divider and enters the lower branch with only a slight delay. When $y_0 = 1.1$ μm, the cell is flattened against the flow divider with one end pulled in each direction. Eventually, the cell enters the lower branch. In the third case, $y_0 = 1.2$ μm, the cell almost fills the entrance to the upper branch as soon as it impinges on the flow divider, and then enters the upper branch.

The numerical method was applied to to carry out a comprehensive study of cell trajectories in bifurcations (Barber *et al.* 2008). Two significant mechanisms were identified leading to deviations from streamlines. As a consequence of their deformability, RBCs tend to migrate toward the centerline upstream of the bifurcation, as already discussed. Cell migration accentuates the uneven partitioning of the hematocrit. However, another mechanism causing the opposite trend, termed daughter vessel obstruction, was identified. The dividing streamline of the underlying flow necessarily lies near the low-flow branch in a bifurcation. A cell traveling near this streamline may therefore intercept a large fraction of the flow entering that branch, and be drawn into it even though its center lies on a streamline that would otherwise enter the high-flow branch. This phenomenon also occurs in the case of rigid particles.

Hematocrit partitioning at a bifurcation is generally analyzed in terms of a Ψ-Φ plot, where Ψ is the volume fraction of the total flow entering a branch and Φ is the fraction of RBCs entering the same branch (e.g., Schmid-Schönbein *et al.* 1980). Lack of phase separation is indicated by a diagonal line with slope equal to 1 passing through the origin. In the configuration described in figure 7.6.2 after Barber *et al.* (2008), cells do not enter the branch labelled 1 for Ψ_1 values below a certain minimum value. Since the low flow branch merely skims plasma from the side of the main flow, the process

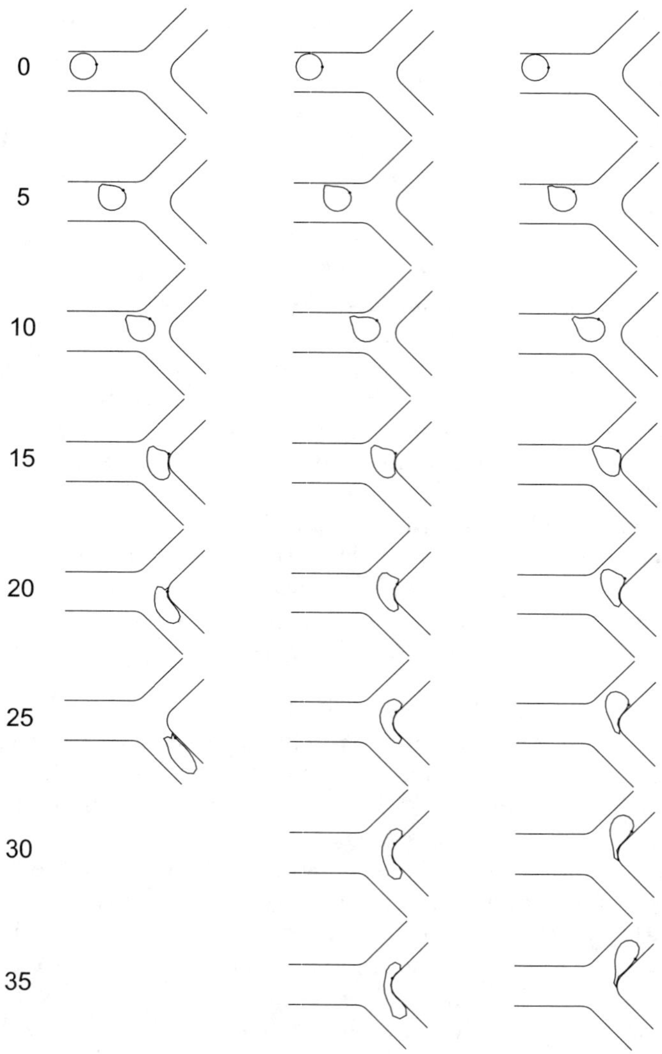

Figure 7.6.1 Time sequences of red blood cell shapes in a diverging bifurcation. Time measured in ms since initiation of motion is shown on the left. The flow direction is from left to right. The flow is divided 25% in the upper branch and 75% in the lower branch. The initial distance of the cell from the centerline is 1.0 μm (left column), 1.1 μm (center column), and 1.2 μm (right column). The suspending medium viscosity is 1 cP.

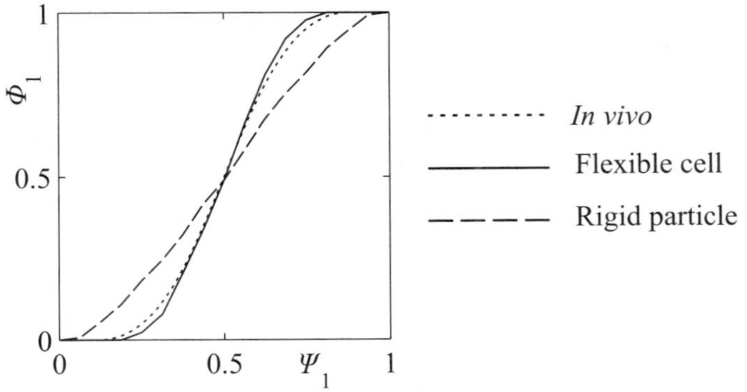

Figure 7.6.2 Partitioning of RBC fluxes in a symmetric diverging bifurcation: Ψ_1 is the volume fraction of the total flow branch 1, and Φ_1 is the fraction of RBCs entering that branch. The width of the parent vessel is 8 μm. The *in vivo* results are taken from Pries *et al.* (1989).

can be described as plasma skimming. Predicted results for flexible RBCs (solid line) are in good agreement with results obtained from an experimental study of hematocrit partitioning in diverging bifurcations of the rat mesentery (Pries *et al.* 1989). In the case of rigid disk-shaped particles, because upstream migration towards the centerline does not take place, a much weaker phase separation is observed.

7.7 Motion of multiple cells

The two-dimensional model is suitable for simulating the motion of multiple interacting cells. An example is the formation of trains of RBCs in capillaries resulting from the presence of white blood cells (WBCs), which are larger and more rigid than RBCs. Because white cells occupy a large fraction of the width of a capillary, they travel with a velocity that is nearly equal to the mean flow velocity. In contrast, RBCs travel near the centerline with a higher velocity. When red cells encounter a white cell in a capillary, they frequently slow down, as in traffic through a single-lane road behind a slow-moving vehicle. The result is the formation of a train of RBCs behind a WBC resulting in increased hematocrit and reduced velocity (Secomb *et al.* 1987).

The results of a simulation are shown in figure 7.7.1. A white blood cell is modeled as a larger and more rigid particle with diameter approximately 7.2 μm. Four RBCs are initially placed behind the WBC, slightly off the centerline. During the initial stage of the motion, the RBCs move closer to the centerline with an increasing streamwise velocity, which causes them to

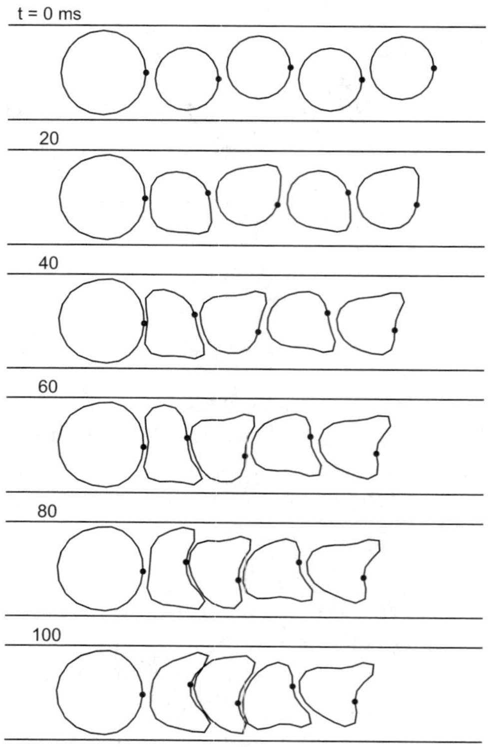

Figure 7.7.1 Train formation in an 8-μm channel. The leading white cell is larger and more rigid than the following red cells. Profiles are shown in a frame of reference moving with the leading cell with a velocity of approximately 0.6 mm/s to the left.

pile up behind the WBC. Accumulation causes the RBCs to spread across the channel, thereby reducing their velocity to match that of the WBC.

The numerical method can be applied to simulate flow in a channel with multi-file formations. By imposing periodic boundary conditions, the motion of an infinite array of cells can be simulated with affordable computational cost. When a cell crosses the domain boundary, it is divided into two non-contiguous parts. An example of the resulting computational domain and finite element mesh is shown in figure 7.7.2 where an array of three cells is periodically repeated along the channel. For computational efficiency, the periodic domain boundaries are realigned at each time step so that they pass through the cell boundary at previously defined cell membrane nodes.

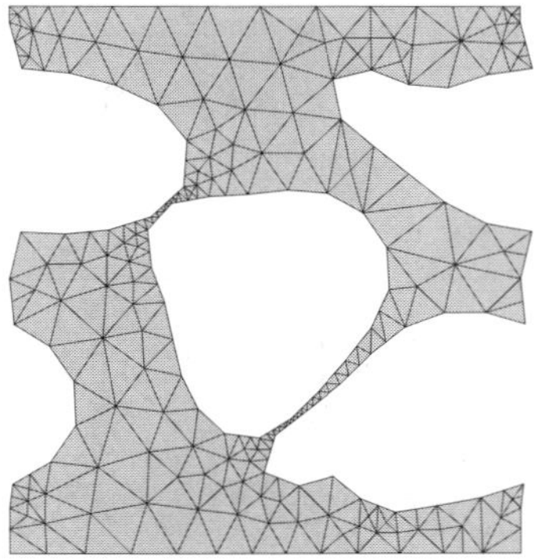

Figure 7.7.2 Typical finite-element mesh used to compute fluid flow with multiple cells in a channel. Periodic boundary conditions are imposed at the left and right boundaries. The boundaries are aligned to conform with nodal points along the cell membranes.

A simulation of a three-file arrangement in a 8-μm channel is shown in figure 7.7.3 The tube hematocrit, defined as the fraction of the domain occupied by cells, is close to 45%. Several interesting phenomena appear in this simulation. During the initial phase of the motion (0 to 40 ms), the cells obtain asymmetric slipper-like shapes while migrating away from the channel walls. Because the central cell is near the center of the channel, it travels faster than the other two cells, occasionally interrupted by interceptions with other cells. Interception brings two participating cells into close proximity, generating lubrication forces that prevent contact. The finite-element software generates a very fine mesh inside small gaps so that these lubrication forces can be adequately resolved, as shown in figure 7.7.2.

Cell-cell interactions tend to drive the outer two cells outwards toward the wall, opposing inward migration. The width of the cell-free layer is determined by the balance between these opposing tendencies (Secomb 2003). The motion is inherently unsteady and nonuniform, leading to spatial and temporal variations in the width of the cell-free layer. In regions where a red cell lies near the wall, the width of the cell-free layer is generally about 1 μm.

In the simulation shown in figure 7.7.3, the dimensionless pressure gradient is 0.2, corresponding to the physical value 2×10^4 dyn/cm^3. In the absence of cells, this pressure gradient would drive a plane Poiseuille flow with mean velocity 2.4 mm/s. During the simulated motion, the velocities of the cells show significant fluctuations. The velocity of the central cell varies in the range 1.5–1.8 mm/s. The two peripheral cells gradually fall behind the

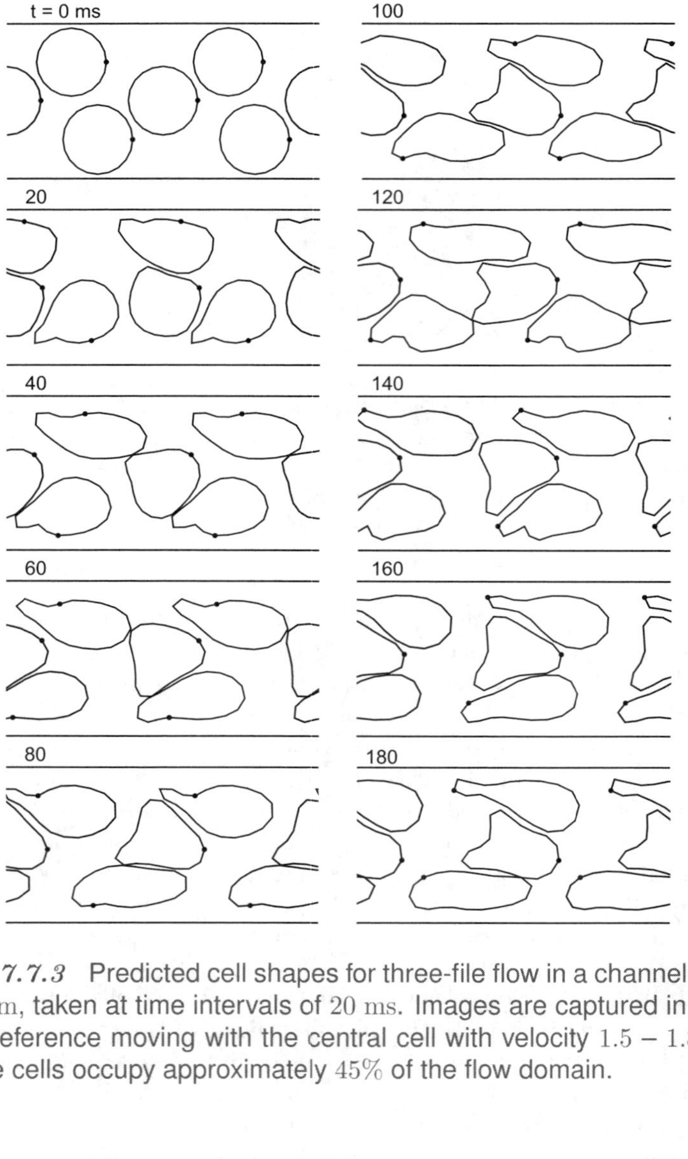

Figure 7.7.3 Predicted cell shapes for three-file flow in a channel of width $8\ \mu\mathrm{m}$, taken at time intervals of $20\ \mathrm{ms}$. Images are captured in a frame of reference moving with the central cell with velocity $1.5 - 1.8\ \mathrm{mm/s}$. The cells occupy approximately 45% of the flow domain.

central cell. Their average velocity relative to the central cell is less than 0.1 mm/s. This small relative velocity is reflected in the blunting of the velocity profile compared to the parabolic profile of the plane Poiseuille flow.

The dot around each RBC contour in figure 7.7.3 marks a material point particle attached to the membrane. The motion of this point particle illustrates the tank-treading of the two peripheral cells. The period of tank-treading motion is approximately 300 ms, corresponding to frequency 3.3 s^{-1}, which is much lower than the shear rate of the surrounding plasma flow at the channel walls, 1200 s^{-1}. The difference underscores the highly viscous constitution of the RBCs.

In the simulations presented in this section, interactions between cells are purely hydrodynamic. The phenomena observed in the simulations arise from the interaction between cellular mechanics and fluid mechanical forces generated by the flow. Real RBC membranes are coated with a layer of macromolecules (glycocalyx) few-tens of nm wide, which affects cell-cell interactions. In the presence of plasma proteins, aggregation of RBCs at low shear stress affects the flow properties (Bagchi *et al.* 2005). The endothelial cells lining living microvessels are coated with a relatively thick layer of macromolecules, on the order of 1 μm, termed the glycocalyx or endothelial surface layer. This layer has a major effect on the radial distribution of RBCs in microvessels and on the resistance to blood flow (Pries *et al.* 2000). Further investigation of these effects in the context of the present model requires further work.

7.8 Discussion

Mammalian red blood cells have a relatively simple mechanical structure in comparison with other types of cells. Lacking a nucleus and an internal cytoskeleton, they resemble capsules filled with a viscous fluid. The physical properties of the cell components have been characterized in considerable detail. It has long been recognized that RBCs undergo continuous deformation as they pass through the circulatory system and, particularly, the microcirculation. The cell mechanical properties are important in allowing blood to flow with a relatively low resistance despite it being a highly concentrated suspension. The formation of cell-free or cell-depleted layers near walls due to the deformability and lateral migration of RBCs is a major reason for the observed reduction in the apparent blood viscosity (Fåhraeus-Lindqvist effect) and tube hematocrit (Fåhraeus effect) in small tubes (Goldsmith *et al.* 1989).

Although the main features of blood flow have been identified in general terms for many decades, a thorough quantitative understanding of blood rheology and blood flow mechanics in the microcirculation remains elusive. For example, a theory that is able to predict the width of the cell-free layer from first principles in terms of the hematocrit and tube diameter is not available. Several factors contribute to this difficulty. The flow of concentrated suspensions is complicated by hydrodynamic interactions among a large number of particles. The specific mechanical properties of the individual cells pose analytical difficulties stemming from large disparities in values of relevant parameters when scaled by typical dimensions, velocities, and forces of interacting cells. Numerical approaches with sufficient resolution to represent hydrodynamic interactions with realistic mechanical properties in fully three-dimensional flow require a prohibitive computational cost.

In this chapter, we have discussed a two-dimensional model. The approach has obvious physical and physiological limitations. Because the fundamental singular solutions of Stokes flow are different in two and three dimensions, the range and decay of particle-particle and particle-wall interactions are only qualitatively described by the two-dimensional model. Even so, the examples presented in this chapter suggest that the two-dimensional approach can be surprisingly effective when applied to the motion of RBCs in confined geometries encountered in the microcirculation. The inclusion of internal viscous elements is critical for incorporating the effect of viscous dissipation. Cell shape and behavior predicted by the two-dimensional model show striking similarities with those observed in the laboratory. Good quantitative agreement is found on the radial migration of RBCs in a specific microvascular geometry (Secomb *et al.* 2007).

A quantitative understanding of blood rheology in the microcirculation ultimately requires a three-dimensional model for the realistic representation of multi-cell interactions. The results presented in this chapter provide guidance and motivation by identifying key cellular and fluid mechanical factors and by illustrating a range of phenomena occurring in the microcirculation where future models could be applied.

Acknowledgment

This research is supported by NIH Grant HL034555.

References

BAGCHI, P., JOHNSON, P. C. & POPEL, A. S. (2005) Computational fluid dynamic simulation of aggregation of deformable cells in a shear flow. *J. Biomech. Eng.* **127**, 1070–1080.

BARBER, J. O., ALBERDING, J. P., RESTREPO, J. M. & SECOMB, T. W. (2008) Simulated two-dimensional red blood cell motion, deformation, and partitioning in microvessel bifurcations. *Ann. Biomed. Eng.* **36**, 1690–1698.

BARNARD, A. C. L., LOPEZ, L. & HELLUMS, J. D. (1968) Basic theory of blood flow in capillaries. *Microvasc. Res.* **1**, 23–34.

CANTAT, I. & MISBAH, C. (1999) Lift force and dynamical unbinding of adhering vesicles under shear flow. *Phys. Rev. Lett.* **83**, 880–883.

EVANS, E. A. (1983) Bending elastic modulus of red blood cell membrane derived from buckling instability in micropipette aspiration tests. *Biophys. J.* **43**, 27–30.

EVANS, E. A. & HOCHMUTH, R. M. (1976) Membrane viscoelasticity. *Biophys. J.* **16**, 1–11.

EVANS, E. A. & SKALAK, R. (1980) *Mechanics and Thermodynamics of Biomembranes*, CRC Press, Boca Raton.

FÅHRAEUS, R. (1928) Die Strömungsverhltnisse und die verteilung der blutzellen im gefäßsystem. Zur frage der bedeutung der intravasculären erythrocytenaggregation. *Klin. Wochenschr.* **7**, 100–106.

FÅHRAEUS, R. & LINDQVIST, T. (1931) The viscosity of the blood in narrow capillary tubes. *Am. J. Physiol.* **96**, 562–568.

FISCHER, T. & SCHMID-SCHÖNBEIN, H. (1977) Tank tread motion of red cell membranes in viscometric flow: Behavior of intracellular and extracellular markers (with film). *Blood Cells* **3**, 351–365.

FISCHER, T. M., STOHR-LISSEN, M. & SCHMID-SCHÖNBEIN, H. (1978) The red cell as a fluid droplet: Tank tread-like motion of the human erythrocyte membrane in shear flow. *Science* **202**, 894–896.

GAEHTGENS, P., DÜHRSSEN, C. & ALBRECHT, K. H. (1980) Motion, deformation, and interaction of blood cells and plasma during flow through narrow capillary tubes. *Blood Cells* **6**, 799–812.

GAEHTGENS, P. & SCHMID-SCHÖNBEIN, H. (1982) Mechanisms of dynamic flow adaptation of mammalian erythrocytes. *Naturwissenschaften* **69**, 294–296.

GOLDSMITH, H. L., COKELET, G. R. & GAEHTGENS, P. (1989) Robin Fåhraeus: Evolution of his concepts in cardiovascular physiology. *Am. J. Physiol.* **257**, H1005–H1015.

HOCHMUTH, R. M. & WAUGH, R. E. (1987) Erythrocyte membrane elasticity and viscosity. *Annu. Rev. Physiol.* **49**, 209–219.

HSU, R. & SECOMB, T. W. (1989) Motion of nonaxisymmetric red blood cells in cylindrical capillaries. *J. Biomech. Eng.* **111**, 147–151.

KELLER, S. R. & SKALAK, R. (1982) Motion of a tank-treading ellipsoidal particle in a shear flow. *J. Fluid Mech.* **120**, 27–47.

LIGHTHILL, M. J. (1968) Pressure-forcing of tightly fitting pellets along fluid-filled elastic tubes. *J. Fluid Mech.* **34**, 113–143.

PRIES, A. R., LEY, K., CLAASSEN, M. & GAEHTGENS, P. (1989) Red cell distribution at microvascular bifurcations. *Microvasc. Res.* **38**, 81–101.

PRIES, A. R., NEUHAUS, D. & GAEHTGENS, P. (1992) Blood viscosity in tube flow: Dependence on diameter and hematocrit. *Am. J. Physiol.* **263**, H1770–H1778.

PRIES, A. R. & SECOMB, T. W. (2008) Blood flow in microvascular networks. In *Handbook of Physiology: Microcirculation*, Second Edition, Tuma, R. F., Duran, W. N. & Ley, K. (Eds.), Academic Press, San Diego, 3–36.

PRIES, A. R., SECOMB, T. W. & GAEHTGENS, P. (2000) The endothelial surface layer. *Pflugers Arch.* **440**, 653–666.

SCHMID-SCHÖNBEIN, G. W., SKALAK, R., USAMI, S. & CHIEN, S. (1980) Cell distribution in capillary networks. *Microvasc. Res.* **19**, 18–44.

SECOMB, T. W. (1987) Flow-dependent rheological properties of blood in capillaries. *Microvasc. Res.* **34**, 46–58.

SECOMB, T. W. (1988) Interaction between bending and tension forces in bilayer membranes. *Biophys. J.* **54**, 743–746.

SECOMB, T. W. (1995) Mechanics of blood flow in the microcirculation. *Symp. Soc. Exp. Biol.* **49**, 305–321.

SECOMB, T. W. (2003) Mechanics of red blood cells and blood flow in narrow tubes. In *Hydrodynamics of Capsules and Cells*, Pozrikidis, C. (Ed.) Chapman & Hall/CRC, Boca Raton, 163–196.

SECOMB, T. W., PRIES, A. R. & GAEHTGENS, P. (1987) Hematocrit fluctuations within capillary tubes and estimation of Fåhraeus effect. *Int. J. Microcirc. Clin. Exp.* **5**, 335–345.

SECOMB, T. W. & SKALAK, R. (1982*a*) A two-dimensional model for capillary flow of an asymmetric cell. *Microvasc. Res.* **24**, 194–203.

SECOMB, T. W. & SKALAK, R. (1982*b*) Surface flow of viscoelastic membranes in viscous fluids. *Quart. J. Mech. Appl. Math.* **35**, 233–247.

SECOMB, T. W., SKALAK, R., OZKAYA, N. & GROSS, J. F. (1986) Flow of axisymmetric red blood cells in narrow capillaries. *J. Fluid Mech.* **163**, 405–423.

SECOMB, T. W., STYP-REKOWSKA, B. & PRIES, A. R. (2007) Two-dimensional simulation of red blood cell deformation and lateral migration in microvessels. *Ann. Biomed. Eng.* **35**, 755–765.

SKALAK, R. & BRANEMARK, P. I. (1969) Deformation of red blood cells in capillaries. *Science* **164**, 717–719.

TIMOSHENKO, S. (1940) *Theory of Plates and Shells*, McGraw–Hill, New York.

TRAN-SON-TAY, R., SUTERA, S. P. & RAO, P. R. (1984) Determination of red blood cell membrane viscosity from rheoscopic observations of tank-treading motion. *Biophys. J.* **46**, 65–72.

ZARDA, P. R., CHIEN, S. & SKALAK, R. (1977) Interaction of viscous incompressible fluid with an elastic body. In *Computational Methods for Fluid-Solid Interaction Problems*, Belytschko, T. & Geers, T. L. (Eds.), American Society of Mechanical Engineers, New York, 65–82.

Multiscale modeling of transport and receptor-mediated adhesion of platelets in the bloodstream 8

N. A. Mody, M. R. King

Department of Biomedical Engineering
Cornell University
Ithaca, New York

Platelet blood cells are microscopic oblate spheroid-shaped particles whose transport and hemostatic functions are strongly affected by the hemodynamic flow environment. A platelet adhesive dynamics (PAD) computational algorithm has been developed to examine the platelet motion and adhesion near a vessel wall at an unprecedented level of spatial and temporal resolution. The numerical model integrates the three-dimensional hydrodynamics of multiple nonspherical cell motion near a wall with the dynamics of receptor–ligand binding. Simulations reveal that a platelet-shaped cell exhibits three distinct types of motion near a wall. Random Brownian motion is found to play a negligible role in the platelet motion, platelet-surface contact frequency and dissociative binding at physiological shear rates. Particle size and proximity of a boundary strongly influence particle collision trajectories, collision times, surface contact areas, and collision frequencies. Computational modeling of platelet aggregation via GPIbα–vWF–GPIbα bond bridges demonstrates the important effects of vWF multimer size, governing receptor–ligand binding kinetics, and nature of cell-cell collisions on the extent of the initial shear-induced thrombus formation. Inter-platelet bond force loading is predicted to be complex and highly nonlinear. The multiscale model is, to date, the most advanced and powerful predictive tool developed for elucidating platelet motion and adhesive phenomena near the vascular wall.

8.1 Introduction

In the language of fluid mechanics, blood flow in the human body is a pressure-driven flow of a concentrated suspension of blood cells, including red blood cells, white blood cells, and platelets, through a series of conduits (blood vessels) and junctions with varying shape and size. The primary role of blood flow is to mediate solute transport, delivery and uptake in the body. Blood flow also plays an important role in immune response, delivering white blood cells to inflamed areas throughout the body.

Many decades of experimental and theoretical work have been devoted to characterizing the constitution and flow properties of blood cells, thereby vastly improving our understanding of the mechanics of blood flow. For example, the distinctive shape of red blood cells (RBCs) allows them to non-reactively aggregate in flow. Red blood cells are highly deformable biconcave discs with diameter 6–8 μm and approximate thickness 2 μm. RBC aggregates are convected along the centerline of blood vessels, pushing leukocytes and platelets to the periphery. Consequently, blood segregates into an RBC-rich core surrounded by a concentric cell-depleted plasma layer near the wall. The tendency of RBCs to migrate toward the centerline is responsible for the Fåhraeus effect, where the apparent hematocrit (volume fraction of RBCs in blood) discharging from a vessel is greater than the overall systemic hematocrit.

The study of multitudinous interactions among blood cells and vessel walls is an interdisciplinary field bridging engineering and medicine. Progress contributes to our understanding of physiological processes, such as oxygen/solute transport, hemostasis, and inflammatory response; it helps determine the root causes of blood disorder, cardiovascular disease, and thrombosis; and it provides us with guidelines for the design of biomedical devices and prosthetics for therapeutic intervention.

Platelets are the smallest formed elements in blood. In their resting unactivated state, platelets are discoidal particles with approximate diameter 2 μm. Unactivated platelets, also called thrombocytes, are the most viscous of the three types of blood cells. Their cytoplasmic viscosity is two orders of magnitude higher than that of leukocytes (Haga *et al.* 1998), and their constitution is ten times more resilient to deformation compared to red blood cells (White 1988). When the integrity of the vascular endothelial wall is compromised and the underlying subendothelial layer is exposed to blood, platelets traveling past injured tissue activate and aggregate at the wound to form a clot.

8.1.1 Role of shear flow in platelet transport and function

Before addressing the role of hemodynamics in platelet function and aggregation, we briefly discuss the medicinal role of platelets. Hemostasis is a natural mechanism for the body to seal a vascular lesion in order to stop blood loss and repair injured tissue. The process of hemostasis involves a succession of events starting with platelet adhesion to an injured vascular endothelium, accumulation of platelets at the injured site, and activation of a coagulation cascade that converts fibrinogen to fibrin and thus permits the solidification and strengthening of a blood clot. Platelets display an abundance of surface receptors mediating platelet adhesion to cell surfaces. In the resting state, most surface adhesion receptors are inactive.

A disk-shaped platelet convected in plasma next to a vessel wall translates over an intact endothelium but rapidly attaches to injured tissue. The initial attachment can be reversible or permanent depending on the activation state of the platelet, the type of participating receptor–ligand bonds, and the prevailing shear rate. In arterioles, when the extracellular matrix components are suddenly exposed to flowing blood, a platelet initiates contact with injured tissue by way of reversible tether bonds developing between the platelet surface receptor GPIbα and the surface-bound plasma protein von Willebrand factor (vWF) (Coller *et al.* 1983, Sakariassen *et al.* 1979).

GPIbα–vWF bonds are weak noncovalent bonds mediating transient tethering of a platelet to a surface and stable rolling of a platelet on a surface in the direction of the flow. The tethering process slows down moving platelets so that they roll or translocate at a lower velocity than the velocity of the local ambient blood (Doggett *et al.* 2002). The formation of multiple transient platelet–surface tethers prolongs the duration of contact between the platelet and the surface and facilitates the binding of other slower forming irreversible bonds.

A captured platelet becomes activated during this process (Yuan *et al.* 1999). Activation causes a change in the platelet shape from ellipsoidal to globular with numerous filopodia-like protrusions. Surface adhesion receptors, such as $\alpha_{IIb}\beta_{III}$, also become activated. The $\alpha_{IIb}\beta_{III}$ receptor participates in the formation of firmly adherent bonds with the subendothelial surface, leading to strong platelet–surface attachment, and the formation of bonds between platelets to facilitate platelet aggregation (Weiss *et al.* 1989). Once a single layer of platelets has adhered to a surface, new layers of platelets accumulate on previously deposited layers to form a platelet thrombus.

The time scale of platelet adherence and accumulation is on the order of milliseconds. Platelet activation is accompanied by the release of granules containing pro-hemostatic chemical factors from the platelet interior, described as degranulation. Activation of the coagulation cascade results in thrombin production and fibrin deposition. Activated globular platelets in a mature thrombus contract their filopodia, flattening themselves to increase the clot density.

Shear flow plays an important role in the dynamics of hemostasis (Wootton & Ku 1999). Transport of platelets to the location of injury, dynamics of platelet attachment to an injured surface, cell activation, and embolization (shearing off of platelet aggregates, called emboli from the thrombus) are shear-linked mechanisms influencing platelet accumulation and thrombus formation. Shear rate and flow streamlines influence platelet trajectories, residence times near a wall, frequency of cell–wall contact, and cell–cell collision frequencies. Fluid shear stress contributes to the forces exerted on the wall and on surface-bound cells and cell aggregates. Shear stress is responsible for domain-level conformation changes of the cell surface receptor proteins and corresponding solution and surface bound ligands (Singh *et al.* 2009, Uff *et al.* 2002). Shear forces are necessary to promote GPIbα–vWF binding. Plausible postulated mechanisms are reviewed by Mody & King (2005).

In certain regions of the vasculature, such as heart valves, branches, stenoses, and aneurysms, complex flows, flow separation, and recirculation zones arise. Platelet deposition depends on the local fluid dynamics and can be correlated with the fluid shear stress, wall drag, and convective fluid motion towards a wall. Cell deposition reaches a maximum in areas of low wall shear stress and stagnation-point flows, and a minimum where high shear and flow separation are encountered (Perktold 1987, Pritchard *et al.* 1995, Schoephoerster *et al.* 1993). Karino & Goldsmith (1979) attributed pronounced platelet deposition to the geometry of the local streamlines. Curved streamlines with large radial velocity components can be partly responsible for flux of blood cells to the walls (Barber *et al.* 1998). Cells (platelets), solutes, and chemical signaling factors tend to accumulate inside recirculation zones. The dynamics of platelet adherence and aggregation proceeds much differently in these regions.

Atherosclerosis is a cardiovascular disease involving the deposition of fatty substances, such as LDL cholesterol and triglycerides, and proteins, such as fibrinogen and albumin, in arterial lesions or dysfunctional endothelium, leading to thickening and hardening of arteriolar wall. The focal occurrence of atherosclerosis in arterial regions of disturbed blood flow, such as branches, bifurcations, and curvatures, suggests a correlation between the location of atherosclerotic lesions and fluid dynamic factors including low mean wall shear stress, oscillatory wall shear stress, flow separation, velocity and velocity gradients, high cell residence time, and three-dimensional flow patterns (Fry 1976). The buildup of plaque on arterial walls leads to the progressive narrowing of the arterial lumen. Abnormally high wall shear stresses in excess of 80 dyn/cm^2 typically encountered in stenoses of arteries induce spontaneous platelet activation and aggregation in the absence of chemical agonists or vascular injury (Goto *et al.* 1998, 1995; Ikeda *et al.* 1993, Konstantopoulos *et al.* 1997, Shankaran *et al.* 2003). Two platelet surface receptors, GPIbα and $\alpha_{IIb}\beta_{III}$, and plasma protein vWF have been implicated in the formation of stable shear-induced platelet aggregates (Alevriadou *et al.* 1993). High shear-induced platelet activation and formation of large suspended platelet thrombi can block blood flow through the narrowed (stenosed) lumen, thereby impairing organ function and causing ischemic organ failure.

Pathological platelet aggregation involving the formation of thrombus in the absence of legitimate vascular injury is called thrombosis. Atherosclerotic lesions are high-risk sites for thrombosis. Rupture of plaque or secretion of inflammatory stimuli leads to the recruitment of platelets to a ruptured site and the formation of thrombus that can obstruct blood flow. Coronary angioplasty, a well-established procedure for widening a stenosis, can cause injuries to the arterial wall. Subsequent platelet deposition and thrombus formation can lead to the dangerous condition of restenosis of the treated vessel. Drug coated stents that supply anti-thrombotic agents at a controlled rate are used to combat restenosis. Exposure of blood to artificial surfaces can also trigger thrombosis. The design of cardiovascular devices, such as blood pumps, cardiopulmonary bypass devices, and prosthetic valves, must account for flow streamlines, cell residence times, and local shear stress in every region inside a device contacting blood. A main challenge in designing prosthetics and extracorporeal blood processing devices and in developing effective treatment routes is to limit thrombus growth during pathological platelet aggregation and prevent circumstances that trigger thrombosis.

By determining and quantifying the influence of the flow environment and the unique particle geometry on the initiation and progression of platelet tethering, activation, and aggregation events, we are better able to understand the physical factors promoting thrombotic events. Experimentally verified computational models can be used to predict platelet adhesive behavior for different types of flow, different cell sizes, and different binding kinetics such as that for a mutant receptor or a normal receptor. Measures can be devised to control or prevent pathological thrombosis and improve the condition of patients suffering from hemorhagic disorders.

8.1.2 Models of transport, deposition, and aggregation

Mathematical models have been developed to describe the dynamics of platelet motion and aggregation. The main purpose is to quantify interactions between flow field, bounding geometry, particulates in blood, and adhesive mechanisms. A common goal is to gain an understanding of how the flow environment can be favorably manipulated. Several investigators have attempted to model particulate blood flow through different vessel geometries in the presence or absence of particle deposition at the wall, neglecting interparticle hydrodynamics (Bluestein *et al.* 1999, Lei *et al.* 1997, Perktold 1987). Other studies have addressed the effect of shear flow on platelet activation and deposition, modeling platelet as infinitesimal point particles or using a species transport model where the platelets are treated as a reactive chemical species (Kuharsky & Fogelson 2001, Wootton *et al.* 2001). Because these models do not account for the finite size and shape of traveling and aggregating particles, they lack the spatial and temporal resolution necessary to study interactions at the cellular level.

In the majority of computational models of cellular (leukocyte and erythrocyte) flow, cell–wall adhesion, and cell–cell aggregation, the particles are idealized as spheres convected in an unbounded fluid, near a plane wall, and inside a tube (Hammer & Apte 1992, Helmke *et al.* 1998, Jadhav *et al.* 2005, King & Hammer 2001, 2001; King *et al.* 2001, 2005; Long *et al.* 1999, Tandon & Diamond 1998). Platelet aggregation has also been modeled by approximating the cells as perfect spheres (Mori *et al.* 2008, Shankaran & Neelamegham 2004, Tandon & Diamond 1997). Activated platelets possess a globular shape with a rough surface and numerous filopodia extending in different directions. On the contrary, resting platelets are disk-like shaped. The unactivated platelet shape can be approximated as an oblate spheroid with diameter 2 μm and aspect ratio 0.25 (Frojmovic *et al.* 1990). The cell shape is expected to influence the nature of interception, frequency of collision, collision contact time, collision contact area, and magnitude of shear and normal forces exerted on platelets and on interplatelet bonds or platelet–surface bonds. Thus, platelet motion, tethering characteristics, and rolling properties on a reactive surface are substantially different than those for leukocytes.

While extensive theoretical and experimental investigations have been conducted to delineate the motion of idealized spherical cells in bounded shear flow, only a limited number of studies are available for nonspherical particle motion in wall-bounded flow. Several studies have modeled platelets and red blood cells as nonspherical particles and investigated their motion in two-dimensional flow (AlMomani *et al.* 2008, Blyth & Pozrikidis 2009, Mody *et al.* 2005).

8.1.3 Motion of oblate spheroids in semi-infinite shear flow

The boundary geometry of a cellular flow problem depends on the specific biomedical problem being investigated. Red blood cell motion near a venular or arterial bifurcation is best addressed in a channel or tubular configuration

(Das *et al.* 1997, Pries *et al.* 1989). The diameter of arterioles is 10–100 μm, whereas the diameter of large blood vessels such as arteries is greater than 100 μm. Since the particle size is much smaller than the vessel diameter, the adhesive behavior of microscopic blood cells, such as platelets convected near an injured vessel wall, can be investigated by approximating a vessel surface with a planar wall. The curvature of the vessel is negligible and the flow can be approximated as simple shear flow with a linear velocity profile near the wall (Gavze & Shapiro 1997). Studies of spheroid motion near a flat wall have examined the effect of geometric parameters including aspect ratio, particle orientation with respect to the surface, and distance from the wall.

Hsu & Ganatos (1994) investigated the motion of a rigid oblate spheroid in linear shear flow near a plane wall. The plane containing the particle axis of symmetry was restricted to be parallel to the direction of the flow and perpendicular to the wall, allowing for plane symmetric flow. Instantaneous translational and rotational velocities of neutrally buoyant spheroids were calculated for zero-drag motion in terms of resistance coefficients encapsulated in a resistance tensor derived in an earlier study using a boundary-integral formulation (Hsu & Ganatos 1989). The resistance coefficients relate the hydrodynamic force and torque exerted on a particle to the particle translational and rotational velocities. The resistance tensor depends on the distance of the particle centroid from the wall, orientation angle of the particle major axis, and particle aspect ratio, λ. The particle velocities were integrated in time to generate trajectories for spheroids with aspect ratio $\lambda = 0.1$ and 0.5 and an initial distance from the wall equal to 1.25 times the spheroid major axis.

The results of Hsu & Ganatos (1994) showed that the centroid of an oblate spheroid dips towards the wall and then moves away from the wall as the platelet tumbles in the direction of the shear flow. The lateral motion becomes more pronounced as the particle aspect ratio becomes higher. Particle trajectories follow a periodic pattern where the particle reaches the same initial distance from the wall after a 180° rotation. In unbounded shear flow, the particle does not experience lateral motion (Jeffery 1922). This behavior should be contrasted with that exhibited by a freely-suspended sphere rotating and translating parallel to a plane wall in the absence of lateral migration (Goldman *et al.* 1967).

Pozrikidis (1994) used a highly accurate boundary-integral method to determine the forces and torques acting on oblate spheroids in the geometrical setting of Hsu & Ganatos (1989, 1994). Calculations were carried out for particle aspect ratios ranging from 1 to 0.1, and for particle–wall separations down to 0.1 times the particle major axis for a spheroid whose major axis is oriented parallel to the wall. The numerical results revealed that the torque experienced by an oblate particle increases significantly as the particle obtains a flatter disk-like shape. An oblate body of revolution undergoes steady translation parallel to the wall without lateral motion or rotation in a narrow range of aspect ratios, $(1 > \lambda > 0.8)$, for specifically two different distances of the spheroid from the wall.

In a more recent investigation of the three-dimensional motion of a prolate spheroid near an infinite plane wall, Pozrikidis (2005) presented a brief study of wall effects on the out-of-plane motion of an oblate spheroid with aspect ratio 0.25. The simulations revealed that, at a critical tilt angle of the platelet axis of revolution about the plane of flow, the platelet rolls on its side when the platelet centroid is below a critical initial height. The effect of the wall on particle motion is greater for oblate than for prolate spheroids.

A limited number of studies have addressed the motion of two or more spheroids in shear flow in low-Reynolds number flow. Pozrikidis (2006) investigated the interception of two prolate spheroids in unbounded linear shear flow using a spectral boundary-element method. An integral equation of the second kind consisting of the double-layer potential was solved to produce the particle translational and angular velocities. The results demonstrated that a net particle displacement in the direction of shear, transverse to the direction of the flow, occurs following interception. Yoon & Kim (1990) developed a least-squares collocation method as a variant of the multipole moment method to solve the mobility problem for two spheroids suspended in linear shear flow. Claeys & Brady (1993) used a Stokesian dynamics method to study the motion of prolate spheroids and particle chains for several configurations.

8.1.4 Preamble

In recent years, we have conducted extensive numerical studies to provide a detailed description of the three-dimensional motion of platelet-shaped spheroids in linear shear flow, particularly concentrating on the following issues: influence of a wall on flow trajectories before and after platelet–surface contact; opportunity for platelet–surface contact; mechanics of interparticle collision; effect of particle size and shape and bounding surfaces on platelet adhesion and aggregation. In this chapter, the latest research developments elucidating the effect of particle size and shape on the mechanics of platelet motion and the physical of platelet-surface interactions and platelet–platelet collisions near a wall in linear shear flow are reviewed.

In Section 8.2, a boundary-integral formulation of three-dimensional low-Reynolds-number flow past oblate spheroidal particles representing unactivated platelets near a plane wall is outlined. The calculation of the particle motion is carried out using the completed double-layer boundary-integral equation method (CDL-BIEM) proposed by Kim & Karilla (1991). We follow Phan-Thien *et al.* (1992) and account for the presence of a wall by using the half-space Green's function automatically satisfying the no-slip boundary condition on the wall.

In section 8.1.3, the behavior of platelet-shaped particles near a flat surface in linear shear flow is reviewed. The role of Brownian motion and the dynamics of platelet-surface adhesion at normal shear rates encountered in blood is discussed in section 8.4. The influence of particle shape and proximity of a wall on particle collision is addressed in section 8.5. A three-dimensional multiscale model is used to explore the mechanics of transient aggregation by GPIbα–vWF–GPIbα bridging of two unactivated platelet-shaped cells in linear shear flow near a wall. The outcome of the reactive collisions is discussed section 8.6.

In our discussion, unless otherwise noted, the terms spheroid, platelet, particle, and cell, all refer to an oblate spheroid.

8.2 Mathematical framework

The hydrodynamic problem considered addresses the motion of neutrally buoyant rigid oblate spheroids in a semi-infinite domain of flow bounded by an infinite plane located at $z = 0$, as shown in figure 8.2.1. The particle size is $2 \times 2 \times 0.5$ μm^3, and the platelet major semi-axis is $a = 1\mu m$. The Reynolds number is estimated to be $\mathrm{Re} = \dot{\gamma}\rho a^2/\mu = O(10^{-3}) < 1$, where $\dot{\gamma} = 1{,}000$ s^{-1} is the shear rate, $\rho = 1.0239$ g/cm^3 is the plasma density and $\mu = 1.2$ cP is the plasma viscosity. Typical shear rates in the arterioles range from 500 to 1600 s^{-1} (Kroll *et al.* 1996).

The motion of the fluid is well within the Stokes flow regime governed by the Stokes equation and the continuity equation,

$$\nabla p = \mu \nabla^2 \mathbf{u}, \qquad \nabla \cdot \mathbf{u} = 0, \tag{8.2.1}$$

where $\mathbf{u}$ is the velocity and p is the pressure. The velocity of the the ambient linear simple shear flow is

$$u_x^\infty = \dot{\gamma} z, \qquad u_y^\infty = 0, \qquad u_z^\infty = 0, \tag{8.2.2}$$

where $\dot{\gamma}$ is the shear rate and z is the distance from the wall (figure 8.2.1). The wall velocity is required to be zero due to the no-slip and no-penetration boundary conditions, $\mathbf{u}(z = 0) = \mathbf{0}$. At the surface of the kth particle, the velocity expresses rigid-body motion,

$$\mathbf{u} = \mathbf{V}^{(k)} + \boldsymbol{\Omega}^{(k)} \times (\mathbf{x} - \mathbf{x}^{(k)}), \tag{8.2.3}$$

where $\mathbf{V}^{(k)}$ is the translational velocity, $\boldsymbol{\Omega}^{(k)}$ is the angular velocity, and $\mathbf{x}^{(k)}$ is the particle centroid. We are interested in the mobility problem where

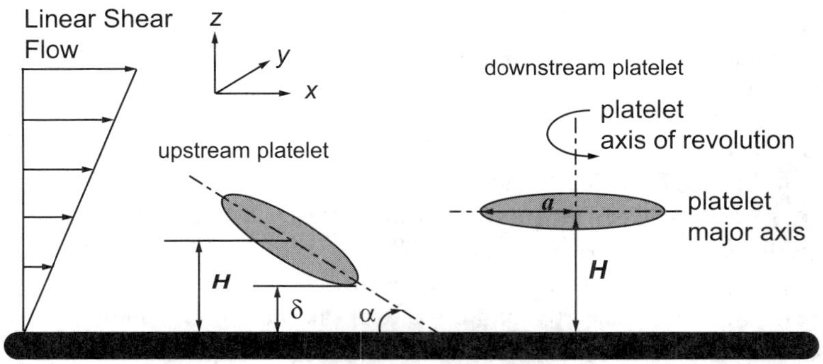

Figure 8.2.1 Schematic illustration of two oblate spheroids with aspect ratio $\lambda = 0.25$ translating and rotating in shear flow near an infinite plane wall. The major axis, a, centroid height H, and closest distance from the wall, δ, are defined in the illustration. At the instant shown in the figure, the major axis of the upstream platelet is oriented at an angle α with respect to the surface measured about the y axis in the clockwise direction. In this illustration, both spheroids have zero tilt about the x axis.

particle forces and torques are specified and the translational and rotational velocities are to be determined as part of the solution. The completed double-layer boundary-integral equation method (CDL-BIEM) is used to solve the governing equations and compute the platelet motion (Kim & Karilla 1991, Phan-Thien *et al.* 1992).

8.2.1 The CDL–BIEM method

The integral representation of Stokes flow provides us with an expression for the velocity at a point inside a selected domain of flow, $\mathbf{x}_0$, bounded by a surface, D,

$$u_j(\mathbf{x}_0) = -\frac{1}{8\pi\mu}\iint_D \mathcal{G}_{ji}(\mathbf{x}_0,\mathbf{x})\, f_i(\mathbf{x})\, \mathrm{d}S(\mathbf{x}) + \frac{1}{8\pi}\iint_D u_i(\mathbf{x})\, \mathcal{T}_{ijk}(\mathbf{x},\mathbf{x}_0)\, n_k(\mathbf{x})\, \mathrm{d}S(\mathbf{x}), \qquad (8.2.4)$$

where $\mathbf{f} = \mathbf{n}\cdot\boldsymbol{\sigma}$ is the boundary traction, $\mathbf{n}$ is the unit normal vector pointing into the ambient fluid, $\boldsymbol{\sigma}$ is the stress tensor, $\mathcal{G}_{ij}(\mathbf{x},\mathbf{x}_0)$ is the Green's function of Stokes flow for the velocity, and $\mathcal{T}_{ijk}(\mathbf{x},\mathbf{x}_0)$ is the corresponding Green's function for the stress. In our case, the integration is carried out over a bounding surface consisting of the union of the plane wall, W, and the surface of the kth particle, $P^{(k)}$, where $k = 1, \ldots, N$ and N is the number of particles. The first and second integrals in the right-hand side of (8.2.4) is the single- and double-layer potential of Stokes flow. By using integral identities, we find that the disturbance flow due to a particle undergoing rigid-body-motion can be described in terms of a single-layer potential alone (e.g., Phan-Thien *et al.* 1992). A Fredholm integral equation of the first kind for the traction arises by placing the point $\mathbf{x}_0$ at the particle contour (Kim & Karilla 1991).

A more efficient integral formulation involves the double-layer potential alone,

$$u_j(\mathbf{x}_0) = u_j^\infty(\mathbf{x}_0) + \frac{1}{8\pi}\iint_D \phi_i(\mathbf{x})\, \mathcal{T}_{ijk}(\mathbf{x}, \mathbf{x}_0)\, n_k(\mathbf{x})\, \mathrm{d}S(\mathbf{x}), \tag{8.2.5}$$

where $\boldsymbol{\phi}$ is an unknown surface density distribution and $\mathbf{u}^\infty$ is the unperturbed velocity prevailing in the absence of particles. However, the double-layer representation, by itself, is not able to mediate a force or torque. Power & Miranda (1987) addressed this shortcoming by completing the deficient range of the double-layer potential. This is done by complementing the integral representation with the velocity field due to point forces and point torques placed at the particle centroids,

$$\begin{aligned} u_j(\mathbf{x}_0) = u_j^\infty(\mathbf{x}_0) &+ \sum_{\alpha=1}^{N}\left(F_i^{(\alpha)} - \frac{1}{2}(\mathbf{T}^{(\alpha)} \times \nabla)\right)\mathcal{G}_{ij}(\mathbf{x}_0, \mathbf{x}_c^{(\alpha)}) \\ &+ \frac{1}{8\pi}\iint_D \phi_i(\mathbf{x})\, \mathcal{T}_{ijk}(\mathbf{x}, \mathbf{x}_0)\, n_k(\mathbf{x})\, \mathrm{d}S(\mathbf{x}), \end{aligned} \tag{8.2.6}$$

where $\mathbf{F}^{(\alpha)}$ is the force and $\mathbf{T}^{(\alpha)}$ is the torque exerted on the α particle at the volume centroid, $\mathbf{x}_c^{(\alpha)}$.

Evaluating (8.2.6) at the surface of the β particle, enforcing the rigid-body-motion boundary condition and rearranging, we obtain an integral equation of the second kind for $\boldsymbol{\phi}$,

$$\begin{aligned} &V_j(\mathbf{x}_0) + \left(\boldsymbol{\Omega}^{(\beta)} \times (\mathbf{x}_0 - \mathbf{x}_c^{(\beta)})\right)_j - u_j^\infty(\mathbf{x}_0) \\ &= \sum_{\alpha=1}^{N}\left(F_i^{(\alpha)} - \frac{1}{2}(\mathbf{T}^{(\alpha)} \times \nabla)\right)\mathcal{G}_{ij}(\mathbf{x}_0, \mathbf{x}_c^{(\alpha)}) \\ &\quad + \frac{1}{2}\phi_j(\mathbf{x}_0) + \frac{1}{8\pi}\iint_D^{PV} \phi_i(\mathbf{x})\, \mathcal{T}_{ijk}(\mathbf{x}, \mathbf{x}_0)\, n_k(\mathbf{x})\, \mathrm{d}S(\mathbf{x}). \end{aligned} \tag{8.2.7}$$

The null space of the linear integral operator comprised of the right-hand side of (8.2.7) contains $6N$ eigenfunctions representing rigid-body-motion,

$\phi_i^{(\alpha)}$, for $\alpha = 1, \ldots, N$ and $i = 1, \ldots, 6$. Accordingly, $6N$ additional linearly independent equations must be introduced to determine the double-layer density and obtain a unique solution. Power & Miranda (1987) coupled the forces and torques acting at the center of the particle with the null solutions at the particle surfaces,

$$F_i^{(\alpha)} = \iint_{P^{(\alpha)}} \boldsymbol{\phi} \cdot \boldsymbol{\phi}_i^{(\alpha)} \, \mathrm{d}S, \qquad T_i^{(\alpha)} = \iint_{P^{(\alpha)}} \boldsymbol{\phi} \cdot \boldsymbol{\phi}_{i+3}^{(\alpha)} \, \mathrm{d}S \tag{8.2.8}$$

for $i = 1, 2, 3$, where $\phi_1^{(\alpha)}$, $\phi_2^{(\alpha)}$, and $\phi_3^{(\alpha)}$ are orthonormalized null solutions expressing translation, and $\phi_4^{(\alpha)}$, $\phi_5^{(\alpha)}$, and $\phi_6^{(\alpha)}$ are orthonormalized null solutions expressing rotation. Substituting (8.2.8) in (8.2.7), we obtain an integral equation of the second kind with a unique solution. Once the double-layer density $\boldsymbol{\phi}$ has been determined, the particle surface velocity can be obtained as

$$\mathbf{u}(\mathbf{x}) = -\sum_{i=1}^{6} \boldsymbol{\phi}_i^{(\alpha)} \iint_{P^{(\alpha)}} \boldsymbol{\phi} \cdot \boldsymbol{\phi}_i^{(\alpha)} \, \mathrm{d}S. \tag{8.2.9}$$

The particle velocity of translation and angular velocity or rotation are extracted by taking the inner product of (8.2.9) with the individual eigensolutions.

The spectrum of the integral operator arising from (8.2.7) and (8.2.8) contains a marginal eigenvalue equal to -1, preventing a solution by the method of successive substitutions. The eigenfunctions of the adjoint integral operator are unit normal vectors over each particle. Following Kim & Karilla (1991), we deflate the spectrum of the integral equation by shifting the marginal eigenvalue to zero. The modified integral equation is then solved by the method of successive substitutions. The presence of the wall is incorporated by employing the half-space Green's function (Phan-Thien *et al.* 1992).

8.2.2 Repulsive contact force

Particle and wall surfaces are coated with a steric layer of thickness 25–50 nm modeling the glycocalyx and natural roughness of the cell surface (Haga *et al.* 1998). Contact or repulsion between the surfaces results from the interaction of the outer edges of these rough layers. In our model, it is assumed that the surface roughness layer allows fluid to easily pass through a surface layer causing a negligible disturbance to the existing flow.

A short-range repulsive force between the surfaces is included to account for nonspecific short-range interactions, namely electrostatic repulsion developing when cell membranes, glycocalices, or glycoproteins come into contact within a distance of less than 20 nm (Lauffenburger & Linderman 1993).

The contact force is expressed by an empirical relationship of the form

$$F_{rep} = F_0 \, \chi \, \frac{e^{-\chi\epsilon}}{1 - e^{-\chi\epsilon}}, \tag{8.2.10}$$

where $F_0 = 500$ pN m, $\chi = 2000\ \mu\text{m}^{-1}$, and ϵ is the surface-to-surface separation based on the distance between the tips of opposing surface roughness layers (King & Hammer 2001). The magnitude of the repulsive force is significant at surface separations smaller than 20 nm. For example, at 15 nm separation, the repulsive force is 0.0374 pN, which is approximately twice the Stokes drag force on a 1 μm diameter sphere sedimenting with velocity 1 μm/s in a fluid with 1 cP viscosity.

Bell *et al.* (1984) also used an exponential relation to empirically model a distance-dependent nonspecific interaction energy causing a net repulsion between cell-cell surfaces. Except for the specific purpose of calculating repulsive forces between two opposing surfaces, the separation distance between the two surfaces is measured from the edges of the real surfaces, not from the outer edges of the steric layers on these surfaces. The short-range repulsive force between two platelets is directed along a line segment connecting two points (nodes) on either one of two opposing particle surfaces in closest proximity.

8.2.3 Numerical implementation

The particle size is fixed at $2 \times 2 \times 0.5\ \mu\text{m}^3$ with platelet major radius $a = 1$ μm. In nonreactive single platelet simulations, the platelet surface is divided into 96 QUAD9 boundary elements. The procedure involves discretizing each face of a cube into sixteen equal square elements, where each element has one node at the center and three nodes along each edge, totaling nine nodes per element. The nodes are then projected radially onto a circumscribing concentric sphere. The discretized sphere is finally compressed along one axis to become an oblate spheroid. In simulations with multiple platelets, or when adhesive interactions are present, the surface mesh is further refined into 384 QUAD9 elements described by 1538 nodes.

The time step is chosen so that the results are virtually independent of the time step. Depending on the shear rate, presence of adhesive reactions determining the formation and breakup of stiff bonds, strength of the flow, and Brownian motion, the time step employed ranges from 10^{-1} to 10^{-10} s.

Unactivated platelets immersed in a linear shear flow near a surface spend most of their time with their major axes aligned parallel to the surface in the horizontal orientation. The platelet angular velocity in linear shear flow is a highly nonlinear function of the orientation. The rate of rotation is the highest when the platelet axis of revolution is nearly parallel to the shear flow. Thus, the platelets remain in nonhorizontal orientations for relatively short periods of time compared to the time spent with their long axis oriented horizontally with respect to the surface (Mody & King 2005). Except when indicated in section 8.2.3, all simulations are initiated with platelet orientation such that the axis of revolution is perpendicular to the surface, $\alpha, \beta = 0$.

Gravitational forces are of the same order of magnitude as Brownian forces. In order to isolate the effect of fluid flow, gravitational forces are not included in the simulations. Since the gap size below which lubrication forces become important is less than the thickness of the surface roughness layer, 25–50 nm, modifications of the basic procedure are not necessary. Other boundary-integral simulations of oblate spheroid motion near a wall have also disregarded lubrication forces for similar reasons (Hsu & Ganatos 1994, Maul *et al.* 1994, Pozrikidis 2005, 2006). Additional information on the numerical implementation will be provided in each section of this chapter, as necessary.

8.2.4 Validation of half-space CDL-BIEM for oblate spheroid motion

Maul *et al.* (1994) carried out experimental and computational studies of the sedimentation of hexagonal aluminum flakes near a wall in low-Reynolds-number flow. Their investigations verified the reliability of the CDL-BIEM formulation in predicting particle-wall hydrodynamic interactions for particle shapes with two disparate length scales. Their numerical predictions are in good agreement with experimental results on platelet sedimentation. We have compared the motion of a platelet with aspect ratio 0.0526 near a wall as predicted by the numerical method described by Phan-Thien *et al.* (1992) with numerical and experimental results by Maul *et al.* (1994) with excellent agreement. Our predictions of the three-dimensional platelet motion near a wall in quiescent fluid and in linear shear flow (Mody & King 2005) compare favorably with solutions by Kim *et al.* (2001) and Jeffery (1922).

8.3 Motion of an oblate spheroid near a wall in shear flow

When a bounding wall is at least several major radii away from the particle surface, the hydrodynamic influence of the wall is minimal and the particle trajectories are similar to those observed in an unbounded fluid. Far from the wall, an oblate spheroid suspended in linear shear flow executes Jeffery (1922) orbits characterized by tumbling motion with a fixed periodicity in the direction of flow. Lateral motion transverse to the flow does not take place. An oblate spheroid moving near a plane wall exhibits three distinct regimes of flow depending on the distance of the particle centroid from the wall (Mody & King 2005).

8.3.1 Regime I: Modified Jeffery orbits

A platelet whose centroid is placed at a distance greater than 1.2 times the platelet radii exhibits periodic rotational motion while translating in the direction of flow, exhibiting a tumbling motion. An important difference between the motion of a spheroid in unbounded shear flow and near a wall is that the proximity of the wall introduces an asymmetry in the flow field, and this causes the particle centroid to periodically move toward and then away from the wall as it translates above the wall.

Figure 8.3.1(a) demonstrates this behavior for initial orientation $\alpha = 0$ and centroid distance from the wall equal to 1.2, 1.4, and 2.4 platelet radii. In this flow regime, a platelet exhibits lateral motion normal to the wall while rotating and translating in the direction of shear flow (Hsu & Ganatos 1994, Yang & Leal 1984). The lateral motion becomes more pronounced as the particle is placed closer to the wall. It is interesting that the platelet centroid remains consistently at or below its initial starting height. Figure 8.3.1(b) shows that the rotational velocity decreases due to presence of the wall compared to that observed in unbounded flow. We call this regime, Regime I: Modified Jeffery orbits.

8.3.2 Regime II: Pole vaulting and periodic tumbling

A platelet initially oriented parallel to the wall, $\alpha = 0$, with its centroid located at a distance between $1.1a$ and $0.75a$ from the wall, is found to initially move toward the wall under the influence of the shear flow. As the platelet approaches the wall while rotating, it makes contact with the wall at an oblique angle. As soon as the platelet contacts the wall, the centroid is rapidly pushed away to a greater final height as in pole vaulting. The occurrence of pole vaulting was first noted by Stover & Cohen (1990) for rod-like particles undergoing tumbling motion near a wall. These authors describe pole vaulting as an irreversible process where the particle centroid is irreversibly shifted to a height above the surface approximately equal to the particle radius along the major axis. In the case of a rod-like particle, the major axis is equal to half the particle length.

Figure 8.3.2(a) shows trajectories of the z position of the platelet centroid for four initial platelet heights from the wall, $0.75a$, $0.8a$, $1.0a$, and $1.1a$. All trajectories collapse onto a single curve after the first contact (pole vaulting) is made. Physically, once pole vaulting has been completed, the subsequent tumbling motion follows a trajectory that is the same, irrespective of the initial particle height and orientation. The pole vaulting motion is the greatest for a platelet with initial height $0.75a$ from the surface. Figure 8.3.2(b) shows the evolution of the closest separation of the platelet from the wall, δ.

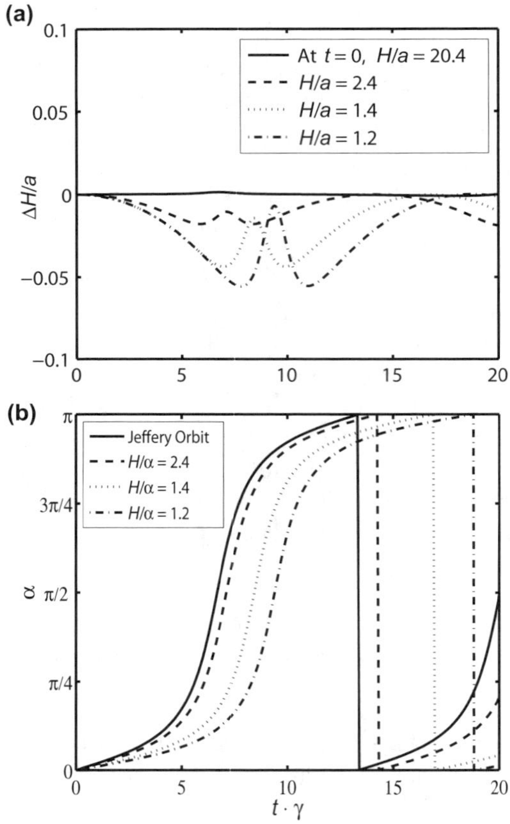

Figure 8.3.1 Plots demonstrating the translational and rotational motion of an oblate spheroid with aspect ratio $\lambda = 0.25$ in Regime I. (a) Evolution of the dimensionless centroid height H/a above the wall scaled by the initial value and (b) rotational trajectory (angular orientation) for varying initial heights of the platelet centroid from the wall. *Reprinted with permission from Mody & King (2005).*

Figure 8.3.3(a) shows graphs of the gap, δ, against the orientation angle about the y axis for a platelet initially placed at a distance $1.0a$ above the wall. The combined thickness of the roughness layer on both surfaces, 0.1 μm, has been subtracted from the closest distance to isolate the clearance between the roughness layers of the two surfaces. The platelet makes contact with the surface at $\alpha \simeq 124°$ and detaches from the surface after crossing the $\alpha = 90°$ angle. The clearance is determined by short-range repulsive forces between the two surfaces. The tip of the particle in contact with the surface

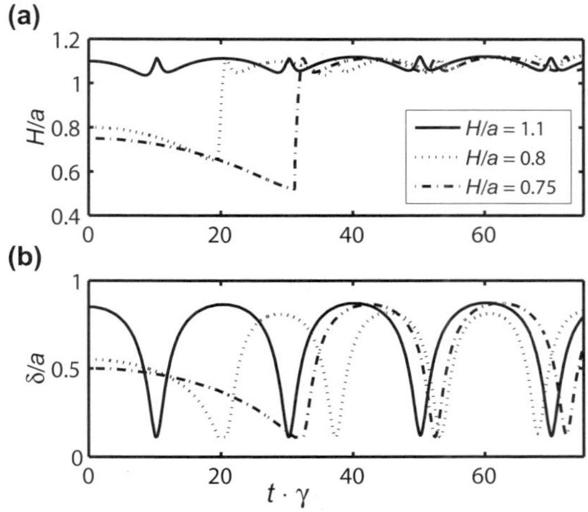

Figure 8.3.2 History of the (a) dimensionless height of the platelet centroid and (b) dimensionless closest separation of the platelet surface from the wall for three different initial platelet centroid heights $H/a =$ 0.75, 0.8, and 1.1. The particle and wall surfaces are coated with a steric layer of thickness 50 nm. *Reprinted with permission from Mody & King (2005).*

is not stationary but constantly translates in the direction of the flow. Three-dimensional views of the platelet starting at height $0.8a$ from the surface are shown in figure 8.3.3(b) during a single rotation.

Figure 8.3.4 illustrates the platelet motion over several tumbling cycles. The platelet repeatedly contacts the surface once every half rotation. After a first contact has been made, the platelet is pushed off the surface so that its centroid is irreversibly elevated to a distance slightly greater than $1.1a$, approximately $1.13a$, from the surface. Following pole vaulting, the platelet undergoes periodic translation and rotation, always attaining the height approximately $1.13a$ when the orientation angle α is zero. Thus, in this regime, spheroid contact with the surface is transient and occurs repeatedly without a net drift away from the wall.

8.3.3 Regime III: Wobble flow

A platelet with initial orientation $\alpha = 0$ and centroid distance less than $0.75a$ from the wall exhibits oscillatory or wobble motion. Instead of contacting the surface as in Regime II flow, the platelet exhibits periodic lateral motion where the major axis oscillates about the horizontal direction. A platelet that

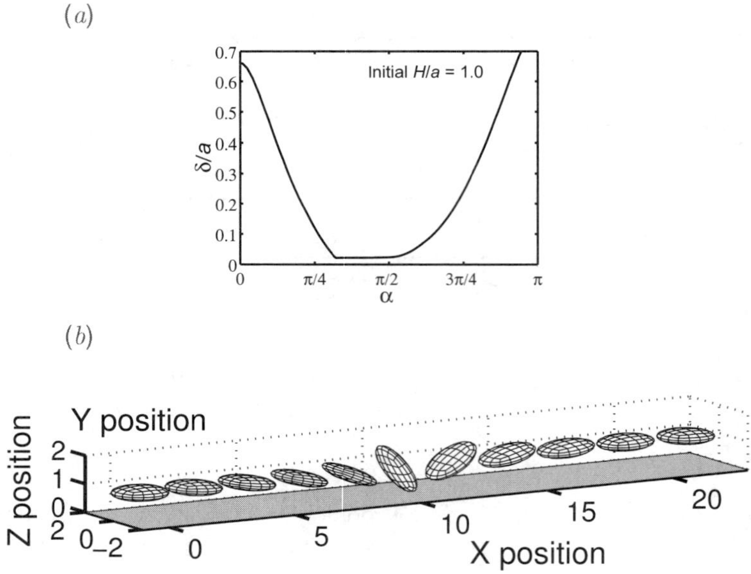

Figure 8.3.3 (a) Graphs of the closest distance of approach δ against the orientation angle α for initial platelet height $H/a = 1.0$. (b) Visualization of three-dimensional platelet motion with initial orientation $\alpha = 0$. Length is scaled by the platelet radius, a. *Reprinted with permission from Mody & King (2005).*

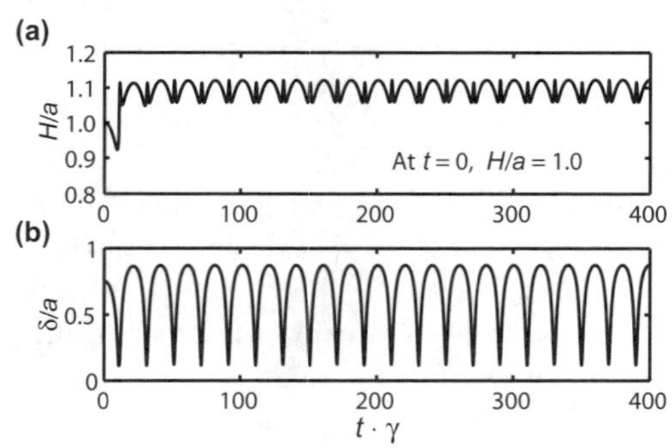

Figure 8.3.4 Evolution of he (a) dimensionless centroid height, and (b) closest distance of the particle from the wall for a platelet with initial height $H/a = 1.0$. *Reprinted with permission from Mody & King (2005).*

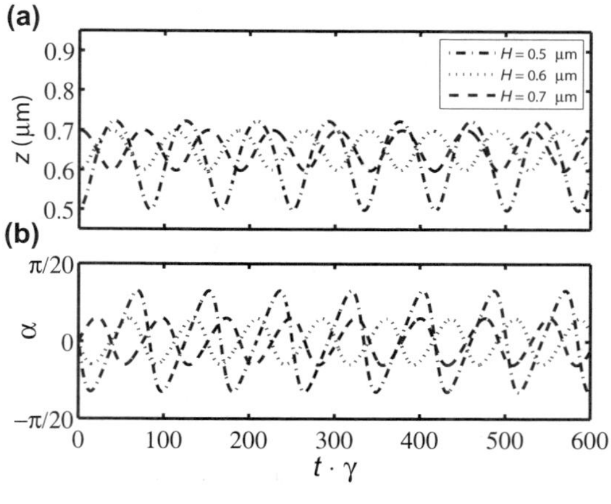

Figure 8.3.5 Periodic oscillatory trajectory of a platelet in wobble flow for three different initial heights. Evolution of the (a) height of the centroid from the surface and (b) platelet orientation about the y axis.

begins moving at a height 0.5 or 0.6 platelet radii above the wall wobbles away from the wall. A platelet that begins moving at a height $0.7a$ above the wall moves toward the wall.

Figure 8.3.5 describes the wobble motion for platelets with initial height $0.5a$, $0.6a$, and $0.7a$. Dampening of the amplitude of the lateral motion or angular velocity does not arise due to reversibility of Stokes flow. Short-range repulsive forces between two opposing surfaces do not enter the hydrodynamic calculations in Regime III flow. Two counteracting hydrodynamic forces act on the platelet at any time: a shear force engaging the platelet into clockwise rotational motion; and a wall drag force pushing the leading platelet edge away from the wall producing an anticlockwise rotational motion.

The dependence of the initial direction of rotation on the particle position was noted by Staben *et al.* (2003) for prolate spheroids with aspect ratio $\lambda = 2.5$ in channel flow. They found that, for small angular orientations with respect to the channel walls $0°– 5°$, and small gaps in the range $0.02 - 0.07$ the channel height or $0.067 - 0.23$ the particle major radius, the particle experiences an instantaneous counterclockwise rotation. At higher angles or larger particle-wall gaps, the particle experiences a clockwise rotation.

The Regime III motion illustrates the interplay between the opposing action of the shear flow and retardation of the wall. Near the wall, retardation forces dominate and the platelet moves away from the wall. As the platelet migrates to a greater height, the effect of the wall diminishes, fluid forces dominate, and the platelet starts rotating in the opposite direction moving toward the surface. The magnitude of the platelet tilt about the horizontal axis is small throughout the motion. Accordingly, the wall exerts a significant drag force on the platelet, precluding platelet-surface contact. In Regime II flow, as the platelet approaches the wall, the tilt angle is steep enough for fluid forces to remain stronger than the wall drag forces throughout the period of surface contact. Within the range of heights pertinent to Regime III flow, a neutrally stable height $0.662a$ is possible. At this position, a horizontal platelet translates parallel to the wall.

For initial heights resulting in Regime III flow and horizontal initial orientations, it is possible that providing an initial platelet tilt about the y axis will cause surface contact. For centroid heights $0.5a$, $0.55a$, $0.6a$, $0.65a$, and $0.7a$, the initial orientation angle was varied from 3° to 15° in the clockwise direction. It was found that the platelet continues to exhibit Regime III flow for a range of tilt angles until a critical tilt angle has been reached. For an initial tilt angle greater than the critical angle, the particle switches to Regime II flow where contact is made with the surface and the platelet flips over and then moves away from the wall. Periodic contact followed by flipping over the surface and rotational motion ensues, typical of Regime II motion.

The critical tilt angle is a function of the initial centroid height H/a, increasing in magnitude with decreasing the initial platelet height. The dependence of the critical angle on H where the flow Regime switches from II to III is expressed by the quadratic function

$$\alpha_{cr} = 114.28H^2 + 107.14H - 14.47, \tag{8.3.1}$$

with an average error 1.14%.

Sugihara-Seki (1996) performed a comprehensive finite-element study of three-dimensional oblate spheroid motion with aspect ratio $\lambda = 0.25$ in Poiseuille flow. Instantaneous particle velocities were calculated for different particle positions and tilts such that the flow is symmetric about the particle midplane. The authors discussed different flow regimes for prolate and oblate spheroids in terms of the initial particle angular orientation and centroid distance from the centerline. The location of neutrally stable points where lateral and rotational motion vanishes depends on the particle radius for a tube of fixed size. The flow pattern of the particle motion is fundamentally changed when the particle size exceeds a critical threshold.

8.3.4 Chaotic motion

The motion of an oblate spheroid is sensitive to the initial angular tilt about the x axis producing a break in symmetry with respect to the xz plane. When a platelet is parallel to the wall, the angular tilt about the x axis is 0°. In the case of nonzero tilt about the x axis, the platelet rotates about the x and y axes and gradually spins about the z axis exhibiting a three-dimensional asymmetric or chaotic motion. Slight out-of-plane motion of the spheroid axis of revolution can suddenly or eventually lead to a transition from wobble flow to pole vaulting to stable rotation with periodic contact.

With only a small tilt angle about the x axis, a platelet starting in Regime III continues to follow Regime III and does not contact the surface. Above a critical tilt angle about the x axis, the platelet displays Regime II motion at the onset of flow. At initial tilt angle above the critical angle, surface contact occurs during the first rotation cycle and the platelet pole vaults over the surface. The tilt of the platelet about the x axis does not remain constant but oscillates about the 0° orientation. The tilt or rotation about the x axis increases with each oscillation.

Because oscillations about the x axis are unstable and increase in time, a platelet that begins moving with an angular tilt about the x axis that is less than the critical value eventually contacts the surface. In such cases, the first dip cycle of the spheroid does not result in contact. However, a second, third, or fourth dip may result in contact and transition into Regime II flow. The critical tilt angle about the x axis resulting in immediate transition from Regime III to II depends inversely on H. Let β be the angular displacement of the platelet about the x axis. The dependence of the critical angle on initial platelet height H is expressed by the quadratic function

$$\beta_{cr} = -57.14H^2 + 32.57H + 8.11, \tag{8.3.2}$$

with an average error of 2.19%.

8.3.5 Oblate spheroids with aspect ratio 0.3–0.5

Particle trajectories were determined for oblate spheroids with other aspect ratios in a range of heights from the surface. At the initial instant, the major axis is parallel to the wall and the axis of revolution is tilted at zero angle about the x and y axes. The motion of a spheroid with aspect ratio $\lambda = 0.42$–0.5 falls in Regime I or II flow. Regime II is observed for spheroid distance from the wall between $1.1a$ and $0.52a$ or $0.6a$, for $\lambda = 0.42$ or 0.5. Below $H = 0.52a$ or $0.6a$, the particle surface and roughness layers overlap for spheroids with $\lambda = 0.42$ and 0.5. Spheroids with aspect ratios 0.41–0.25 exhibit all three flow regimes. The range of heights where Regime III prevails decreases with increasing λ. Figure 8.3.6 documents limiting heights where Regime III prevails for different aspect ratios.

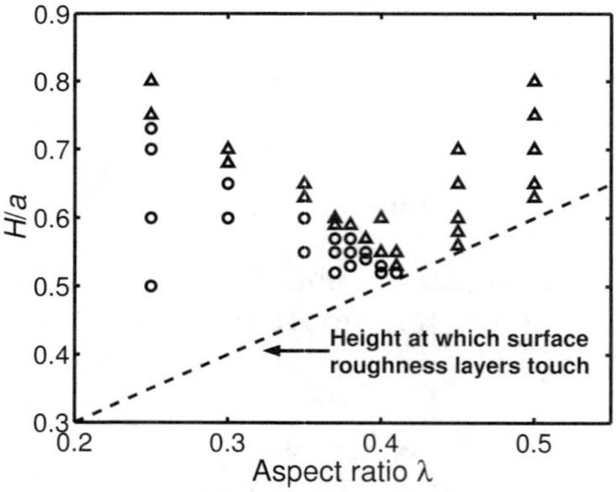

Figure 8.3.6 Centroid height, H, where transition from Regime II to III occurs for particle aspect ratios ranging from 0.25 to 0.5; $\triangle$ denotes Regime II flow and $\circ$ denotes Regime III flow. Regime III flow appears for aspect ratios up to 0.41. Above this aspect ratio, Regime II prevails for all spheroid heights ranging from $1.1a$ down to the height where the spheroid contacts the surface. *Reprinted with permission from Mody & King (2005).*

8.4 Brownian motion

Platelets are colloidal particles due to their submicron size, ~ 0.25 μm (Frojmovic *et al.* 1990). We wish to explore the role of Brownian motion in flow trajectories and dynamics of platelet adhesion. In this section, we address three pertinent questions: What are the characteristics of Brownian motion of a platelet-shaped particle near a wall? What is the relevance of platelet Brownian motion in the presence of shear flow? What is the influence of Brownian motion on the rupture dynamics of platelet-surface molecular bonds?

Previous authors have studied the rheology of suspensions of spherical and nonspherical Brownian particles (mainly solids of revolution) in linear shear or extensional flow (Asokan & Ramamohan 2004, Brenner 1974, Yamamoto *et al.* 2005). A few experimental and theoretical studies have incorporated the effects of a proximal wall (Garnier & Ostrowsky 1991, Unni & Yang 2005). Theoretical investigations of Brownian motion near a wall employ a single characteristic particle length scale. Our goal is to understand the effect of platelet size and shape and document the influence of Brownian motion on particle trajectories in shear flow. The ultimate objective is to determine the physiological role of Brownian motion.

We have simulated translational and rotational Brownian motion of platelets with dimensions $2 \times 2 \times 0.5$ μm^3 near an infinite plane wall (Mody & King 2007). Since Brownian motion is an unbiased stochastic process with normal distribution, the Brownian forces and torques are calculated using a Gaussian random distribution with zero mean, variance $F_{Br} = k_B T/\hat{a}$ for the force and $T_{Br} = k_B T$ for the torque, where $k_B T$ is the product of the Boltzmann constant and the absolute temperature, and $\hat{a}$ is the characteristic radius of the projected area of the platelet cast onto the plane normal to the direction of motion (Carpen & Brady 2005, Kim & Karilla 1991). Brownian forces and torques were calculated in three directions, x, y, and z (figure 8.2.1). The characteristic radii of the projected area of the platelet in the yz, xz, and xy plane, a_x, a_y, and a_z, were determined at each case.

The Reynolds number is $\mathrm{Re} = O(10^{-4}) << 1$, where $\dot{\gamma} = 100$ s^{-1} is the shear rate. In the absence of a linear shear flow field, the Reynolds number is $\mathrm{Re} = U\rho a/\mu = O(10^{-7}) << 1$ in the case of translation, where $U \sim 0.62$ μm/s is the translational velocity of a horizontal platelet whose centroid is located at a distance 0.8 μm above the wall due to Brownian force $k_B T/\hat{a}$ (one standard deviation of the zero mean normal distribution that defines the translational Brownian motion) in the x, y, or z direction. In the case of rotation, the Reynolds number is $\mathrm{Re} = \Omega\rho a^2/\mu = O(10^{-7}) << 1$, where $\Omega = 0.27$ rad/s is the angular velocity of a platelet whose centroid is placed at a distance 0.8 μm above the wall, due to Brownian torque $k_B T$ (one standard deviation of the zero mean normal distribution that defines the rotational Brownian motion) acting on the platelet about the z direction.

The platelet shape has a much shorter length in one dimension than in the other two dimensions. This results in asymmetric Brownian motion where the largest displacement occurs in the dimension(s) where the projected (viewable) area of the particle is the smallest. The net Brownian motion of the particle is calculated from the sum of displacements due to Brownian forces acting in all three directions and rotational motions due to Brownian torque acting in directions parallel to the x, y, and z axis passing through the platelet centroid.

8.4.1 Brownian motion near a wall in a quiescent fluid

Brownian motion is retarded by the presence of a neighboring boundary. While wall drag reduces particle Brownian velocities in all directions, the effect is greatest in the direction normal to the wall (Mody & King 2007). Figure 8.4.1 illustrates the time it takes for a platelet undergoing translational and rotational Brownian motion to contact the surface in the absence of an external flow for a range of starting heights. Platelet–surface contact occurs when the platelet approaches the surface within a distance that allows

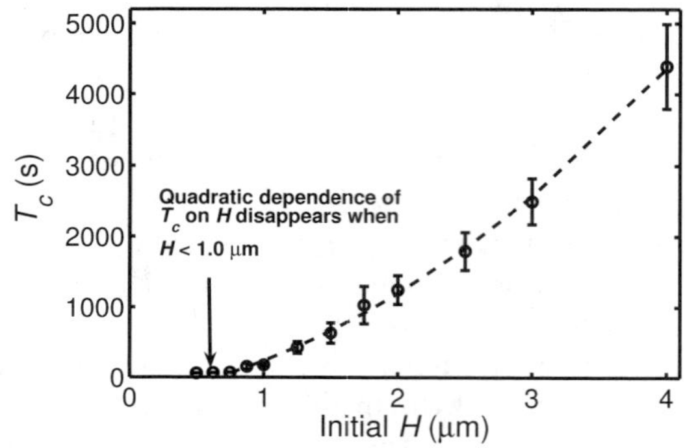

Figure 8.4.1 Average time for a platelet at an initial height H to contact the surface while undergoing translational and rotational Brownian motion in a quiescent fluid. The best-fit broken curve depicts the quadratic dependence of the average time T_c on the initial platelet height, $T_c = 207.1H^2 + 339.6H - 310.6$, with an 8.2% average error, for initial height $H \geq 0.75$ μm. The fit does not include data at initial heights $H = 0.5$ and 0.625 μm. *Reprinted with permission from Mody & King (2007).*

reactions to occur between the molecules on both surfaces. We specifically consider binding reactions between the platelet surface receptor GPIbα, and surface-bound vWF. The GPIbα–vWF bond length is estimated by adding the respective sizes of the two participating molecules to an approximate length of 70 nm, (Fox *et al.* 1988, Siedlecki *et al.* 1996). Physical contact between the edges of the surfaces is not required for binding of GPIbα and vWF (Li *et al.* 2002). A surface-to-surface separation distance within 120 nm, equal to the sum of the glycocalyx roughness layers on both surfaces and the 70 nm bond length of GPIbα–vWF is set as the threshold defining a contact event.

In all simulations, the platelet major axis is initially horizontal. Because the time for contact with the surface T_c is random, at least 200 surface-contact events are sampled to obtain a meaningful average for each initial height, H. On average, it takes 57 seconds before platelet–surface contact is observed when the platelet centroid is initially situated at a distance 0.5 μm above the surface. This time interval is three orders of magnitude greater than the convective time scale $1/\dot{\gamma}$ at a shear rate of 50 s^{-1}. For initial height 2.0 or 4.0 μm, the average time for contact, T_c, is 21 or 73 minutes, which is much higher than the time scale relevant to physiological blood clotting, on the order of seconds. Based on these large contact times, we expect that convective processes dominate the transport of platelets from the blood stream to an injured vascular surface. Consistent with the diffusive nature of Brownian motion, the time to contact the surface, T_c, depends quadratically on the initial centroid height, H.

For initial height $H < 1.0$ μm, the quadratic dependence of the diffusion time on distance no longer applies. When the platelet is situated near the surface, two simultaneous mechanisms come into play to initiate cell–surface contact, including translation and rotation of the cell body. At larger initial distances from the surface, translational motion is the dominant mechanism for bringing the platelet within a reactive distance from the wall.

For platelet starting heights greater than 1 μm, not all platelets are able to reach the surface. In this range of starting heights, some platelets drift away from the surface beyond the critical distance 25 μm, chosen as the cut-off height for terminating numerical simulations. This threshold is comparable to the average inner diameter of arterioles, 50 μm (Berne & Levy 1997). A platelet that has reached a height of 25 μm is considered to have crossed the vessel centerline. The probability that a platelet reaches the wall decreases as the platelet is situated farther from the wall. The percentage of platelets that drift away and do not return to the wall is 1.5%, 5.5%, and 12% for initial heights 1.5, 2.0 and 4.0 μm. Platelets positioned far from the wall enjoy a greater freedom in translating and rotating in all directions without contacting the wall. For smaller initial heights, $H < 1.0$ μm, all platelets make contact with the wall.

8.4.2 Convective and diffusive transport

At time scales relevant to blood flow, platelet Brownian motion is barely able to alter the platelet trajectory. For example, a platelet initially at 1.3 μm above a surface in linear shear flow with shear rate 100 s^{-1} undergoes 14 rotations during the first two seconds of flow, as shown in figure 8.4.2. Enabling platelet Brownian motion changes the z position of the centroid by 0.002 μm and the y position of the centroid by 0.01 μm at the end of the simulation, after the platelet has traveled a distance of 250 μm along the x axis.

We see that Brownian motion has a negligible effect on the platelet trajectory along the x and z axes and on the platelet frequency of rotation, and a significant effect on the platelet trajectory along the y axis. However, over the entire length of the simulation, the distance traveled by the platelet in the y direction is only 0.004% that traveled in the x direction, which is 1% the platelet radius. Zero translation of the platelet in the y direction in the absence of Brownian motion is an idealization. In reality, deviations occur due

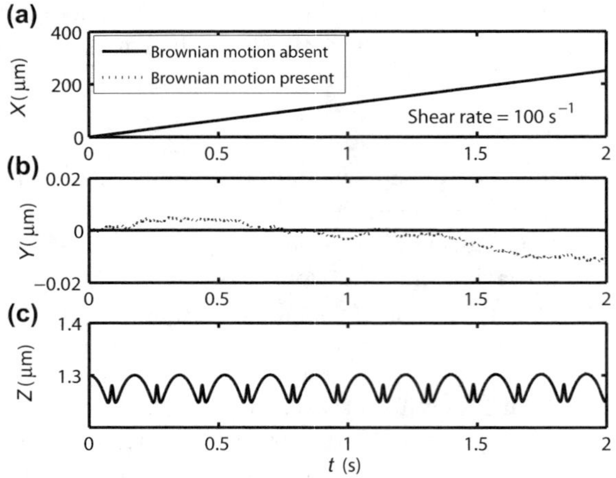

Figure 8.4.2 Graphs of the (*a*) x, (*b*) y, and (*c*) z position of a platelet centroid in linear shear flow with shear rate 100 s^{-1}. The platelet is initially placed at a distance 1.3 μm above the wall parallel to the wall. The solid line depicts the path of a non-Brownian platelet and the dotted line depicts the path of a Brownian platelet. *Reprinted with permission from Mody & King (2007).*

to presence of other particles, nonuniformities in the flow field, and presence of bounding walls.

Next, we consider the influence of Brownian motion on the frequency of contact of a platelet with a surface in the presence of fluid flow. As noted in section 8.2, cell–surface contact occurs only in Regime II flow. For motion with initial height below 0.75 μm and an approximately horizontal initial spheroid orientation, the platelet moves nearly parallel to the surface for a long distance without making contact with the wall. A transition from Regime III (wobble flow very close to the surface) to Regime II flow can occur only if the particle motion is sufficiently chaotic, characterized by particle rotation about all three axes. To determine if Brownian motion can speed up the average time to contact the surface, the Brownian platelet was placed at a distance 0.8 μm above the wall with a horizontal initial orientation. This initial height lies inside the window of platelet centroid distance resulting in transient platelet contact with the surface in simple shear flow, 0.75–1.1 μm.

In the absence of particle Brownian motion, the dimensionless time necessary for the platelet to touch the surface scaled by the inverse shear rate is independent of the shear rate. This is expected because of the linearity

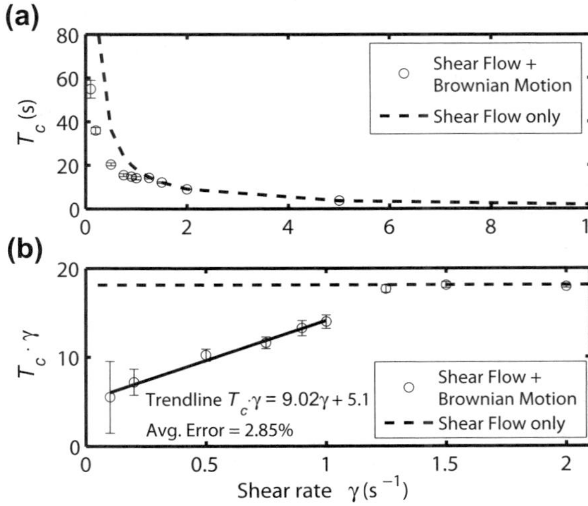

Figure 8.4.3 Graphs of the time until surface contact, T_c, against the shear rate for a Brownian platelet with initial centroid height 0.8 μm in linear shear flow. In frame (*b*), the time to contact the surface is non-dimensionalized by the inverse shear rate. A best-fit curve is plotted to illustrate the linear dependence of the scaled time for contact on the shear rate over the range of shear rates where Brownian motion (diffusion) accelerates cell transport to the surface. In simulations including Brownian motion, 200 data points are collected for each shear rate to determine the average time to contact. *Reprinted with permission from Mody & King (2007).*

of the Stokes equation and continuity equation. As a result, creeping flow streamlines are independent of the shear rate. Figure 8.4.3 shows the time it takes for a Brownian platelet to contact the surface in the presence of linear shear flow for a range of shear rates. For $\dot{\gamma} \leq 1$ s^{-1}, the effect of Brownian motion on T_c is significant. For $\dot{\gamma} > 5$ s^{-1}, the contact time for shear flow and Brownian motion is indistinguishable from the contact time for shear flow alone. In the range of shear rates 1–5 s^{-1}, a transition from dominant diffusive Brownian motion to dominant convective motion takes place.

The shear rates considered are extremely low compared to physiological shear rates. For example, wall shear rates in the veins where the flow is the slowest ranges from 20 to 200 s^{-1}. In large arteries, the wall shear rate generally ranges from 300 to 800 s^{-1} (Kroll *et al.* 1996). Arteriolar blood flow experiences even higher shear rates ranging from 500 to 1600 s^{-1} (Kroll *et al.* 1996). The Péclet number, $\mathrm{Pe} = LU/D_{AB}$, provides us with a measure of the relative time scales for convection and diffusion. In our problem, Pe depends on the distance of a platelet from the wall. For a platelet undergoing Brownian motion in linear shear flow, Pe is given by the following expression deduced from the computational model,

$$\mathrm{Pe} = \dot{\gamma}\,(1.56H + 0.66), \tag{8.4.1}$$

for $H > 0.3$ μm, with an average error of 0.51%. This empirical formula expresses a linear dependence of Pe on H obtained by assuming that the platelet is horizontal and considering translation due to Brownian motion in the x direction. The Brownian force is equal to the standard deviation of the normal distribution, $k_B T/\hat{a}$. For $H = 0.8$ μm, the platelet Péclet number for shear rates in the range 0.1–100 s^{-1} varies in the range 0.19–191.42. We may conclude that, at time scales relevant to shear flow in blood, Brownian motion plays a negligible role in influencing platelet motion and creating further opportunities for platelet–surface contact.

8.4.3 Influence on surface adhesive dynamics

Bond formation of GPIbα with the A1 domain of surface-bound von Willebrand factor (vWF) plays a critical role in prolonging platelet surface contact. Bond formation has been found to occur over a wide range of shear rates up to 6000 s^{-1} and higher (Savage *et al.* 1996). We have studied the effect of Brownian motion on the breakup dynamics of platelet–surface GPIbα–vWF–A1 bonds. The bond between a GPIbα receptor on the platelet surface and a surface–bound vWF molecule was modeled as a linear spring with end points located at the cell periphery and the wall, as shown in figure 8.4.4. The platelet is bound to the surface by a single preassigned cell-surface GPIbα–vWF–A1 bond with unstressed bond length.

The bond is initially placed at the center of the cell surface facing the wall and is aligned vertically with bond length 70 nm. Because of Brownian motion, the platelet-surface bond is stressed (pulled and pushed), increasing the probability of dissociation. When the bond length shifts away from its equilibrium (unstressed) length, a force and torque develops on the cell to restore equilibrium. The kinetics of bond dissociation of GPIbα and vWF–A1 follow the Bell model describing the force-dependent dissociation rate of weak noncovalent bonds according to the equation

$$k_{off}(F) = k^0_{off} \exp\left[\frac{\gamma F}{k_B T}\right], \tag{8.4.2}$$

where $k_{off}(F)$ is the bond dissociation (off) rate, k^0_{off} is the unstressed off-rate, γ is the reactive compliance, F is the force applied on the bond, and

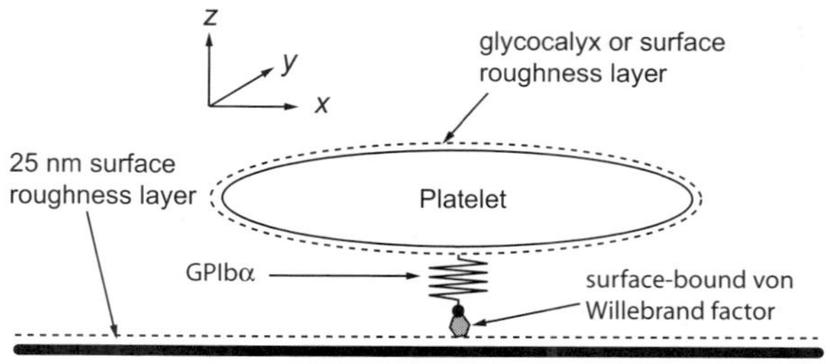

Figure 8.4.4 Schematic illustration of a horizontal platelet bound to a surface by way of a single GPIbα–vWF bond. A steric layer with thickness 25 nm coats the platelet and wall emulating the glycocalyx layer and surface roughness.

$k_B T$ is the product of Boltzmann constant and absolute temperature (Bell 1978). The reactive compliance has dimensions of length, and its magnitude is comparable to the atomic radius, which is on the order of Ångströms. The Bell model parameters $k^0_{off} = 5.47$ s^{-1} and $\gamma = 0.71$ nm used in this study for GPIbα–vWF dissociation kinetics were obtained from optical tweezer experiments by Arya *et al.* (2005).

Figure 8.4.5 illustrates the average bond lifetimes and bond rupture forces due to platelet translational and rotational Brownian motion in the absence of flow. Figure 8.4.5(a) shows bond lifetimes and rupture forces obtained for a range of temperatures, 273–323 K. The viscosity of the ambient medium is varied with temperature to assess the effect on the particle drag force, and hence particle mobility, simultaneously with the effect of temperature-dependent Brownian motion on bond dissociation dynamics.

The dependence of the bond dissociation rate on temperature is twofold: the Bell model predicts that the bond half-life (inverse of the dissociation rate constant) increases with temperature; and the Brownian motion intensifies with an increase in temperature. A regular trend of the bond lifetime, t_b, with respect to the ambient temperature is not observed in figure 8.4.5(a). An expected monotonic increase of F_{rup} with T is seen in figure 8.4.5(b). Physically, lower solution viscosities and higher temperatures lead to more vigorous Brownian motion and therefore higher bond stress. The unstressed bond lifetime is the inverse of the unstressed off rate, $\simeq 1/5.47 = 0.183$ s.

Figure 8.4.5 demonstrates that the lifetime of a single cell–surface bond subjected to perturbations due to the platelet Brownian motion is similar to the unstressed bond lifetime. The average of the bond rupture forces plotted in figure 8.4.5(b) is 0.0034 pN. At shear rate 100 s^{-1}, platelet adhesive dynamics

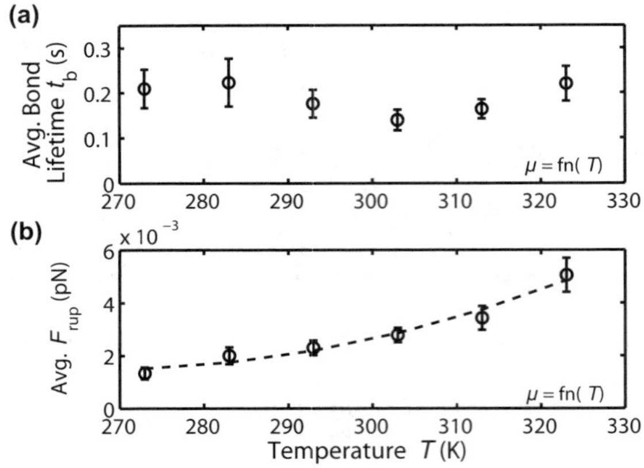

Figure 8.4.5 GPIbα–vWF-A1 bond lifetimes and bond rupture forces for a platelet undergoing Brownian motion in a quiescent fluid. (*a*) Bond lifetimes and (*b*) bond rupture forces in water. Thirty-seven observations are collected at each temperature. The quadratic best-fit curve in (*b*) carries a 7.8% average error. *Reprinted with permission from Mody & King (2007).*

simulations predict that bond rupture due to fluid forces acting on the platelet occur within 0.3 ms, accompanied by bond rupture forces of 30–100 pN. The bond rupture force in shear flow at $\dot{\gamma} = 50\ \mathrm{s}^{-1}$ is found from calculations based on experimental studies to be approximately 36 pN (Doggett *et al.* 2002). This force is roughly 10^4 times greater than that calculated for platelet Brownian motion. The insensitivity of the GPIbα–vWF bond lifetimes and the negligible stressing of bonds due to Brownian forces and torques exerted on a platelet indicate that Brownian motion has a minor effect on the dissociation dynamics of these receptor–ligand bonds.

8.5 Shape and wall effects on hydrodynamic collision

The mechanics of cell–cell encounter plays an important role in determining the outcome of cell aggregation. In this section, we discuss the hydrodynamic effect of a planar wall and the influence of particle shape–oblate versus spherical–on platelet collision in linear shear flow (Mody & King 2008*a*).

To quantify the influence of the system geometry, several important collision metrics are introduced, including the collision frequency, the collision contact time, and the cell surface area in contact with another cell during collision. The interception of two particles with equal volume is considered in the simulations. Platelet–platelet interactions of interest are confined to the region of flow adjacent to a vessel wall inside the cell-depleted plasma layer. Since red blood cells are primarily absent in this layer, the motion of erythrocytes and its influence on platelet–platelet encounters is not considered.

A hydrodynamic encounter of two platelets is classified as a collision when the platelets approach by less than a small distance that permits reaction between molecules on the two surfaces to take place. During collision, the platelet surfaces do not necessarily touch. The nomenclature for the two-platelet systems is defined in figure 8.5.1. Unlike sphere–sphere collisions where the faster moving sphere rolls over the slower moving sphere, platelet–platelet collision occurs by several mechanisms.

8.5.1 Collision mechanisms

Figure 8.5.1 illustrates three collision mechanisms between two oblate spheroids with the same shape and size, including the following:

Glide-over: The upstream (faster) platelet moves above the downstream or slower platelet making reactive contact. The two platelets rotate together and then move apart, as shown in figure 8.5.1(a).

Glide-under: The upstream platelet moves by a small distance under the downstream platelet. The two platelets rotate together and then separate, as shown in figure 8.5.1(b).

Intermittent contact: A short duration of contact is followed by a brief period of separation lasting for two or more rotations, as shown in figure 8.5.1(c). This mechanism is observed when the platelets are introduced at approximately the same height.

These three mechanisms are representative of collisions that are symmetric about the plane of the flow. Other collision mechanisms are possible. The prevailing mechanism depends on the initial platelet separation in all three directions.

8.5.2 Collision frequency

The collision frequency of two particles depends on the relative particle velocity. If a homogeneous suspension of particles with concentration (number density) of $\dot{n}$ particles per volume is sheared, the rate at which one particle collides with another particle is

$$\dot{R} = \dot{n} \iint_{\Omega} v_{rel} \, \mathrm{d}S \tag{8.5.1}$$

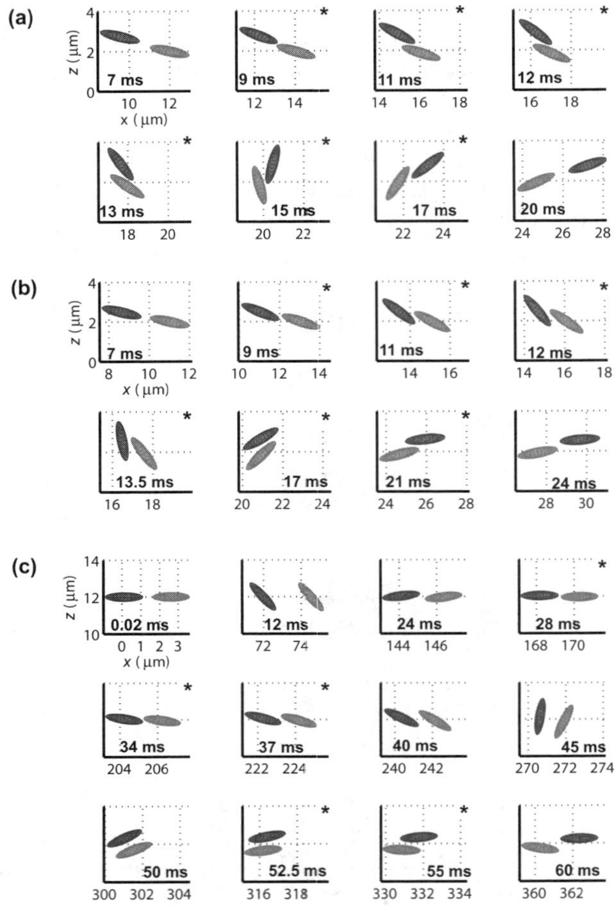

Figure 8.5.1 Illustration of different collision mechanisms between two platelets when the flow is symmetric about the xz plane: (a) glide-over, (b) glide-under, and (c) intermittent contact. In frames containing a star in the right-hand upper corner, the two platelets depicted are within reactive distance. The simulation time corresponding to the shown particle orientations are indicated in the respective frames. The fluid shear rate is $500\ \mathrm{s}^{-1}$ and the reactive contact distance between the particle edges is set to $260\ \mathrm{nm}$. A steric layer with thickness $40\ \mathrm{nm}$ is coated on each surface. *Reprinted from Mody & King (2008).*

where v_{rel} is the relative particle velocity and Ω is the upstream cross-sectional area defined such that all particles convected through this area collide with a particle of interest located downstream. The surface Ω lies in the plane containing the velocity gradient (z plane) and the vorticity (y plane). The rate of collision per unit volume is

$$\dot{R}_v = \dot{n}^2 \iint_\Omega v_{rel}\, \mathrm{d}S. \tag{8.5.2}$$

The relative velocity of two remote particles in the absence of hydrodynamic interaction in linear shear flow is $v_{rel} = \dot{\gamma} z_{rel}$. Discretizing the surface integral in (8.5.2), we obtain

$$\dot{R}_v = 2\dot{n}^2 \dot{\gamma}\, \Delta S \sum z_{rel} \tag{8.5.3}$$

where ΔS is an element of the upstream cross-sectional area resulting in collision. Since the rate of collision is symmetric about the y axis (vorticity direction), a factor of two has been introduced so that only summation in the positive y direction is carried out.

To compare the collision rates of two different geometries or particle shapes, we consider the ratio of collision rates,

$$\frac{(\dot{R}_v)_1}{(\dot{R}_v)_2} = \frac{\sum (z_{rel})_1}{\sum (z_{rel})_2}. \tag{8.5.4}$$

One way to quantify the effect of particle shape and wall proximity is to determine the collision frequencies of platelets and spherical particles suspended in shear flow, near and far from a wall. Figure 8.5.2 maps the regions of upstream cross-sectional area through which convected particles collide with a downstream particle for four different geometries. The collision maps were generated for two platelets (figure 8.5.2(*a*, *b*)), two spherical particles whose radius is equal to the equivalent platelet radius 0.63 μm (figure 8.5.2(*c*, *d*)), and a platelet–sphere pair (figure 8.5.2(*e*, *f*)). A platelet–sphere collision approximates a physical encounter between a freely convected unactivated platelet and a freely convected activated (globular) platelet. A darkened square in the grid indicates that the corresponding initial relative particle position results in collision.

The maximum reactive distance between two particles was set equal to the length of the trimolecular noncovalent bond between two GPIbα receptors and a vWF ligand, 128 nm (see section 8.6). For each initial relative position depicted in figure 8.5.2(*a*–*f*), the two particles are initially separated by distance 10 μm in the flow direction along the x axis (10 times the platelet

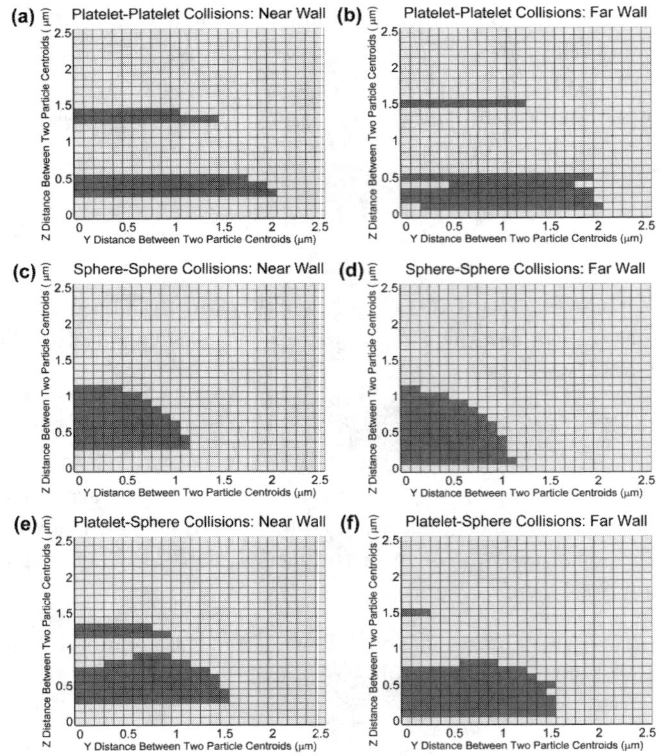

Figure 8.5.2 Grid maps depicting the initial relative particle position resulting in collision based on reactive particle gap 128 nm for different particle shapes and distances from the wall. (*a–f*) Two-dimensional grids depicting incremental relative separations in the y and z direction between two particles with initial distance $10.0 \ \mu\text{m}$ apart in the x direction. Dark gray shading represents initial positions that result in collision. Each two-dimensional grid depicts 676 different initial particle configurations. (a) Collision pattern of two platelets near a surface; the downstream platelet is introduced at a height $1.5 \ \mu\text{m}$. (b) Collision pattern of two platelets far from the surface; the downstream platelet is introduced at a height $12.0 \ \mu\text{m}$. Panels (c) and (d) show collision patterns of two spheres with radius $0.63 \ \mu\text{m}$ with starting initial configurations as in (a) and (b). Panels (e) and (f) show collision patterns for a downstream platelet with major axis $1 \ \mu\text{m}$ and an upstream sphere with radius $0.63 \ \mu\text{m}$ with starting initial configurations as in (a) and (b), respectively. *Reprinted with permission from Mody & King (2008).*

radius), so that hydrodynamic interactions are weak. In the near-wall simulations, the initial height of the downstream particle is 1.5 μm. In far-wall simulations conducted to ascertain the influence of particle shape on particle collisions in the absence of a wall, the downstream particle is initially placed at a distance 12 μm above the wall. Maps were generated for particle collisions in similar geometries where the reactive distance defining the collision takes a different value (Mody & King 2008*a*).

The initial platelet position resulting in collision between two particles with the same size and shape are very different for spheres and oblate spheroids. In the case of spheroids, the upstream collisional cross-sectional area consists of two separated regions. The lower cross-sectional region represents all collisions where the platelets initiate contact when their major axes are nearly parallel to the plane. The smaller upper cross-sectional area is associated with collisions observed when the platelet tips graze each other. This occurs when the axis of revolution of at least one platelet is approximately parallel to the plane when contact is initiated. The contact time during collision represented by the upper cross-sectional region is at least 2 to 4 and up to 37 times less than the corresponding contact time for collisions represented by the lower cross-sectional region at the same initial particle center-to-center separation in the y direction. Note that the contact time during collision is a strong function of the separation between the platelet centers in the y direction.

The collision map patterns shown in figure 8.5.2 are altered when the particles are brought closer to the wall. The change in the pattern of the upstream cross-sectional area is more pronounced for platelet-shaped particles than for spherical particles. The effect of the wall and particle shape on the overall collision frequency is illustrated in figure 8.5.3. The graphs in this figure compare the collision frequency of two particles for six geometries whose collision maps are shown in figure 8.5.2. The collision frequencies described in 8.5.3 have been normalized by the collision rate for 0.63 μm spheres described in figure 8.5.2(d), whose hydrodynamic encounters are essentially unaffected by the presence of boundaries.

The presence of a wall promotes the collision rate for particles of any shape. In unbounded flow, the discoidal shape of the platelets reduces the collision frequency. Physically, platelets separated farther in the z direction pass each other without colliding when their axes of revolution are approximately normal to their plane. The wall enhances the collision rates by approximately 25% in the case of platelets and by 13% in the case of spheres. It is interesting that platelet-shaped particles have a greater collision frequency near the wall compared to spheres in bounded and unbounded flow. The wall has a greater effect on the frequency of platelet–platelet rather than sphere–sphere

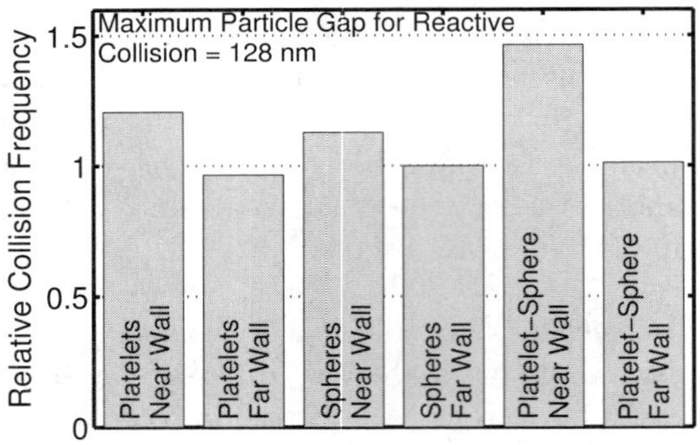

Figure 8.5.3 Collision frequency for the six geometries described in figure 8.5.2(a–f). The collision frequency has been normalizing by that for far-wall spheres; see equation (8.5.4). *Reprinted with permission from Mody & King (2008).*

collisions. If the concentration of spherical and platelet particles in the system geometry shown in figure 8.5.2(e, f) is the same as that for the platelet–platelet and sphere–sphere systems described in figure 8.5.2(a–d), their respective collision frequencies can also be compared. The wall augments the platelet-sphere collision rate by 44%.

Figure 8.5.2 shows that fewer particle-particle trajectories or a smaller upstream cross-sectional area result in particle–particle collisions closer to the wall compared to those far from the wall. However, the net effect of the wall is to shift the upstream collision cross-sectional area upward, away from the wall. As a result, on average, near-wall collisions involve higher relative velocity differences between particles, accounting for increased collision frequency near the wall.

The collision trajectories and associated diagnostics, such as collision time and collision contact area, were determined for all 85 observed initial positions resulting in collision between two platelets near the wall. The corresponding upstream collision cross-section is described in figure 8.5.2(a). In Stokes flow, the contact area and collision trajectories are independent of the shear rate and the duration of collision is inversely proportional to the shear rate. Because a short-range repulsive force is present, the simulated particle collision is not precisely scalable with the inverse shear rate. Simulations were performed at two shear rates, 1500 and 8000 s^{-1}.

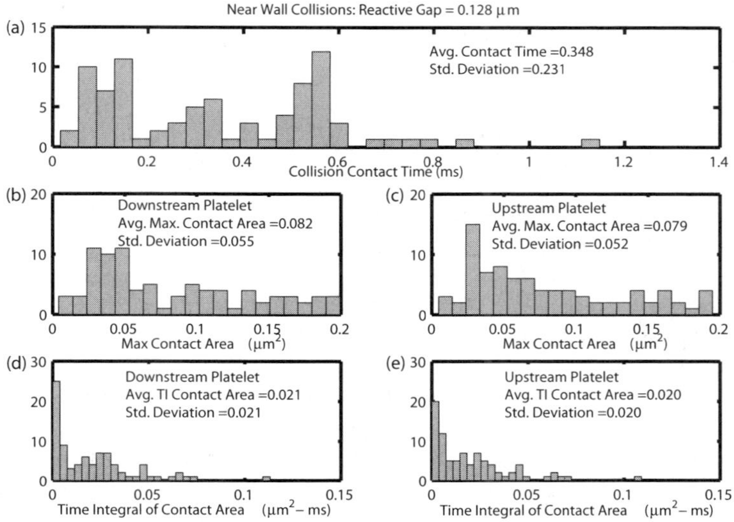

Figure 8.5.4 Histograms of characteristics of 85 platelet–platelet collisions depicted in figure 8.5.2(a) occurring close to a wall at shear rate $8000\ \mathrm{s}^{-1}$. The reactive collision gap is set to 128 nm. (a) Contact time between platelets during collision. (b, c) Maximum instantaneous surface contact area of the downstream and upstream platelet during collision. (d, e) Time-integral of contact area for the downstream and upstream platelet, respectively. *Reprinted with permission from Mody & King (2008).*

The collision properties depend weakly on the shear rate (Mody & King 2008a). Accordingly, we discuss the nature of the collisions irrespective of the shear rate. The collision contact time and maximum instantaneous contact surface area of colliding platelets are presented in figure 8.5.4(a–c) for each initial configuration shown in figure 8.5.2(a) resulting in collision. Due to the asymmetry of the interactions between nonspherical particles, the maximal and average contact area are slightly different for each platelet participating in the collision.

The duration of collision, t_c, and contact area during collision, A_c, can be combined to produce a single metric called the time-integral of contact area, A_{TI},

$$A_{TI} = \int_{t_i}^{t_i+t_c} A_c^{platelet}\, dt, \tag{8.5.5}$$

where t_i is the time when the two platelets first come into contact. A_{TI} is a measure of the probability of adhesion between the two blood cells that briefly encounter each other during flow.

This third characteristic of a two-particle collision is plotted in figure 8.5.4(*d, e*). The histograms of the collision times and maximum contact areas show a fairly uniform distribution with few data points near zero and multiple peaks inside the range of observation. In contrast, the lone peak of the histogram for the time-integral of contact area is located at the leftmost end.

The spreading of the time-integral of contact area over the range of observation does not mirror the shape of the histograms for collision contact time and maximum area of contact. Physically, a collision trajectory with a long contact time does not necessarily involve platelet orientations favoring maximum particle surface areas in the contact zone. The instantaneous contact area depends on the relative platelet orientation and position during interception. Relative positions and orientations resulting in significant overlap of the platelet surfaces correspond to nonzero distances between the platelet centroids in the y direction. Consequently, numerous collisions occur for a small time-integral of the contact area, and fewer collisions occur with a sizeable time-integral of the contact area.

To further illustrate the hydrodynamic significance of the wall on the platelet motion, we consider individual platelet trajectories, duration of cell–cell contact during collision, maximum instantaneous area of contact, and the time-integral of the contact area for each platelet in the case of near-wall and far-wall platelet–platelet interactions. Figure 8.5.5 illustrates particle trajectories for two colliding platelets near and far from the wall. The platelets are initially separated by 2.0 μm in the x direction and by 0.25 μm in the z direction. The mechanism of collision is different near and far from the wall.

In the near-wall collision, the upstream platelet rotates as though it were hinged at the left end of the downstream platelet. The downstream platelet is convected without rotation until the upstream platelet has completed its rotation. After rotation, the upstream platelet slides over the downstream platelet and moves away, while the second platelet enters a rotation cycle. In the far-wall collision, the two platelets rotate together with a small overlap and then separate with little relative sliding motion.

The influence of the wall on the collision outcome is found significant for certain initial configurations and less important for others. Comparing the contact time and maximum instantaneous contact area for near- and far-wall encounters with the same initial relative particle positions, we observe that the influence of the wall is greater for initial configurations where the relative distance between the two platelet centroids in the z direction is small.

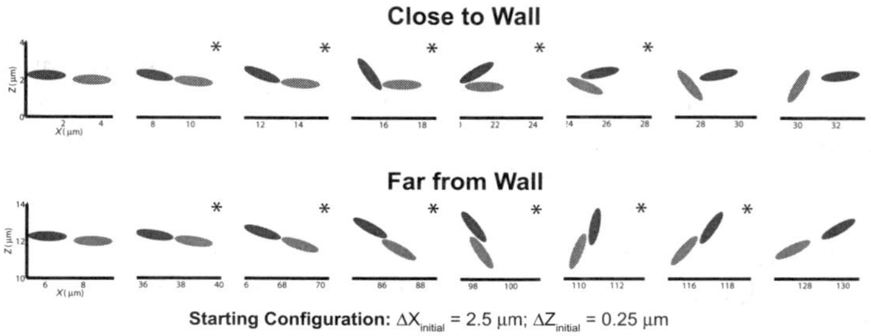

Figure 8.5.5 Sequential frames showing platelet trajectories during collision near and far from the wall. The initial distance between the platelet centroids is $2.5\ \mu\mathrm{m}$ in the x direction and $0.25\ \mu\mathrm{m}$ in the z direction. *Reprinted with permission from Mody & King (2008).*

Collision metric	*Near-wall*	*Far-wall*
Contact time t_c ($\dot{\gamma} = 500\ \mathrm{s}^{-1}$)	$22.84\ \mathrm{ms}$	$17.39\ \mathrm{ms}$
Max instantaneous contact area	$1.27\ \mu\mathrm{m}^2$	$0.85\ \mu\mathrm{m}^2$
Time integral of contact area (upstream, downstream platelet)	$10.0, 9.7\ \mu\mathrm{m}^2 - \mathrm{ms}$	$6.1, 6.0\ \mu\mathrm{m}^2 - \mathrm{ms}$

Table 8.5.1 Comparison of characteristics of near-wall and far-wall collisions presented in figure 8.5.5.

Wall effects are less profound for collisions between particles separated by larger heights. In all instances surveyed, the wall is found to increase the time-integral of the contact area.

8.6 Transient aggregation of two platelets near a wall

The hydrodynamic model of unactivated platelet motion near a wall can be coupled with an adhesive dynamics model to account for noncovalent binding reactions between the sticky surfaces of adhesive cells (Hammer & Apte 1992). We have developed a multiscale numerical simulation method termed platelet adhesive dynamics (PAD) (Mody & King 2008*b*). The algorithm incorporates stochastic kinetic equations to predict binding and unbinding reactions between GPIb receptors at the platelet surface and the A1 domain of multivalent vWF protein molecules in solution or bound to a platelet. The attachment of circulating vWF multimers to platelet GPIb receptors and the formation of cross-links or multiple bonds between vWF and the receptors is modeled as an ongoing process during runtime. The multiscale model was used to analyze the transient formation of GPIbα–vWF–GPIbα bond bridges between two platelets convected in linear shear flow at high shear rates, greater than 4000 s^{-1}.

Because GPIbα–vWF bonds are short-lived, they do not support the formation of stable platelet aggregates (Goto *et al.* 1995). However, interplatelet binding mediated by this bond pair is the sole medium possessing the mechanical strength to promote initial interplatelet contact in high-shear flow (Goto *et al.* 1998). The cytoplasmic domain of GPIbα possesses connections with the cytoskeleton attached to the plasma membrane. Following bond formation of the GPIbα surface receptor with vWF, if sufficient mechanical force on the receptor is generated, the binding event is signaled to the cell via a mechanotransduction pathway. This signal partially activates the platelet and key platelet receptors responsible for mediating stronger interplatelet connections. Platelets already activated by exogenous agonists (e.g. ADP, thrombin) can form aggregates with other platelets at high shear rates only after platelet–platelet bridging via GPIbα–vWF–GPIbα bonds is established (Goto *et al.* 1998).

The kinetics of the GPIbα–vWF–A1 bond is characterized by fast association and dissociation rates, weakening of the bond when acted upon by a tensile or compressive force, and the requirement of a critical level of shear stress to promote adhesive interactions (Arya *et al.* 2005, Doggett *et al.* 2002, Kumar *et al.* 2003). Mutations in the protein structures may alter the kinetics of binding and become responsible for hereditary bleeding disorders such as von Willebrand disease (vWD). Type 2-B vWD disease is caused by a gain-of-function mutation in the A1 domain or vWF. Platelet-type vWD disease is due to a mutation in the GPIb receptor. The altered binding kinetics is characterized by longer bond lifetimes, enhanced bond formation rates, or both (Arya *et al.* 2005, Doggett *et al.* 2003, Doggett *et al.* 2002, Dumas *et al.* 2004, Huizinga *et al.* 2002, Kumar *et al.* 2003). The strong interactions between the mutant GPIb receptor and vWF permit the spontaneous formation of non-permanent platelet aggregates in blood at normal flow rates. This reduces the availability of vWF in plasma, especially large vWF multimers, for binding to subendothelial components at the time of injury. The enhanced GPIbα–vWF binding activity promotes platelet-vWF association in the blood rather than at an injured surface, thereby increasing bleeding time (Ruggeri 2004).

Von Willebrand factor multimers are comprised of repeated identical subunits exhibiting a wide spectrum of molecular weights in blood, from 500 KDa:vWF dimer to 20000 KDa:Ultra-large-vWF (ULVWF) (Arya *et al.* 2002, Fischer *et al.* 1996, Li *et al.* 2004). The binding affinity of vWF to platelet surface receptors GPIbα and $\alpha_{IIb}\beta_3$ increases with an increase in the molecular weight of the multimers (Federici *et al.* 1989, Fischer *et al.* 1996, Furlan 1996, Kumar *et al.* 2006, Li *et al.* 2004, Moake 1995, Moake *et al.* 1988). In the absence of vascular injury, convected platelets interact minimally with circulating plasma vWF. Newly secreted vWF by platelets and endothelial cells at the time of injury contain ultra large vWF (ULVWF) normally not found in blood. When the regulation of ULVWF size is compromised, the persistence of ULVWF in blood can cause the formation of shear-induced platelet aggregates at normal blood flow rates. Thrombi in blood can occlude smaller blood vessels and thereby cause reduced oxygen supply and organ damage.

The multiscale PAD model has been used to explore the influence of binding kinetics and ligand size on the adhesive dynamics of platelet–platelet binding in shear flow. In this section, we discuss the efficiency of interplatelet binding for three different cases, the influence of interplatelet binding on hydrodynamic collision between platelets, and the force mechanics of bond rupture as a function of the collision properties with reference to our system geometry and receptor–ligand pair.

8.6.1 Adhesive dynamics model

The adhesive dynamics model pertinent to transient platelet aggregation is briefly presented in this section (Mody & King 2008*b*). Each platelet is assigned a uniform distribution of GPIb receptors deployed over the platelet surface. Areas with sparse or dense receptor concentrations are not allowed. The exact location of the receptors is determined at runtime and is unique for each simulation. The individual vWF molecules are modeled as virtual prolate spheroids. While their size is taken into consideration in the adhesive dynamics calculations, these molecules are infinitesimal particles in the context of hydrodynamics.

Two scenarios of vWF distribution in plasma are considered. With regard to n-vWF, to simulate normal conditions, a vWF molecule is sized as a 175×28 nm spheroid with 18 A1 binding sites connecting up to 8 receptors on any platelet (Siedlecki *et al.* 1996, Singh *et al.* 2006). With regard to L-vWF, to represent pathological conditions, vWF is sized as a 400×28 nm spheroid with 34 A1 binding sites (Slayter *et al.* 1985). The spheroid can bind with up to 17 receptors on any platelet. The location and number of receptors that a vWF molecule can bind is restricted by spatial limitations, including the location of receptors on the platelet surface, the length of the vWF molecule, and the number of binding sites present on the molecule. In the simulations, the entire population of vWF molecules in solution that are attached to platelets is monodisperse. The solution concentration of the ligand vWF was fixed at 2 nM.

When sheared at high flow rates in the presence of vWF multimers, a limited number of platelets are observed to bind vWF on their surface (Konstantopoulos *et al.* 1997). In keeping with this observation, every simulation is started with one platelet devoid of vWF on its surface and the other platelet with surface-bound vWF. The number and extent of ligation of the multimers on the platelet surface is determined by equilibrium calculations using the equivalent site hypothesis assumption (Perelson 1981). According to this hypothesis, the kinetic parameters for cross-linking (two-dimensional association and dissociation rates) are the same for every potential receptor-ligand bond formed on the platelet surface. Note that cross-linking kinetics is spatially restricted to a two-dimensional (surface) region and is therefore different from three-dimensional binding kinetics of a circulating ligand attaching to a surface-bound receptor.

The equilibrium concentration of ligand over the platelet surface is a strong function of the size of the ligand and binding kinetics expressed by the dissociation constant, K_D. We have studied platelet aggregation for two different dissociation constants, one fivefold smaller than the other, $K_D = 7.73 \times 10^{-5}$ M and 1.55×10^{-5} M (Goto *et al.* 1995, Mody & King 2008*b*). The fivefold difference, typical of platelet-type vWD, can be due to a decrease in dissociation rate, increase in association rate, or both (Doggett *et al.* 2003, Miura *et al.* 2000).

The three-dimensional on-rate for association of solution vWF with GPIbα receptors, k_{on}, is calculated by combining the intrinsic forward reaction rate constant with the diffusion-limited forward rate constant to obtain an overall forward rate constant. Bond breakup between the A1 domain of vWF multimers and their corresponding receptors is governed by the Bell model (8.4.2) with parameters listed in Section 8.4. Bond formation and breakup occur based on their respective probabilities calculated from the expressions

$$P_f = 1 - \exp(-k_f \Delta t), \qquad P_r = 1 - \exp(-k_r \Delta t), \tag{8.6.1}$$

where the subscripts f and r denote the forward and reverse rates (Hammer & Apte 1992).

Monte Carlo simulations are performed to test the formation of new bonds. Bonds between unoccupied platelet surface receptors and surface-attached vWF molecules are described by a two-dimensional cross-linking kinetics where k_f is the two-dimensional cross-linking forward rate constant, $k^0_{f,2D}$. In the case of bonds between unoccupied surface receptors and vWF molecules in solution, $k_f = k_{on} L$, where L is the solution ligand concentration. Free GPIbα receptors on one platelet located close to a vWF molecule bound to the opposing platelet surface, are tested for the formation of interplatelet bonds when two platelets come within binding distance of each other.

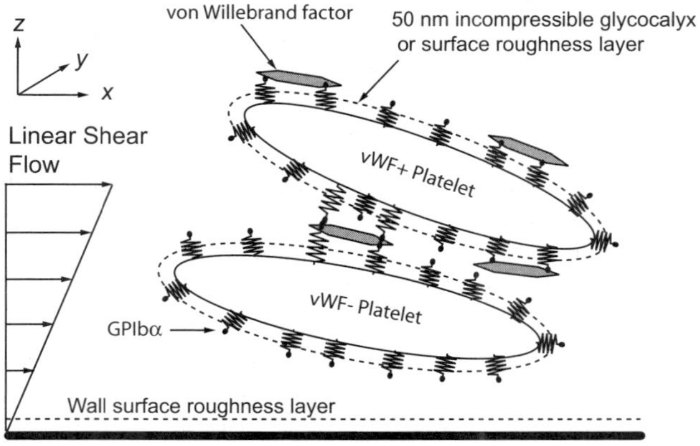

Figure 8.6.1 Schematic illustration of two platelets transiently aggregating in linear shear flow via GPIbα–vWF–GPIbα bond formation. Each platelet is coated with a 50 nm incompressible surface roughness layer whose thickness is equal to the height of the GPIb surface receptors. The various components in the diagram have not been drawn to scale.

All GPIbα–vWF–A1 bonds involved in the bridging of two platelets are treated as individual linear springs, as shown in figure 8.6.1. The dependence of the bond formation rate constant k_f on the deviation bond length, $|x_b - l_b|$, is described by the equation

$$k_f = k^0_{f,2D} \exp\left(\sigma\, |x_b - l_b|\, \frac{\gamma - 0.5|x_b - l_b|}{k_B T}\right), \tag{8.6.2}$$

where $k^0_{f,2D}$ is the intrinsic cross-linking forward rate constant, σ is the spring constant, l_b is the equilibrium bond length, x_b is the distance between the end points of the GPIbα receptor on the platelet surface and the vWF-A1 binding site, and γ is the bond reactive compliance (Bell *et al.* 1984). Because the two-dimensional cross-linking forward rate constant is not well determined experimentally, it is the single adjustable parameter. If a bond spring deviates from its equilibrium length, an instantaneous force and torque are transmitted to the bound cells depending on the deviation length and orientation of the stretched or compressed bond with respect to the participating bound surfaces, with spring constant $\sigma = 10$ pN/nm (Chtcheglova *et al.* 2004).

Monte Carlo simulations were performed to examine the breakup of bonds between vWF and receptors on either platelet in the case of interplatelet bridging, and bonds between vWF and one or more receptors on a platelet surface where it is bound. The bond length used for determining the bond force, $F_b = \sigma|x_b - l_b|$ and $2l_b = 0.128$ μm, for the two receptor-ligand bonds forming the trimolecular bridge is half the total distance between the linked receptor nodes on either platelet.

In performing simulations of platelet aggregation, it is not possible to consider all infinite possibilities of collision trajectories. Narrowing down the set of unique collisions to 85 different collision trajectories depicted in figure 8.5.2(a) is still prohibitive in terms of computational resources. Because adhesive dynamics is probabilistic, time-consuming simulations must be performed to produce an adequate sample size and obtain meaningful averages. To adequately represent the characteristics of platelet-platelet collisions close to a bounding surface, three representative initial platelet positions were chosen from a multitude of initial positions resulting in cell interception.

The three initial platelet positions considered produce collision trajectories that possess three time integrals of collision contact area, referred to as representative collisions type I, II, type III, respectively: average time-integral contact area reduced by 0.5 times the standard deviation; average time-integral contact area; average time-integral contact area enhanced by 0.5 times the standard deviation. Average and standard deviation values are listed in figure 8.5.4. The cross-linking forward rate constant was estimated by matching the simulation predictions with experimentally observed platelet aggregation behavior by trial and error. Since binding molecules must undergo reversible shear-stress-induced activation to participate in binding, the two-dimensional forward rate constant depends on the prevailing shear stress (Doggett *et al.* 2002). Bond formation is promoted by increasing the shear rate (Konstantopoulos *et al.* 1997).

Huang & Hellums (1993ab) calculated the overall collision efficiency η_c of platelets at high shear rates. A population balance was used to describe the aggregation dynamics of platelets in unbounded flow typically observed in a cone and plate viscometer. The parameters of their mathematical model were obtained by fitting equations to experimental data of shear-induced platelet aggregation. Platelets were approximated as rigid spheres and the effect of hydrodynamic interactions on the particle trajectories was ignored. Such interactions between approaching particles tend to push the particles away from each other during interception. This interaction prevents collision that would have occurred if the particles had followed straight-line trajectories. We have determined the collision frequency of spherical particles influenced by hydrodynamic interactions to be 0.53 times that for spheres following linear trajectories. Accordingly, we can extract the binding efficiency from the overall efficiency determined by Huang & Hellums (1993ab) for a range of shear rates.

The binding efficiency is the probability of an adhesive interaction between two platelets in brief contact. The number of platelets in a platelet-rich mixture that bind vWF under high-shear conditions is small compared to the total platelet population sheared in the presence of vWF multimers (Goto *et al.* 1995, Konstantopoulos *et al.* 1997). Using the experimental data of Konstantopoulos *et al.* (1997), we can estimate the binding efficiency between two platelets, one functionalized with L-vWF molecules (L-vWF+platelet) and the other devoid of L-vWF molecules (L-vWF-platelet). In our model, the fraction of collisions participating in interplatelet GPIbα–vWF binding is assumed to be the same as the fraction of collisions that immediately or eventually form stable platelet aggregates via $\alpha_{IIb}\beta_3$–vWF bonds. This assumption is supported by the observation that all vWF-positive platelets generated in a high shear environment participate in aggregate formation (Goto *et al.* 1995).

In our study of transient platelet aggregation, the binding efficiency is defined as the probability of formation of at least one cell-cell bond bridge between two colliding cell surfaces. Binding efficiency depends on the receptor distribution over the platelet surface, ligand distribution over the platelet surface, cross-linking forward rate constant, and hydrodynamics of platelet collisions with reference to the time integral of the contact area and separation of surfaces during collision. Values producing a good match between theoretically predicted two-platelet binding efficiencies and experimentally determined platelet binding efficiencies were obtained for six shear rates, 4500, 5400, 6300, 7200, 7700, and 8000 s^{-1} (Huang & Hellums 1993*ab*, Konstantopoulos *et al.* 1997). We found that the GPIb–vWF–A1 bond formation rate exhibits a piecewise linear dependence on the shear rate, and the rate of association is a strong function of the shear rate for shear rates greater than 7200 s^{-1} and a weak function of the shear rate otherwise. This prediction was confirmed experimentally by small angle neutron scattering (SANS) experiments with vWF in solution (Singh *et al.* 2009).

8.6.2 Binding efficiency for GPIbα–vWF kinetics

Estimates for $k^0_{f,2D}$ were obtained by matching the predicted binding efficiencies for L-vWF mediated platelet–platelet binding with values obtained experimentally by Huang & Hellums (1993*ab*). In the absence of further evidence, the forward cross-linking rate constant can be assumed to be independent of the vWF size and GPIbα-vWF dissociation constant. With known values of the two-dimensional rate constant as a function of shear rate, the binding efficiency can be determined for n-vWF mediated interplatelet binding with a fivefold lower K_D. A smaller equilibrium dissociation constant for the GPIbα–vWF-A1 bond increases the equilibrium number and ligation of vWF multimers on the surface of a n-vWF+ platelet.

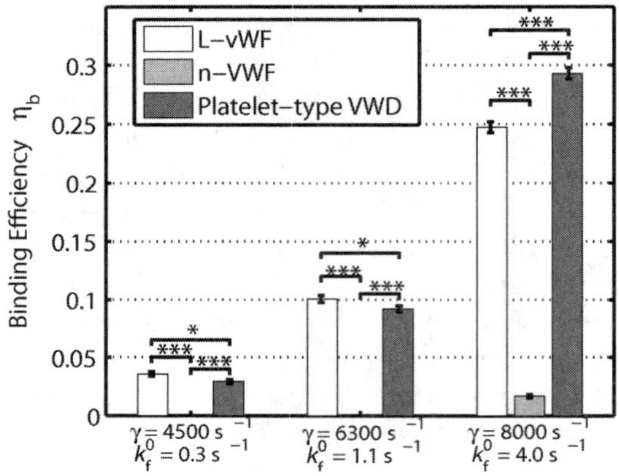

Figure 8.6.2 Comparison of two-platelet binding efficiencies for three aggregation scenarios at three high shear rates. The error bars indicate the standard error of the mean (SEM); $n = 300$ for each average value. *Reprinted with permission from Mody & King (2008).*

The bar graph in figure 8.6.2 compares the two-platelet binding efficiency for all three cases at shear rates, 4500, 6300, and 8000 s^{-1}. Platelets undergoing collisions under normal conditions (normal binding kinetics) where n-vWF is present in solution do not form interplatelet bonds at shear rates 4500 and 6300 s^{-1}. This agrees with published reports stating that von Willebrand factor-mediated shear-induced platelet aggregation occurs only at shear stress higher than 80 dyn/cm^2 (Ikeda *et al.* 1991, Shankaran *et al.* 2003). Binding efficiency for n-vWF-mediated interplatelet binding is significantly different from binding efficiency for L-vWF mediated platelet binding and binding under conditions of platelet-type vWD (unpaired student t-test, $p < 0.001$). The binding efficiencies for the three binding scenarios described in figure 8.6.2 at a particular shear rate differ due to the dissimilar concentration of vWF multimers on the surface of one of the two colliding platelets, which, in turn, is a function of the vWF multimer size and binding kinetics of the GPIbα–vWF-A1 bond.

8.6.3 Effect of interplatelet binding on collision

Depending on the number and location of transient GPIbα–vWF–GPIbα bond bridges formed between two colliding platelets, collision properties, such as the time-integral of contact area, are likely to be altered subtly or dramatically. Table 8.6.1 summarizes the average percentage enhancement of the time-integral of contact area, A_{TI}, due to interplatelet bond formation for

L-vWF

Collision type	*Avg. % change*	*Max % increase*	*Max % decrease*
I	0.99	9.75	6.05
II	2.25	11.48	1.64
III	1.32	20.44	3.21

Platelet-type vWD

Collision type	*Avg % change*	*Max % increase*	*Max % decrease*
I	5.00	24.11	15.05
II	5.86	40.70	3.38
III	−1.52	8.43	31.18

Table 8.6.1 Effect of interplatelet binding on collision properties for L-vWF mediated platelet aggregation and (n-vWF + platelet-type vWD)–mediated platelet aggregation at shear rate $8000\ \mathrm{s}^{-1}$. The table shows variations in the A_{TI}. *Reprinted with permission from Mody & King (2008).*

L-vWF-mediated binding and n-vWF-mediated binding with platelet-type vWD kinetics. In the case of L-vWF, the number of successful collisions is $n = 30, 35$, and 35, respectively, for collision type I, II, and III. In the case of vWD, $n = 38, 35$, and 29, respectively, for collision type I, II, and III.

The results in table 8.6.1 show that, on average, short-lived interplatelet bonds increase the extent of cell-cell contact during collision only marginally at the high shear rate, 8000 s^{-1}. Inter-platelet binding decreases the time-integral of the collision contact area. The majority of the 202 collisions described in table 8.6.1 resulting in interplatelet binding involve the formation of only one bond. Of 202 successful collisions, 160, 37 and 5 collisions involve the formation of one, two, and three interplatelet bonds, respectively. Of 160 single-bond aggregation events, 69% have A_{TI} greater than that in the non-reactive case. Of the 37 double-bond aggregation events, 81% involved an increase in A_{TI}.

It appears that, when two or more bonds are formed, the reduction in the time-integral of the contact area due to changes in the collision trajectory is dominated by the contact time gained from the bonds constraining the cells from separating. All triple-bond aggregation events experience a cell–

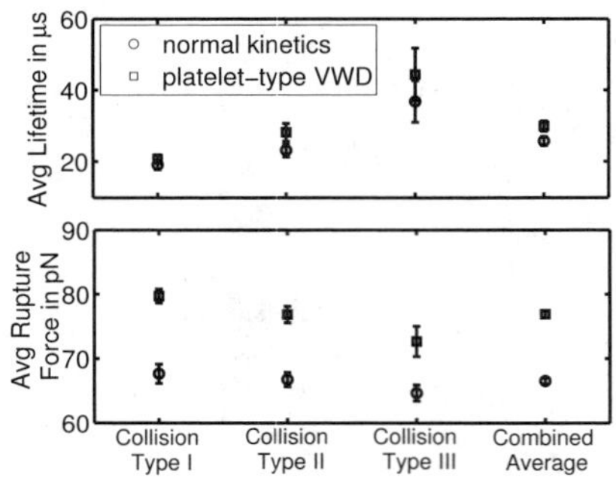

Figure 8.6.3 Average bond lifetimes and average bond rupture forces of interplatelet GPIbα–vWF–A1 bonds at shear rate 8000 s^{-1} for normal binding kinetics and mutant binding kinetics corresponding to platelet-type vWD. The number of binding events is $n = 33$, 42, and 46, respectively, for collision types I, II and III in the first case, and $n = 45$, 43, and 40, respectively, for collision types I, II and III in the second case. The error bars indicate the standard error of the mean (SEM). *Reprinted with permission from Mody & King (2008).*

cell contact that is greater than that during purely hydrodynamic collision. Of the 42 collisions resulting in two or more binding events, 14% involve two or more bonds coexisting for some time. Thus, the majority of binding events occur independently during platelet collision. That is, a bond bridge forms only after a previous bond bridge has been ruptured, and the formation of the first bond does not seem to assist the formation of additional bonds. The maximum number of interplatelet bonds formed during any collision at shear rate 8000 s^{-1} was three.

8.6.4 Force mechanics of bond rupture

The average bond lifetime and average force experienced by bonds at the time of rupture are compared in figure 8.6.3 for three representative collisions of type I, II, and III, at shear rate 8000 s^{-1}. Figure 8.6.3 shows that the average bond lifetime increases as the duration of contact and the surface area available for contact between the two colliding platelet surfaces increase during collision. Inter-platelet bond characteristics are shown for normal binding kinetics and platelet-type vWD binding kinetics. A combined average of the bond lifetime or bond rupture force is obtained by taking a weighted average of the average bond lifetimes and the average bond rupture forces for the three representative collisions, as discussed by Mody & King (2008*b*).

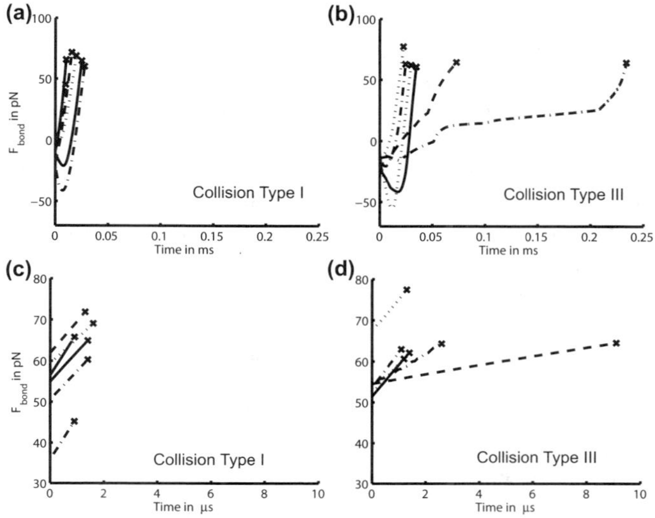

Figure 8.6.4 Bond force-loading histories of twelve representative interplatelet bonds with normal binding kinetics. Complete force loading histories are plotted for interplatelet bonds created during (*a*) type I and (*b*) type III collisions. Partial force-loading histories are plotted for the last 10 pN of force prior to rupture experienced by bonds formed during (*b*) type I and (*d*) type III collisions. *Reprinted with permission from Mody & King (2008).*

As expected, the fivefold smaller dissociation rate constant of platelet type-vWD binding kinetics produces larger average GPIbα–vWF–A1 bond rupture forces and bond lifetimes than those observed under normal kinetics (no gain-of-function mutation). Average bond rupture forces for normal and mutant binding kinetics are significantly different for all three representative collision types ($p < 0.005$ for all three cases, unpaired student t-test – one-tailed). The combined average bond lifetimes for normal and mutant binding kinetics are also found to be significantly different ($p < 0.05$). The combined average bond lifetimes for the two cases differ by 4.2 μs. The difference in the combined average bond rupture forces for the two cases is 10.4 pN.

Figure 8.6.4 displays bond force–loading histories for six representative platelet–platelet bond bridges developing during type I collision and type III collision when platelet aggregation is governed by normal binding kinetics. Note that the time integral of the contact area for type III collision is threefold higher than that for type I collision. The nature of the bond force histories for type II collisions is intermediate of type I and III. A highly nonlinear behavior is observed. Bond forces switch from compressive stress to tensile stress with failure almost always occurring under tension. Except for three rupture events, the remaining 246 bond breakage events at shear rate 8000 s^{-1} result from tensile failure. The bonds fail under a large linear force ramp.

Figure 8.6.4(*c, d*) depicts the linear tensile force ramp exerted on each bridging bond over the final 10 pN of the life of the bonds prior to rupture. At rupture, the bonds experience on average a linear force ramp of 8.1 pN/μs with standard deviation 1.82 for the binding events shown in figure 8.6.4(*c*), and on average 5.9 pN/μs with standard deviation 2.89 for the binding events shown in figure 8.6.4(*d*). Similar magnitudes of the force ramp at breakage are observed for bonds governed by mutant binding kinetics.

The bond force-loading histories described in figure 8.6.4 explain the differences observed in the combined average bond lifetimes and average rupture forces of the normal and mutant bonds shown in figure 8.6.3. The mutant bonds are capable of withstanding a bond force that is, on average, greater by 10 pN than that of normal bonds. Near rupture, bonds experience a linear ramp of bond force in the range 6–8 pN/μs. The additional lifetime experienced by mutant bonds is of the same order of magnitude as the time period where the force on the bond increases by approximately 10 pN.

8.7 Conclusions and future directions

Platelet motion, interaction with a vascular wall, and platelet–surface and platelet–platelet adhesive binding can be studied at an unprecedented level of spatial and temporal detail and resolution using the three-dimensional platelet adhesive dynamics simulation method. In this chapter, we have discussed the effects of hydrodynamics, nature of interactions between platelet-shaped particles among themselves and with bounding surfaces, and the significance of complex transport phenomena arising from Brownian motion on the platelet trajectories and adhesive behavior.

Platelet hydrodynamics near a surface

The motion of platelets is significantly different than that of leukocytes and spherical cells in linear shear flow. Platelets spend most of their time in the horizontal orientation. A platelet modeled as an oblate spheroid with aspect ratio 0.25 convected near a surface exhibits three distinct modes of motion depending on the initial position of the platelet centroid (Mody & King 2005). In Regime I, platelets rotate and translate in the direction of the shear flow while exhibiting periodic lateral motion normal to the wall. In Regime II, platelets contact the surface at an oblique angle and flip over the surface, effectively performing a pole vaulting. In this regime, contact with the surface is transient but occurs repeatedly with no net platelet drift away from the wall.

When platelets are situated near a wall at a distance less than $0.75a$ with their axis of revolution nearly normal to the wall, Regime III prevails. The platelets glide over the surface exhibiting periodic oscillation about the vorticity axis accompanied by lateral motion in the direction of the velocity gradient. Contact between the platelets and the surface is not made in this case. Oscillatory and lateral motion of a spheroid completely disappears at height $0.662a$ where the platelet moves along a straight line with no rotation about any axis. Thus, platelet–surface encounters are observed only in Regime II flow where contact with the surface is transient but recurring. Destabilizing external forces inducing angular tilt about the direction of the flow or velocity gradients can cause a transition from Regime III to II motion, often observed in fully three-dimensional flow. The range of heights where Regime III is observed depends on the spheroid aspect ratio. Wobble flow completely disappears for oblate spheroids with aspect ratio greater than 0.41.

In Regime II, once a platelet has contacted a surface at an oblique angle (always less than 90°) about the y axis (the vorticity direction), it remains in contact with the surface until the angle between the platelet major axis and the surface becomes less than 90°. A two-dimensional analytical model for calculating platelet flipping motion on a vWF-coated surface also predicts platelet–surface contact for certain angular orientations such that $\alpha < 90°$ (Mody *et al.* 2005). The calculated hydrodynamic radial forces are found to be compressive, directed along the platelet major axis towards the wall, only in these orientations. The radial forces are found to be tensile, directed away from the wall, when the platelet is oriented at angles greater than 90° with respect to the wall. Thus, as a platelet rotates or flips over a surface, an alternating hydrodynamic radial force acts along the platelet major axis.

Experimental observations of platelet tethering to a surface corroborate the theoretical predictions of the two-dimensional analytical model (Mody *et al.* 2005) and three-dimensional computational model (Mody & King 2005). Platelets are found to consistently attach to surface-bound vWF only in orientations where a radial compressive fluid force arises. Detachment of the platelet from the surface occurs only for platelet orientations greater than 90°. Fluid forces thus govern the frequency and orientation of a platelet during platelet–surface encounters and thereby influence the ability of a platelet to tether to a reactive surface.

Brownian motion

The influence of unbiased Brownian motion of a microscopic blood platelet on the platelet trajectory, orientation and frequency of contact with a planar surface was discussed in section 8.4. At physiological shear rates, Brownian motion has a negligible effect on platelet paths. Platelet motion near a surface, the frequency of platelet–surface contact and platelet–surface bond dissociation dynamics measured by bond lifetimes and bond rupture forces remain virtually unchanged when Brownian motion is incorporated into the computational model. Brownian motion is a slow process on a physiological time scale. At shear rates typical of normal blood flow, fluid convective motion is the dominant mechanism of cell transport.

The Péclet number of a Brownian platelet calculated for shear rates of interest greater than 100 s^{-1} in a range of heights from the surface is found to be greater than 200, and this corroborates the weak effect of the random motion. Brownian motion becomes important when the Péclet number is of order unity. For linear shear flow, this requires a low shear rate less than 5 s^{-1}. Brownian motion could play a role in influencing platelet motion under certain pathological flow conditions, such as in regions of recirculating flow proximal or distal to a stenosis where the local flow velocities are small. In addition, Brownian motion may play an important role in cell–surface adhesion of micron-sized spherical particles. Because spheres do not migrate laterally in Stokes flow, small-amplitude random motion could be instrumental in promoting cell-surface contact followed by adhesive interactions (Yago *et al.* 2007).

Collision

The detailed characterization of particle collision in the proximity of a wall discussed in section 8.5 illustrates the significant effect of particle shape and distance from the wall on the particle trajectories, collision time, collision contact area on each cell, and collision frequency. Wall effects are evidently more pronounced for nonspherical particles (platelet–platelet and platelet–sphere encounters) than spherical particles. Homogeneous platelet collision frequency increases by 25% when platelets are brought into close proximity to a wall for a reactive gap of 128 nm. The size and shape of individual particles participating in an encounter greatly influence the collision characteristics, including the opportunity for collision. Heterogeneous platelet–sphere (equal volume) collision frequency is 21% greater than homogeneous platelet collision frequency and 30% greater than homogeneous sphere collision frequency near a bounding wall for reactive gap 128 nm. While platelet–platelet collisions are less probable in unbounded fluid compared to sphere–sphere collisions, the presence of a wall significantly increases the collision rate. Platelet collision frequency near a wall is greater by a factor of 1.106 than that for spheres. We expect that particle shape as well as the hydrodynamic effects of the bounding wall will strongly influence the aggregation characteristics of reactive cells.

Adhesion and aggregation

In Section 8.6, we presented an adhesive dynamics model of GPIbα–vWF–GPIbα mediated transient aggregation of two resting platelets. The stochastic adhesion model was coupled with the three-dimensional multi-particle hydrodynamic flow model discussed in section 8.2. This is the first fully three-dimensional study of cell adhesion and aggregation incorporating hydrodynamic effects for nonspherical cell shapes and accounting for the presence of a wall. To date, the platelet adhesive dynamics method is the most advanced and realistic method for modeling platelet adhesion at the cellular level.

Our simulations have shown that interplatelet GPIbα–vWF–GPIbα bond bridges are short-lived and cannot withstand the dominating shearing forces that quickly separate platelets captured in an aggregate. According to the model, an increase in the average size of vWF in plasma or a reduction in the dissociation constant K_D of the GPIbα–vWF–A1 bond strongly influence the equilibrium binding kinetics and correspondingly increase the interplatelet binding efficiency. These abnormal conditions have pathological consequences by promoting shear-induced platelet aggregation at shear rates greater than 5000 s^{-1}, as shown in figure 8.6.2. GPIbα–vWF–A1 bond lifetimes and bond rupture forces at shear rate $\dot{\gamma} = 8000$ s^{-1} were found to be 20–40 μs and approximately 70 pN for normal binding kinetics.

The time-integral of the contact area during collision is a metric of the adhesion probability between two surfaces. Bond lifetimes increase as the time-integral becomes higher. The bonds linking two platelets in shear flow are subjected to highly nonlinear bond forces during their lifetime. Bonds fail primarily under tension while experiencing a sharp linear increase in bond force. All GPIbα–vWF–A1 bonds rupture at shear rate $\dot{\gamma} = 8000$ s^{-1} under a high linear force ramp of $6 - 8$ pN/μs. Cell collisions involving more intimate contact tend to form a larger number of interplatelet bonds. These bonds exhibit less predictable and more chaotic bond force-loading histories. The insights gained from these studies underscore the importance of the simulation method as a powerful predictive model for elucidating platelet adhesive under flow.

Adhesive dynamics simulations

Finally, we briefly touch upon the future scope and potential applications of the multiscale platelet adhesive dynamics method for modeling platelet motion and adhesive interactions in a hemodynamic environment. The current algorithm can be applied with straightforward modifications to simulate single platelet tethering and adhesive rolling on a flat subendothelial surface. The objective is to study the influence of the platelet shape and size and of experimentally determined healthy and diseased bond kinetics on the mechanics of platelet attachment to and detachment from a surface with regard to local hydrodynamic flow conditions and concentration of adhesive molecules on the surfaces. Such simulations are useful in elucidating the etiology of disease states caused by irregular platelet shape and size, abnormal binding kinetics, and the lack or preponderance of a particular reactive molecule on the platelet surface at the subendothelial surface or within plasma.

A related and more challenging application involves simulating the early stages of the hemostatic cascade where multiple platelets make a first contact with an injured vascular surface, accompanied by tethering bond formation and stable rolling on surface-bound vWF. A major contribution to the numerical model would be the addition of a deterministic or probabilistic algorithm mimicking the signaling cascade downstream of the transduction of an GPIbα–vWF–A1 binding event that leads to calcium influx and the transmission of a platelet activating signal. A comprehensive algorithm incorporating a predictive platelet activation (signal transduction) model will allow us to simulate the sequential progression of hemostatic and thrombotic events, from early to late stages, including a variety of receptor–ligand interactions and elliptical to globular platelet shape transformations.

Platelet adhesive dynamics can be adapted to simulate homotypic and heterotypic aggregation of blood cells, such as high shear-induced pathological platelet aggregation resulting in thrombotic occlusion of blood vessels, or low shear-induced platelet–leukocyte binding responsible for a circulation disorder known as venous stasis (deep vein thrombosis). Pathological aggregation can be studied as a function of hemodynamic factors including different flow fields, concentration of activated blood cells, cell shape and size, and the presence of pro-thrombotic mediators. The three-dimensional model lends itself to computer parallelization that allows for substantial speedup in simulations with multiple particles (Fuentes & Kim 1992). The development of more sophisticated and advanced models that can predict the physiological and pathological behavior of platelets in specified hemodynamic environments with increasing accuracy is closely tied with the availability of quantitative experimental data to characterize the individual processes.

Acknowledgment

This material is based upon work supported by the National Science Foundation under Grant No. CBET-0935889. Any opinions, findings and conclusions or recommendations expressed in this material are those of the authors and do not necessarily reflect the views of the National Science Foundation (NSF).

References

ALEVRIADOU, B. R., MOAKE, J. L., TURNER, N. A., RUGGERI, Z. M., FOLIE, B. J., PHILLIPS, M. D., SCHREIBER, A. B., HRINDA, M. E. & MCINTIRE, L. V. (1993) Real-time analysis of shear-dependent thrombus formation and its blockade by inhibitors of von Willebrand factor binding to platelets. *Blood* **81**, 1263–1276.

ALMOMANI, T., UDAYKUMAR, H. S., MARSHALL, J. S. & CHANDRAN, K. B. (2008) Micro-scale dynamic simulation of erythrocyte-platelet interaction in blood flow. *Ann. Biomed. Eng.* **36**, 905–920.

ARYA, M., ANVARI, B., ROMO, G. M., CRUZ, M. A., DONG, J. F., MCINTIRE, L. V., MOAKE, J. L. & LOPEZ, J. A. (2002) Ultra-large multimers of von Willebrand factor form spontaneous high-strength bonds with the platelet glycoprotein Ib-Ix complex: Studies using optical tweezers. *Blood* **99**, 3971–3977.

ARYA, M., KOLOMEISKY, A. B., ROMO, G. M., CRUZ, M. A., LOPEZ, J. A. & ANVARI, B. (2005) Dynamic force spectroscopy of glycoprotein Ib-Ix and von Willebrand factor. *Biophys. J.* **88**, 4391–4401.

ASOKAN, K. & RAMAMOHAN, T. R. (2004) The rheology of a dilute suspension of Brownian dipolar spheroids in a simple shear flow under the action of an external force. *Phys. Fluids* **16**, 433–444.

BARBER, K. M., PINERO, A. & TRUSKEY, G. A. (1998). Effects of recirculating flow on U-937 cell adhesion to human umbilical vein endothelial cells. *Am. J. Physiol.* **275**, H591-9.

BELL, G. I. (1978) Models for the specific adhesion of cells to cells. *Science* **200**, 618–627.

BELL, G. I., DEMBO, M. & BONGRAND, P. (1984) Cell adhesion. Competition between non-specific repulsion and specific bonding. *Biophys. J.* **45**, 1051–1064.

BERNE, R. M. & LEVY, M. N. (1997) *Cardiovascular Physiology.* Mosby, St. Louis.

BLUESTEIN, D., GUTIERREZ, C., LONDONO, M. & SCHOEPHOERSTER, R. T. (1999) Vortex shedding in steady flow through a model of an arterial stenosis and its relevance to mural platelet deposition. *Ann. Biomed. Eng.* **27**, 763–73.

BLYTH, M. G. & POZRIKIDIS, C. (2009) Adhesion of a blood platelet to injured tissue. *Eng. Anal. Bound. Elem.* **33**, 695–703.

BRENNER, H. (1974) Rheology of a dilute suspension of axisymmetric Brownian particles. *Int. J. Multiph. Flow* **1**, 195–341.

CARPEN, I. C. & BRADY, J. F. (2005) Microrheology of colloidal dispersions by Brownian dynamics simulations. *J. Rheol.* **49**, 1483–1502.

CHTCHEGLOVA, L. A., SHUBEITA, G. T., SEKATSKII, S. K. & DIETLER, G. (2004) Force spectroscopy with a small dithering of AFM tip: A method of direct and continuous measurement of the spring constant of single molecules and molecular complexes. *Biophys. J.* **86**, 1177-1184.

CLAEYS, I. L. & BRADY, J. F. (1993) Suspensions of prolate spheroids in Stokes flow. 1. Dynamics of a finite number of particles in an unbounded fluid. *J. Fluid Mech.* **251**, 411–442.

COLLER, B. S., PEERSCHKE, E. I., SCUDDER, L. E. & SULLIVAN, C. A. (1983) Studies with a murine monoclonal antibody that abolishes ristocetin-induced binding of von Willebrand factor to platelets: Additional evidence in support of GPIb as a platelet receptor for von Willebrand factor. *Blood* **61**, 99–110.

DAS, B., ENDEN, G. & POPEL, A. S. (1997) Stratified multiphase model for blood flow in a venular bifurcation. *Ann. Biomed. Eng.* **25**, 135–153.

DOGGETT, T. A., GIRDHAR, G., LAWSHE, A., MILLER, J. L., LAURENZI, I. J., DIAMOND, S. L. & DIACOVO, T. G. (2003) Alterations in the intrinsic properties of the GPIb*alpha*–vWF tether bond define the kinetics of the platelet-type von Willebrand disease mutation, Gly233Val. *Blood* **102**, 152–160.

DOGGETT, T. A., GIRDHAR, G., LAWSHE, A., SCHMIDTKE, D. W., LAURENZI, I. J., DIAMOND, S. L. & DIACOVO, T. G. (2002). Selectin-like kinetics and biomechanics promote rapid platelet adhesion in flow: The GPibα–vWF tether bond. *Biophys. J.* **83**, 194–205.

DUMAS, J. J., KUMAR, R., MCDONAGH, T., SULLIVAN, F., STAHL, M. L., SOMERS, W. S. & MOSYAK, L. (2004) Crystal structure of the wild-type von Willebrand factor A1-glycoprotein Ibα complex reveals conformation differences with a complex bearing von Willebrand disease mutations. *J. Biol. Chem.* **279**, 23327–23334.

FEDERICI, A. B., BADER, R., PAGANI, S., COLIBRETTI, M. L., DE MARCO, L. & MANNUCCI, P. M. (1989) Binding of von Willebrand factor to glycoproteins Ib and Iib/Iiia complex: Affinity is related to multimeric size. *Br. J. Haemat.* **73**, 93–99.

FISCHER, B. E., KRAMER, G., MITTERER, A., GRILLBERGER, L., REITER, M., MUNDT, W., DORNER, F. & EIBL, J. (1996) Effect of multimerization of human and recombinant von Willebrand factor on platelet aggregation, binding to collagen and binding of coagulation Factor Viii. *Thromb. Res.* **84**, 55–66.

FOX, J. E., AGGERBECK, L. P. & BERNDT, M. C. (1988) Structure of the glycoprotein Ib.Ix complex from platelet membranes. *J. Biol. Chem.* **263**, 4882–4890.

FROJMOVIC, M., LONGMIRE, K. & VAN DE VEN, T. G. (1990) Long-range interactions in mammalian platelet aggregation. Ii. The role of platelet pseudopod number and length. *Biophys. J.* **58**, 309–318.

FUENTES, Y. O. & KIM, S. (1992) Parallel computational microhydrodynamics–Communication scheduling strategies. *AIChE J.* **38**, 1059–1078.

FURLAN, M. (1996) Von Willebrand factor: Molecular size and functional activity. *Ann. Hematol.* **72**, 341–348.

GARNIER, N. & OSTROWSKY, N. (1991) Brownian dynamics in a confined geometry–Experiments and numerical simulations. *J. Phys. II* **1**, 1221–1232.

GAVZE, E. & SHAPIRO, M. (1997) Particles in a shear flow near a solid wall: Effect of nonsphericity on forces and velocities. *Int. J. Multiph. Flow* **23**, 155–182.

GOLDMAN, A. J., COX, R. G. & BRENNER, H. (1967) Slow viscous motion of a sphere parallel to a plane wall. I. Motion through a quiescent fluid. *Chem. Eng. Sci.* **22**, 637–651.

GOTO, S., IKEDA, Y., SALDIVAR, E. & RUGGERI, Z. M. (1998) Distinct mechanisms of platelet aggregation as a consequence of different shearing flow conditions. *J. Clin. Invest.* **101**, 479–86.

GOTO, S., SALOMON, D. R., IKEDA, Y. & RUGGERI, Z. M. (1995) Characterization of the unique mechanism mediating the shear-dependent binding of soluble von Willebrand factor to platelets. *J. Biolog. Chem.* **270**, 23352–23361.

HAGA, J. H., BEAUDOIN, A. J., WHITE, J. G. & STRONY, J. (1998) Quantification of the passive mechanical properties of the resting platelet. *Ann. Biomed. Eng.* **26**, 268–277.

HAMMER, D. A. & APTE, S. M. (1992) Simulation of cell rolling and adhesion on surfaces in shear flow: General results and analysis of selectin-mediated neutrophil adhesion. *Biophys. J.* **63**, 35–57.

HELMKE, B. P., SUGIHARA-SEKI, M., SKALAK, R. & SCHMID-SCHÖNBEIN, G. W. (1998) A mechanism for erythrocyte-mediated elevation of apparent viscosity by leukocytes in vivo without adhesion to the endothelium. *Biorheology* **35**, 437–448.

HSU, R. & GANATOS, P. (1989) The motion of a rigid body in viscous fluid bounded by a plane wall. *J. Fluid Mech.* **207**, 29–72.

HSU, R. & GANATOS, P. (1994) Gravitational and zero-drag motion of a spheroid adjacent to an inclined plane at low Reynolds number. *J. Fluid Mech.* **268**, 267–292.

HUANG, P. Y. & HELLUMS, J. D. (1993*a*) Aggregation and disaggregation kinetics of human blood platelets: Part I. Development and validation of a population balance method. *Biophys. J.* **65**, 334–343.

HUANG, P. Y. & HELLUMS, J. D. (1993*b*) Aggregation and disaggregation kinetics of human blood platelets: Part II. Shear-induced platelet aggregation. *Biophys. J.* **65**, 344–353.

HUIZINGA, E. G., TSUJI, S., ROMIJN, R. A., SCHIPHORST, M. E., DE GROOT, P. G., SIXMA, J. J. & GROS, P. (2002) Structures of glycoprotein Ibα and its complex with von Willebrand factor A1 domain. *Science* **297**, 1176–1179.

IKEDA, Y., HANDA, M., KAMATA, T., KAWANO, K., KAWAI, Y., WATANABE, K., KAWAKAMI, K., SAKAI, K., FUKUYAMA, M., ITAGAKI, I., YOSHIOKA, A. & RUGGERI, Z. M. (1993) Transmembrane calcium influx associated with von Willebrand factor binding to GPIb in the initiation of shear-induced platelet aggregation. *Thromb. Haemost.* **69**, 496–502.

IKEDA, Y., HANDA, M., KAWANO, K., KAMATA, T., MURATA, M., ARAKI, Y., ANBO, H., KAWAI, Y., WATANABE, K., ITAGAKI, I., SAKAI, K. & RUGGERI, Z. M. (1991) The role of von Willebrand factor and fibrinogen in platelet aggregation under varying shear stress. *J. Clin. Invest.* **87**, 1234–1240.

JADHAV, S., EGGLETON, C. D. & KONSTANTOPOULOS, K. (2005) A three-dimensional computational model predicts that cell deformation affects selectin-mediated leukocyte rolling. *Biophys. J.* **88**, 96–104.

JEFFERY, G.B. (1922) The motion of ellipsoidal particles immersed in a viscous fluid. *Proc. R. Soc. Lond. A* **102**, 161–179.

KARINO, T. & GOLDSMITH, H. L. (1979) Aggregation of human platelets in an annular vortex distal to a tubular expansion. *Microvasc. Res.* **17**, 217–237.

KIM, M.-U., KIM, K. W., CHO, Y.-H. & KWAK, B. M. (2001) Hydrodynamic force on a plate near the plane wall. I. Plate in sliding motion. *Fluid Dyn. Res.* **29**, 137–170.

KIM, S. & KARILLA, S. J. (1991) *Microhydrodynamics: Principles and Selected Applications.* Butterworth-Heinemann, Stoneham.

KING, M. R. & HAMMER, D. A. (2001) Multiparticle adhesive dynamics. Interactions between stably rolling cells. *Biophys. J.* **81**, 799–813.

KING, M. R. & HAMMER, D. A. (2001) Multiparticle adhesive dynamics: Hydrodynamic recruitment of rolling leukocytes. *Proc. Nat. Acad. Sci.* **98** 14919–14924.

KING, M. R., HEINRICH, V., EVANS, E. & HAMMER, D. A. (2005) Nano-to-micro scale dynamics of P-selectin detachment from leukocyte interfaces. III. Numerical simulation of tethering under flow. *Biophys. J.* **88**, 1676–1683.

KING, M. R., RODGERS, S. D. & HAMMER, D. A. (2001) Hydrodynamic collisions suppress fluctuations in the rolling velocity of adhesive blood cells. *Langmuir* **17**, 4139–4143.

KONSTANTOPOULOS, K., CHOW, T. W., TURNER, N. A., HELLUMS, J. D. & MOAKE, J. L. (1997) Shear stress-induced binding of von Willebrand factor to platelets. *Biorheology* **34**, 57–71.

KROLL, M. H., HELLUMS, J. D., MCINTIRE, L. V., SCHAFER, A. I. & MOAKE, J. L. (1996) Platelets and shear stress. *Blood* **88**, 1525–1541.

KUHARSKY, A. L. & FOGELSON, A. L. (2001). Surface-mediated control of blood coagulation: The role of binding site densities and platelet deposition. *Biophys. J.* **80**, 1050–1074.

KUMAR, R. A., DONG, J. F., THAGGARD, J. A., CRUZ, M. A., LOPEZ, J. A. & MCINTIRE, L. V. (2003) Kinetics of GPIbα–Vwf–A1 tether bond under flow: Effect of GPIbα mutations on the association and dissociation rates. *Biophys. J.* **85**, 4099–4109.

KUMAR, R. A., MOAKE, J. L., NOLASCO, L., BERGERON, A. L., SUN, C., DONG, J. F. & MCINTIRE, L. V. (2006) Enhanced platelet adhesion and aggregation by endothelial cell-derived unusually large multimers of von Willebrand factor. *Biorheology* **43**, 681–691.

LAUFFENBURGER, D. A. & LINDERMAN, J. J. (1993) *Receptors: Models for Binding, Trafficking and Signaling.* Oxford University Press.

LEI, M., KLEINSTREUER, C. & ARCHIE, J. P. (1997) Hemodynamic simulations and computer-aided designs of graft–artery junctions. *J. Biomech. Eng.* **119**, 343–348.

LI, C. Q., DONG, J. F. & LOPEZ, J. A. (2002) The mucin-like macroglycopeptide region of glycoprotein Ibα is required for cell adhesion to immobilized von Willebrand factor (vWF) under flow but not for static vWF binding. *Thromb. Haemost.* **88**, 673-677.

LI, F., LI, C. Q., MOAKE, J. L., LOPEZ, J. A. & MCINTIRE, L. V. (2004) Shear stress-induced binding of large and unusually large von Willebrand factor to human platelet glycoprotein Ibα. *Ann. Biomed. Eng.* **32**, 961–969.

LONG, M., GOLDSMITH, H. L., TEES, D. F. & ZHU, C. (1999) Probabilistic modeling of shear-induced formation and breakage of doublets cross-linked by receptor-ligand bonds. *Biophys. J.* **76**, 1112–1128.

MAUL, C., KIM, S. T., ILIC, V., TULLOCK, D. & NHAN, P. T. (1994) Sedimentation of hexagonal flakes in a half-space–Numerical predictions and experiments in Stokes flow. *J. Imag. Sci. Tech.* **38**, 241–248.

MIURA, S., LI, C. Q., CAO, Z., WANG, H., WARDELL, M. R. & SADLER, J. E. (2000) Interaction of von Willebrand factor domain A1 with platelet glycoprotein Ibα-(1-289). Slow intrinsic binding kinetics mediate rapid platelet adhesion. *J. Biol. Chem.* **275**, 7539–7546.

MOAKE, J. L. (1995) Thrombotic thrombocytopenic purpura. *Thromb. Haemost.* **74**, 240-245.

MOAKE, J. L., TURNER, N. A., STATHOPOULOS, N. A., NOLASCO, L. & HELLUMS, J. D. (1988) Shear-induced platelet aggregation can be mediated by Vwf released from platelets, as well as by exogenous large or unusually large Vwf multimers, requires adenosine diphosphate, and is resistant to aspirin. *Blood* **71**, 1366–1374.

MODY, N. A. & KING, M. R. (2005) Three-dimensional simulations of a platelet-shaped spheroid near a wall in shear flow. *Phys. Fluids* **17**, 1432–1443.

MODY, N. A. & KING, M. R. (2007) Influence of Brownian motion on blood platelet flow behavior and adhesive dynamics near a plane wall. *Langmuir* **23**, 6321–6328.

MODY, N. A. & KING, M. R. (2008*a*) Platelet adhesive dynamics. Part 1: Characterization of platelet hydrodynamic collisions and wall effects. *Biophys. J.* **95**, 2539–55.

MODY, N. A. & KING, M. R. (2008*b*) Platelet adhesive dynamics. Part 2: High shear-induced transient aggregation via GPIbα-vWF-GPIbα bridging. *Biophys. J.* **95**, 2556–2574.

MODY, N. A., LOMAKIN, O., DOGGETT, T. A., DIACOVO, T. G. & KING, M. R. (2005) Mechanics of transient platelet adhesion to von Willebrand factor under flow. *Biophys. J.* **88**, 1432–1443.

MORI, D., YANO, K., TSUBOTA, K., ISHIKAWA, T., WADA, S. & YAMAGUCHI, T. (2008) Simulation of platelet adhesion and aggregation regulated by fibrinogen and von Willebrand factor. *Thromb. Haemost.* **99**, 108–115.

PERELSON, A. S. (1981) Receptor clustering on a cell surface. 3. Theory of receptor cross-linking by multivalent ligands–Description by ligand states. *Math. Biosci.* **53**, 1–39.

PERKTOLD, K. (1987) On the paths of fluid particles in an axisymmetrical aneurysm. *J. Biomech.* **20**, 311–317.

PHAN-THIEN, N., TULLOCK, D. & KIM, S. (1992) Completed double-layer in half-space: A boundary element method. *Comput. Mech.* **9**, 121–135.

POWER, H. & MIRANDA, G. (1987) Second kind integral-equation formulation of Stokes flows past a particle of arbitrary shape. *SIAM J. Appl. Math.* **47**, 689–698.

POZRIKIDIS, C. (1994) The motion of particles in the Hele-Shaw cell. *J. Fluid Mech.* **261**, 199–222.

POZRIKIDIS, C. (2005) Orbiting motion of a freely suspended spheroid near a plane wall. *J. Fluid Mech.* **541**, 105–114.

POZRIKIDIS, C. (2006) Flipping of an adherent blood platelet over a substrate. *J. Fluid Mech.* **568**, 161–172.

POZRIKIDIS, C. (2006) Interception of two spheroidal particles in shear flow. *J. Non-Newt. Fluid Mech.* **136** 50–63.

PRIES, A. R., LEY, K., CLAASSEN, M. & GAEHTGENS, P. (1989) Red cell distribution at microvascular bifurcations. *Microvasc. Res.* **38**, 81–101.

PRITCHARD, W. F., DAVIES, P. F., DERAFSHI, Z., POLACEK, D. C., TSAO, R., DULL, R. O., JONES, S. A. & GIDDENS, D. P. (1995) Effects of wall shear stress and fluid recirculation on the localization of circulating monocytes in a three-dimensional flow model. *J. Biomech.* **28**, 1459–1469.

RUGGERI, Z. M. (2004) Type Iib von Willebrand disease: A paradox explains how von Willebrand factor works. *Thromb. Haemost.* **2**, 2–6.

SAKARIASSEN, K. S., BOLHUIS, P. A. & SIXMA, J. J. (1979) Human blood platelet adhesion to artery subendothelium is mediated by factor Viii-von Willebrand factor bound to the subendothelium. *Nature* **279**, 636–638.

SAVAGE, B., SALDIVAR, E. & RUGGERI, Z. M. (1996) Initiation of platelet adhesion by arrest onto fibrinogen or translocation on von Willebrand factor. *Cell* **84** 289–97.

SCHOEPHOERSTER, R. T., OYNES, F., NUNEZ, G., KAPADVANJWALA, M. & DEWANJEE, M. K. (1993) Effects of local geometry and fluid dynamics on regional platelet deposition on artificial surfaces. *Arterioscl. Thromb.* **13**, 1806–1813.

SHANKARAN, H., ALEXANDRIDIS, P. & NEELAMEGHAM, S. (2003) Aspects of hydrodynamic shear regulating shear-induced platelet activation and self-association of von Willebrand factor in suspension. *Blood* **101**, 2637–2645.

SHANKARAN, H. & NEELAMEGHAM, S. (2004) Hydrodynamic forces applied on intercellular bonds, soluble molecules, and cell-surface receptors. *Biophys. J.* **86**, 576–88.

SIEDLECKI, C. A., LESTINI, B. J., KOTTKE-MARCHANT, K. K., EPPELL, S. J., WILSON, D. L. & MARCHANT, R.E. (1996) Shear-dependent changes in the three-dimensional structure of human von Willebrand factor. *Blood* **88**, 2939–2950.

SINGH, I., SHANKARAN, H., BEAUHARNOIS, M. E., XIAO, Z., ALEXANDRIDIS, P. & NEELAMEGHAM, S. (2006) Solution structure of human von Willebrand factor studied using small angle neutron scattering. *J. Biol. Chem.* **281**, 38266–38275.

SINGH, I., THEMISTOU, E., PORCAR, L. & NEELAMEGHAM, S. (2009) Fluid shear induces conformation change in human blood protein von Willebrand factor in solution. *Biophys. J.* **96**, 2313–2320.

SLAYTER, H., LOSCALZO, J., BOCKENSTEDT, P. & HANDIN, R. I. (1985) Native conformation of human von Willebrand protein. Analysis by electron microscopy and quasi-elastic light scattering. *J. Biol. Chem.* **260**, 8559–8563.

STABEN, M. E., ZINCHENKO, A. Z. & DAVIS, R. H. (2003) Motion of a particle between two parallel plane walls in low-Reynolds-number Poiseuille flow. *Phys. Fluids A* **15**, 1711–1733.

STOVER, C. A. & COHEN, C. (1990) The motion of rod-like particles in the pressure-driven flow between two flat plates. *Rheol. Acta* **29**, 192–203.

SUGIHARA-SEKI, M. (1996) The motion of an ellipsoid in tube flow at low Reynolds numbers. *J. Fluid Mech.* **324**, 287–308.

TANDON, P. & DIAMOND, S. L. (1997). Hydrodynamic effects and receptor interactions of platelets and their aggregates in linear shear flow. *Biophys. J.* **73**, 2819–2835.

TANDON, P. & DIAMOND, S. L. (1998) Kinetics of β2-integrin and L-selectin bonding during neutrophil aggregation in shear flow. *Biophys. J.* **75**, 3163–3178.

UFF, S., CLEMETSON, J. M., HARRISON, T., CLEMETSON, K. J. & EMSLEY, J. (2002) Crystal structure of the platelet glycoprotein Ib(Alpha) N-terminal domain reveals an unmasking mechanism for receptor activation. *J. Biol. Chem.* **277**, 35657–35663.

UNNI, H. N. & YANG, C. (2005) Brownian dynamics simulation and experimental study of colloidal particle deposition in a microchannel flow. *J. Coll. Interf. Sci.* **291**, 28–36.

WEISS, H. J., HAWIGER, J., RUGGERI, Z. M., TURITTO, V. T., THIAGARAJAN, P. & HOFFMANN, T. (1989) Fibrinogen-independent platelet adhesion and thrombus formation on subendothelium mediated by glycoprotein IIb-IIIa complex at high shear rate. *J. Clin. Invest.* **83**, 288–297.

WOOTTON, D. M. & KU, D. N. (1999) Fluid mechanics of vascular systems, diseases, and thrombosis. *Ann. Rev. Biomed. Eng.* **1**, 299–329.

WOOTTON, D. M., MARKOU, C. P., HANSON, S. R. & KU, D. N. (2001) A mechanistic model of acute platelet accumulation in thrombogenic stenoses. *Ann. Biomed. Eng.* **29**, 321–329.

YAGO, T., ZARNITSYNA, V. I., KLOPOCKI, A. G., MCEVER, R. P. & ZHU, C. (2007) Transport governs flow-enhanced cell tethering through L-Selectin at threshold shear. *Biophys. J.* **92**, 330–342.

YAMAMOTO, T., SUGA, T. & MORI, N. (2005) Brownian dynamics simulation of orientational behavior, flow-induced structure, and rheological properties of a suspension of oblate spheroid particles under simple shear. *Phys. Rev. E*, **72**, 021509.

YANG, S. M. & LEAL, L. G. (1984) Particle motion in Stokes flow near a plane fluid-fluid interface. 2. Linear shear and axisymmetric straining flows. *J. Fluid Mech.* **149**, 275–304.

YOON, B. J. & KIM, S. (1990) A boundary collocation method for the motion of two spheroids in Stokes flow–hydrodynamic and colloidal interactions. *Int. J. Multiph. Flow* **16**, 639–649.

YUAN, Y., KULKARNI, S., ULSEMER, P., CRANMER, S. L., YAP, C. L., NESBITT, W. S., HARPER, I., MISTRY, N., DOPHEIDE, S. M., HUGHAN, S. C., WILLIAMSON, D., DE LA SALLE, C., SALEM, H. H., LANZA, F. & JACKSON, S. P. (1999) The von Willebrand factor-glycoprotein Ib/V/Ix interaction induces actin polymerization and cytoskeletal reorganization in rolling platelets and glycoprotein Ib/V/Ix-transfected cells. *J. Biol. Chem.* **274**, 36241–51.

Index